AF821996

Manhattan Review®

Test Prep & Admissions Consulting

Turbocharge Your GMAT: Quantitative Reasoning Guide

Published on August 20th, 2024

Version 416

- ☐ *Mapped according to the GMAT Focus Edition*
- ☐ *300 GMAT-like practice questions*
- ☐ *Comprehensive, Intuitive, and graphical explanations of concepts*
  - · *Great collection of 700+ level questions*
  - · *Ample questions with Alternate Approaches*
- ☐ *Detailed explanations of questions*

www.manhattanreview.com

Copyright and Terms of Use

International Standard Book Number (ISBN): 978-1-62926-109-6

Published on August 20th, 2024.

Manhattan Review, 2 Park Avenue, Suite 2010, New York, NY 10016.
Phone: +1 (212) 316-2000. E-Mail: info@manhattanreview.com. Web: www.manhattanreview.com

About the Turbocharge your GMAT Series (Focus Edition)

The Turbocharge Your GMAT Series is carefully designed to be clear, comprehensive, and content-driven. Long regarded as the gold standard in GMAT prep worldwide, Manhattan Review's GMAT prep books offer professional GMAT instruction for dramatic score improvement. Now in its updated 7th edition, the full series is designed to provide GMAT Focus test-takers with complete guidance for highly successful outcomes. As many students have discovered, Manhattan Review's GMAT books break down the different test sections in a coherent, concise, and accessible manner. We delve deeply into the content of every single testing area and zero in on exactly what you need to know to raise your score. The full series covers concepts in mathematics from the most basic through the most advanced levels, making them a great study resource for all stages of GMAT preparation. Students who work through all of our books benefit from a substantial boost to their GMAT knowledge and develop a thorough and strategic approach to taking the GMAT.

- ☐ **GMAT Focus Math Essentials (ISBN: 978-1-62926-108-9)**
- ■ **GMAT Focus Quantitative Reasoning Guide (ISBN: 978-1-62926-109-6)**
- ☐ **GMAT Focus Reading Comprehension Guide (ISBN: 978-1-62926-110-2)**
- ☐ **GMAT Focus Critical Reasoning Guide (ISBN: 978-1-62926-111-9)**
- ☐ **GMAT Focus Data Insights Guide (ISBN: 978-1-62926-112-6)**
- ☐ **GMAT Focus Quantitative Question Bank (ISBN: 978-1-62926-113-3)**
- ☐ **GMAC Business Writing Assessment Guide (ISBN: 978-1-62926-114-0)**

About the Company

Manhattan Review's origin can be traced directly back to an Ivy League MBA classroom in 1999. While teaching advanced quantitative subjects to MBAs at Columbia Business School in New York City, Professor Dr. Joern Meissner developed a reputation for explaining complicated concepts in an understandable way. Remembering their own less-than-optimal experiences preparing for the GMAT, Prof. Meissner's students challenged him to assist their friends, who were frustrated with conventional GMAT preparation options. In response, Prof. Meissner created original lectures that focused on presenting GMAT content in a simplified and intelligible manner, a method vastly different from the voluminous memorization and so-called tricks commonly offered by others. The new approach immediately proved highly popular with GMAT students, inspiring the birth of Manhattan Review.

Since its founding, Manhattan Review has grown into a multi-national educational services firm, focusing on GMAT preparation, MBA admissions consulting, and application advisory services, with thousands of highly satisfied students all over the world. The original lectures have been continuously expanded and updated by the Manhattan Review team, an enthusiastic group of master GMAT professionals and senior academics. Our team ensures that Manhattan Review offers the most time-efficient and cost-effective preparation available for the GMAT. Please visit www.ManhattanReview.com for further details.

About the Founder

Professor Dr. Joern Meissner has more than 25 years of teaching experience at the graduate and undergraduate levels. He is the founder of Manhattan Review, a worldwide leader in test prep services, and he created the original lectures for its first GMAT preparation class. Prof. Meissner is a graduate of Columbia Business School in New York City, where he received a PhD in Management Science. He has since served on the faculties of prestigious business schools in the United Kingdom and Germany. He is a recognized authority in the areas of supply chain management, logistics, and pricing strategy. Prof. Meissner thoroughly enjoys his research, but he believes that grasping an idea is only half of the fun. Conveying knowledge to others is even more fulfilling. This philosophy was crucial to the establishment of Manhattan Review, and remains its most cherished principle.

The Advantages of Using Manhattan Review

- **Time efficiency and cost effectiveness.**
  - For most people, the most limiting factor of test preparation is time.
  - It takes significantly more teaching experience to prepare a student in less time.
  - Our test preparation approach is tailored for busy professionals. We will teach you what you need to know in the least amount of time.
- **Our high-quality and dedicated instructors are committed to helping every student reach her/his goals.**

International Phone Numbers and Official Manhattan Review Websites

Manhattan Headquarters	+1-212-316-2000	www.manhattanreview.com
USA & Canada	+1-800-246-4600	www.manhattanreview.com
Argentina	+1-212-316-2000	www.review.com.ar
Australia	+61-3-9001-6618	www.manhattanreview.com
Austria	+43-720-115-549	www.review.at
Belgium	+32-2-808-5163	www.manhattanreview.be
Brazil	+1-212-316-2000	www.manhattanreview.com.br
Chile	+1-212-316-2000	www.manhattanreview.cl
China	+86-20-2910-1913	www.manhattanreview.cn
Czech Republic	+1-212-316-2000	www.review.cz
France	+33-1-8488-4204	www.review.fr
Germany	+49-89-3803-8856	www.review.de
Greece	+1-212-316-2000	www.review.com.gr
Hong Kong	+852-5808-2704	www.review.hk
Hungary	+1-212-316-2000	www.review.co.hu
India	+1-212-316-2000	www.review.in
Indonesia	+1-212-316-2000	www.manhattanreview.id
Ireland	+1-212-316-2000	www.gmat.ie
Italy	+39-06-9338-7617	www.manhattanreview.it
Japan	+81-3-4589-5125	www.manhattanreview.jp
Malaysia	+1-212-316-2000	www.review.my
Mexico	+1-212-316-2000	www.manhattanreview.mx
Netherlands	+31-20-808-4399	www.manhattanreview.nl
New Zealand	+1-212-316-2000	www.review.co.nz
Philippines	+1-212-316-2000	www.review.ph
Poland	+1-212-316-2000	www.review.pl
Portugal	+1-212-316-2000	www.review.pt
Qatar	+1-212-316-2000	www.review.qa
Russia	+1-212-316-2000	www.manhattanreview.ru
Singapore	+65-3158-2571	www.gmat.sg
South Africa	+1-212-316-2000	www.manhattanreview.co.za
South Korea	+1-212-316-2000	www.manhattanreview.kr
Sweden	+1-212-316-2000	www.gmat.se
Spain	+34-911-876-504	www.review.es
Switzerland	+41-435-080-991	www.review.ch
Taiwan	+1-212-316-2000	www.gmat.tw
Thailand	+66-6-0003-5529	www.manhattanreview.com
Turkey	+1-212-316-2000	www.review.com.tr
United Arab Emirates	+1-212-316-2000	www.manhattanreview.ae
United Kingdom	+44-20-7060-9800	www.manhattanreview.co.uk
Rest of World	+1-212-316-2000	www.manhattanreview.com

Contents

1 **Welcome** 1

2 **Introduction** 3

3 **Number Properties** 5

3.1 Understanding numbers 6
3.1.1 Classification of numbers 6
3.1.2 Number line 8
3.1.3 Basic operations on numbers 9
3.1.4 Face value and place value 10
3.1.5 Even and Odd numbers 10
3.1.6 Consecutive numbers 12
3.1.7 Prime numbers 12
3.1.8 Co-Prime numbers 13
3.1.9 Rounding off 14
3.1.10 Order of operations—PEMDAS 15
3.1.11 Divisibility 16
3.1.12 Factors 17
3.1.13 Perfect Square numbers 18
3.1.14 Multiples 19
3.1.15 Lowest Common Multiple—LCM 19
3.1.16 Highest Common Factor—HCF OR Greatest Common Divisor—GCD 20
3.2 Exponents 23
3.2.1 Unit digits of numbers with exponents 24
3.2.2 Deduction of unit digit 25
3.2.3 Roots and Surds 25
3.2.3.1 Surds 26
3.2.3.2 Rationalization of surds 27
3.3 Multiplication and Division of decimals with 10^x 27
3.4 Common Binomial expressions 30
3.5 Some important summation of series 31
3.6 Fractions 32
3.6.1 Concept of fraction 32
3.6.2 Mathematical operations on fractions 33
3.6.3 Comparing fractions 35
3.6.4 More on fractions 36
3.6.5 Fractions and their percent equivalents 37

3.6.6 See-Saw Approach 38
3.7 Absolute numbers 41
3.7.1 Inequality 42
3.7.2 Inequalities with absolute numbers 44
3.7.3 Compound Inequality 46
3.7.4 Mathematical operations on inequalities 47

4 Practice Questions: Number Properties 49

5 Answer Key: Number Properties 61

6 Solution: Number Properties Practice Questions 63

7 Algebra 95
7.1 Identities 96
7.2 Linear Equations 98
7.3 Quadratic Equations 107
7.4 Inequalities 113
7.5 Functions and graphs of functions 123

8 Practice Questions: Algebra 133

9 Answer Key: Algebra 147

10 Solution: Algebra Practice Questions 149

11 Word Problems 179
11.1 Percents 180
11.2 Profit, Loss, and Discount 187
11.3 Ratio, Proportion and Variation 193
11.4 Mixtures 199
11.5 Computation 207
11.6 Averages 210
11.7 Speed/Time/Distance 214
11.8 Time and Work 226
11.9 Simple & Compound Interest 232
11.10 Combinatorics 239
11.11 Permutation 239
11.12 Combination 239
11.13 What's the difference? 240
11.14 Permutation 241
11.14.1 Permutation with repetition of objects 243
11.14.2 Circular permutation 244
11.15 Combination 246
11.16 Concepts Revisited 248
11.17 Permutation of multiple indistinguishable objects 249
11.18 Restricted Permutation 249
11.19 Restricted Combination 250
11.20 Rule of Sum 251

11.21 Rule of Product 251
11.22 Probability 252
11.23 Chance vs. Probability 252
11.24 A few definitions 252
11.24.1 Experiment 252
11.24.2 Event 252
11.24.3 Equally likely events 253
11.24.4 Sample set 253
11.25 An experiment with two dice 254
11.26 A few more definitions 255
11.26.1 Mutually Exclusive Events 255
11.26.2 Independent Events 256
11.26.3 Exhaustive event 256
11.27 AND vs. OR rule 257
11.28 Conditional Probability 257
11.29 Independent and Dependent Events 258
11.30 Complement of an Event 259
11.31 Application of combination 260
11.32 Sets 262

12 Practice Questions: Word Problems 269

13 Answer Key: Word Problems 311

14 Solutions: Word Problems Practice Questions 315

15 Statistics 399
15.1 Arithmetic mean: 400
15.2 Median: 403
15.3 Range: 405
15.4 Standard Deviation: 405

16 Practice Questions: Statistics 407

17 Answer Key: Statistics 419

18 Solution: Statistics Practice Questions 421

19 Talk to Us 455

Chapter 1

Welcome

Dear Students,

Welcome to the exciting, resource-filled world of Manhattan Review—a place where we relentlessly pursue excellence in all we do, which includes providing you with unrivaled educational content for your standardized test preparation. Get ready to study for the GMAT with a renewed vigor because we understand exactly how critical this life-changing exam can be!

The GMAT is an ever-changing exam that has most recently morphed into the GMAT Focus but have no fear...we've already successfully navigated the transition! This book has been meticulously crafted to reflect the updated GMAT Focus edition, making your preparation journey smoother than ever and ensuring you will be prepared and confident on the day of your exam.Choosing the best test-prep guide from a sea of options can be incredibly daunting. Manhattan Review is here to support your preparation by ensuring you possess the most effective test-taking skills, setting yourself up to achieve your highest score while also helping you master important time management skills.

You might wonder: what sets our GMAT Quantitative Reasoning book apart from all the other test prep books? See for yourself how our proprietary, engaging approach makes us the preferred resource for GMAT preparation. This book teaches you how to answer Quantitative Reasoning questions using efficient and easy-to-remember techniques. With a blend of text and illustrative graphics, the concepts are explained so vividly you might actually enjoy studying for your GMAT! Get ready to explore a wide array of problems and discover alternative strategies to tackle those particularly tough questions. We've painstakingly analyzed the tricks and traps a GMAT test-maker might employ, which is how we can teach you to identify the right answers. You'll find a treasure trove of 300 questions in this information-packed manual.

Comprehensive and detailed, Manhattan Review's "Quantitative Reasoning" book will help you stay on track with your GMAT preparation while teaching you to break down even the most challenging questions and avoid common mistakes. Still have questions? Don't hesitate! We're just an email away at *info@manhattanreview.com.*

Happy Learning!

Professor Dr. Joern Meissner
& The Manhattan Review Team

Chapter 2

Introduction

The GMAT Focus Edition: A Snapshot

GMAC launched GMAT Focus

The Graduate Management Admissions Council (GMAC) has launched the GMAT Focus Edition, a new version of the GMAT test. The GMAT Focus Edition brings big and important changes to how the test is set up, what it covers, and even how it is scored. As these changes will have an impact on both your GMAT preparation and your testing experience, it is important to educate yourself on the details of the GMAT Focus Edition.

What is the GMAT Focus Edition?

This newer version of the GMAT?GMAT Focus Edition?is, according to GMAC, "a more focused test-taking experience that hones in on the higher-order critical reasoning and data literacy skills that are especially relevant and applicable in the business environment of tomorrow." With this test, there is a heightened focus on efficiency, flexibility, and providing test-takers with greater insights into their performance, allowing those who take the test to be as informed as possible about their abilities and educational options.

How is the GMAT Focus Edition different from the earlier GMAT Edition?

Let's examine the changes in format and content between the GMAT Focus Edition and the earlier GMAT.

Format Changes

With the upcoming Focus Edition, the GMAT's format will undergo significant changes, specifically in the areas of test time, section order choice, adaptability, question reviewing and editing, scoring, score reporting, and enhanced score reports.

Test Time

One of the most significant changes to the GMAT is the duration of the test. The GMAT Focus Edition is just 2 hours and 15 minutes long (excluding breaks). Thus, the GMAT Focus Edition is approximately 1 hour shorter than the earlier GMAT.

The Quantitative Reasoning section will only have Problem-Solving questions; Data Sufficiency is moved to the data Insights section. Moreover, only 21 Problem-Solving questions will be asked, as opposed to 31 questions in the earlier GMAT.

Question Review and Edit

One especially welcome feature of the GMAT Focus Edition is that you can go back and change your answers to questions in each section.

Unfortunately, this functionality is not unlimited, since the GMAT Focus Edition continues to be a question-adaptive test. For the scoring system to function, the majority of your responses in a section must stay unaffected. Because of this, you can change up to three responses in each section, which is still an improvement over the current GMAT, which doesn?t allow candidates to change any answers.

In addition, you may review your responses to as many questions as you want in each of the three sections; the bookmarking option allows you to mark as many questions as you like so that you can simply return to them later.

So, let's say you bookmarked six questions in a section because you were uncertain of your responses. Understanding that you can only change responses to 3 questions, you could, if time allows, revisit and review them, then decide which 3 responses need to be changed. It is important to note that this is time-dependent, and if you run out of time, you will not be able to go back and change any of your answers.

Changes in Content

The majority of the GMAT Focus Edition's content will be the same as what the earlier GMAT covered. Nonetheless, there are significant changes to be aware of in all sections of the test.

Quants

All topics that were covered in the earlier GMAT quant section are retained in the Focus Edition, with one important exception: Geometry. It will no longer be tested in the GMAT, and the GMAT Focus Edition will only focus on Algebra and Arithmetic.

This book on Quantitative Reasoning covers all major topics: Number Properties, Algebra, Word Problems, and Statistics. Each topic starts with detailed concepts and is followed by a number of GMAT-like questions. The book has 300 Problem-Solving questions. The book provides detailed solutions to each question. Many questions have alternate approaches, too. Whenever needed, the book provides text-cum-graphic solutions. In a nutshell, the Manhattan Review's "Quantitative Reasoning" book is holistic and comprehensive in all respects.

Chapter 3

Number Properties

3.1 Understanding numbers

Numbers—the backbone of GMAT Quant; we all are familiar with numbers, but the devil lies in the detail!

Let us understand numbers from the GMAT perspective.

3.1.1 Classification of numbers

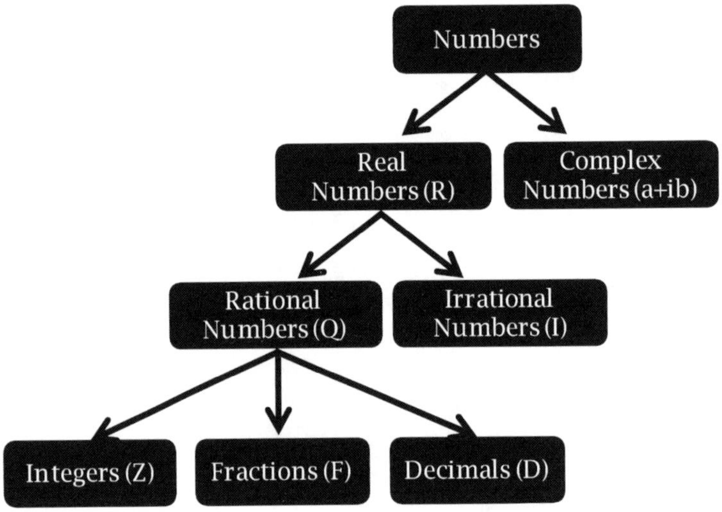

(A) **Real numbers:** Almost all the numbers you can imagine are Real numbers. Any number that can be represented on a number line is a real number. These include $0, 1, -3, 4/3, -0.3457, 48.87, \sqrt{3}, \pi$, and many more!

If most numbers are real numbers, then what are not real numbers?

The numbers such as $\sqrt{-1}, (2 + \sqrt{-5})$, etc. are not real numbers. They are Imaginary number and Complex numbers, respectively. Fortunately, these are beyond the scope of the GMAT.

Real numbers can further be classified into Rational and Irrational numbers.

(a) **Rational numbers:** Numbers that can be expressed in the form of $\frac{p}{q}$ where p and q are integers, and $q \neq 0$. Remember that p may be 0.

The above definition may look a little scary to those who are not very familiar with the concepts of numbers; however, the good news for them is that you need not know the above definition to master the GMAT Quant.

So what do rational numbers include? Simply put, these include all integers, decimal, and fractions.

So what do rational numbers exclude? These exclude roots of non-perfect square numbers, roots of non-perfect cube numbers, special numbers such as $\pi = 3.14, e = 2.718$, and many others, basically, non terminating & non recurring decimals.

Rational numbers can be broadly classified into three categories—Integers, Fractions, and Decimals.

i. **Integers:** All counting numbers are integers whether negative, positive or zero. **Example:** {1, 2, 3, 4, 5,...} are called positive integers; $\{-1, -2, -3, -4, -5, ...\}$ are called negative integers. Note that "0" is also an integer.

   You may come across terms, such as non-negative integers or non-positive integers. A non-negative integer would be one among the set of {0, 1, 2, 3, 4, 5,...}; note that "0" is included; whereas, a non-positive integer would be one among the set of $\{0, -1, -2, -3, -4, -5, ...\}$.

ii. **Fractions:** Any number that can be expressed with an integer numerator and an integer denominator is called fraction. In other words, a number that can be expressed in the form of x/y is a fraction such that $y \neq 0$. **Example:** $1/3, 5/4, -3/8$ etc.

iii. **Decimal numbers:** Decimal numbers are another way of expressing fractions. The decimal numbers are written with the use of a decimal ("."). The left of the decimal point (.) has place values of digits for units, tens, hundreds, thousands, and more; whereas, the right of the decimal point (.) has place values of digits for tenths, hundredths, thousandths, and more.

   Only those decimals which are terminating, for example $2.4, 4.35, -2.1$, etc. or recurring, for example 2.3333, $-4.151515..$, 2.3404040, etc. are rational.

   This is because only these can be expressed in the form p/q, as discussed earlier. For example: A terminating decimal 2.35 is simply $235/100 = 47/20$. A recurring decimal, say 2.343434... can be analyzed as follows:

   Say $x = 2.343434...$
   => $100x = 234.343434$
   Subtracting: $99x = 234 - 2 = 232$
   => $x = \frac{232}{99}$, and hence, is a rational number.

   We will discuss place value in the following pages.

(b) **Irrational numbers:** Numbers that cannot be expressed in the form of $\frac{p}{q}$, where p and q are integers, and $q \neq 0$ are irrational numbers or in other words, all real numbers that are not rational are irrational numbers. As stated earlier that these include roots of non-perfect squared numbers: 2, 5 and others; roots of non-perfect cube numbers, special numbers such as $\pi = 3.14, e = 2.718$, and many others.

By default any number mentioned in the quantitative section is a **real number**; so, you must NOT assume that it is an integer unless stated as such.

(B) **Complex numbers:** As stated earlier that he numbers such as $\sqrt{-1}$, $(2 + \sqrt{-5})$, etc. are not real numbers. They are Imaginary number and Complex numbers, respectively. Fortunately, these are beyond the scope of the GMAT.

3.1.2 Number line

A number line is a line with "0" as its center. Numbers on the right side of "0" are positive and those on the left side are negative. Number line helps define the direction of measurement.

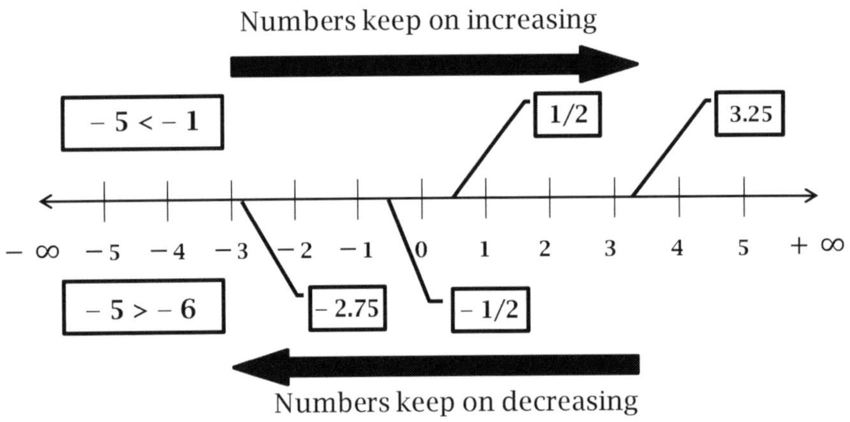

3.1.3 Basic operations on numbers

Symbol	Mathematical operation	Application	Examples
$+$	Addition	Sum, Add, Plus, Total, Increase	• $2 + 3 = 5$; (Two positives is added) • $(-2) + (-3) = -5$; (Two negatives is added) • $2 + (-3) = -1$; (A positive & a negative is subtracted) • $(-2) + 3 = 1$; (A negative and a positive is subtracted)
$-$	Subtraction	Deduct, Minus, Decrease, Less, Reduce, Difference	• $2 - 3 = -1$; (Two positives is subtracted) • $(-2) - (-3) = 1$; (Two negatives is subtracted) • $2 - (-3) = 5$; (A positive & a negative is added) • $(-2) - 3 = -5$; (A negative & a positive is added)
$\times$	Multiplication	Multiply, Product, Times, $\times$, By	• $2 \times 3 = 6$; (Positive $\times$ Positive = Positive) • $(-2) \times (-3) = 6$; (Negative $\times$ Negative = Positive) • $2 \times (-3) = -6$; (Positive $\times$ Negative = Negative) • $(-2) \times 3 = -6$; (Negative $\times$ Positive = Negative)
$\div$	Division	Divide, Quotient, Part of, Per unit calculation	• $6 \div 3 = 2$; (Positive $\div$ Positive = Positive) • $(-6) \div (-3) = 2$; (Negative $\div$ Negative = Positive) • $6 \div (-3) = -2$; (Positive $\div$ Negative = Negative) • $(-6) \div 3 = -2$; (Negative $\div$ Positive = Negative)

3.1.4 Face value and place value

The number system we use is set up from ten symbols only $\{0, 1, 2, 3, 4, 5, 6, 7, 8, 9\}$. Each symbol is called a digit and each has its value called **Face value.** So, the face value of "9" is nine times the face value of the digit "1" or three times the face value of the digit "3." For example, for numbers 243, & 234, the face values of digits "2," "3," and "4" do not vary upon their placement.

Place value of a digit varies depending on where it is placed in a number. For example, for numbers 234, and 243, the place values of digits "3," and "4" are different.

If a digit placed on the left of another digit, the place value of the first digit is 10 times of the other digit. For a number 234, the place value of digit "2" is 200, that of digit "3" is 30, and that of "4" is 4. Similarly, the place values of digits placed after the decimal point $(\cdot)$ reduces in its value by 1/10 times. This can be better understood this way.

- $243 = 2 \times 100 + 4 \times 10 + 3 \times 1$;
- $234 = 2 \times 100 + 3 \times 10 + 4 \times 1$.

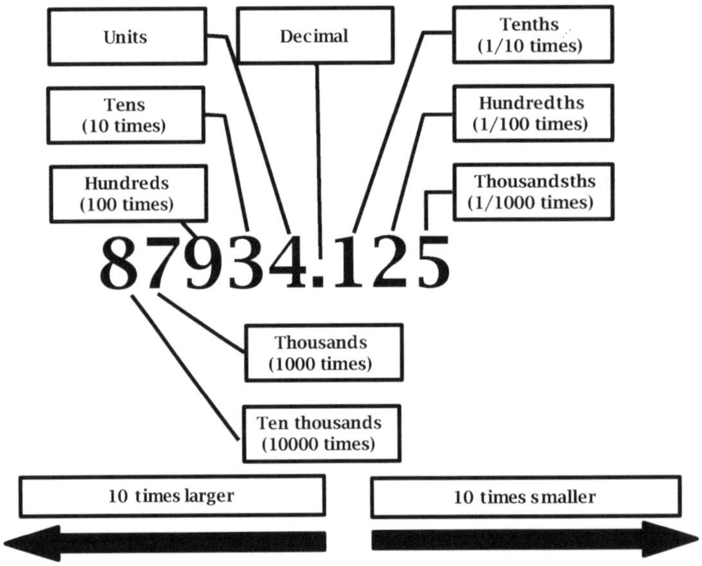

3.1.5 Even and Odd numbers

Even number: An integer completely divisible by "2," without leaving a remainder is called an **Even number.** In other words, an even number can be expressed as "$2m$," where "m" is an integer. As per this definition, "0" is an even integer. **Example:** $\{0, 2, 4, 6, 8, ...\}$, $\{-8, -4, 10, 18, ...\}$.

Odd number: An integer when divided by "2" leaves a remainder "1" is called **Odd number.** In other words, an odd number can be expressed as $(2m + 1)$, where "m" is an integer. All the integers are either even or odd integers. **Example:** $\{1, 3, 5, 7, ...\}$, $\{-3, -1, 9, 11...\}$.

Basic operations on Even and Odd numbers:

Mathematical operation	Application	Examples
Addition/ Subtraction	Even ± Even = Even; Odd ± Odd = Even; Even ± Odd = Odd Any number of even numbers can make an even number; however, only an even number of odd numbers can make an even number; e.g.: Two odd numbers on addition give even, four odd numbers on addition also give even; however, three odd numbers on addition give odd number.	• 4 + 2 = 6; 4 − 2 = 2; Even • 7 + 3 = 10; 3 − 7 = − 4; Even • 5 + 2 = 7; 5 − 2 = 3; Odd
Multiplication	Even × Even = Even; Odd × Odd = Odd; Even × Odd = Even Multiplication of any number of odd numbers always results in another odd number. If the product of two integers is Even and one of them is an Even integer, we cannot comment on the nature of the other number, i.e. whether it is Even or Odd.	• 4 × 2 = 8; Even • 3 × 3 = 9; Odd • 2 × 3 = 6; Even
Division	Not every operation of Division is completely divisible. We show you the results if the division leaves no remainder. Odd ÷ Odd = Odd; Even ÷ Odd = Even; Odd ÷ Even = Not completely divisible; Even ÷ Even = Even or Odd	• 9 ÷ 3 = 3; Odd • 10 ÷ 5 = 2; Even • 11 ÷ 2 = ? Not divisible • 4 ÷2 = 2; Even • 10 ÷ 2 = 5; Odd

3.1.6 Consecutive numbers

An ordered set of continuous integers written in ascending order are called **Consecutive numbers.**

Example: $\{23, 24, 25\}, \{-2, -1, 0, 1, 2, 3\}$

Example of consecutive Even numbers: $\{12, 14, 16\}$

Example of consecutive Odd numbers: $\{11, 13, 15\}$

3.1.7 Prime numbers

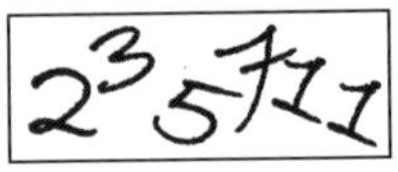

A number greater than "1" which has no factors other than "1" and the number itself, is called **Prime number**. A number with exactly two distinct factors is called prime. Hence, "1" cannot be considered as prime since it has only one factor, "1" itself.

Example: $\{2, 3, 5, 7, 11, 13, 17, 19, 23, 29, ...\}$. There are infinitely many Prime numbers.

First 25 prime numbers are:

$\{2, 3, 5, 7, 11, 13, 17, 19, 23, 29, 31, 37, 41, 43, 47, 53, 59, 61, 67, 71, 73, 79, 83, 89, 97\}$

"2" is the only even prime number or all the prime numbers are odd except "2."

Also, the difference between any two primes greater than "2" is always even.

Another point to note is that any prime, other than "2" or "3," if divided by "6," will always leave a remainder of "1" or "5" (or -1). Thus, any prime can be expressed as "$6k \pm 1$," where k is some positive positive integer.

However, you should note that not all numbers which leave a remainder of 1 or 5 when divided by 6 are prime, for example, 25, 35, etc.

How to verify whether a number "n" is a prime number:

Step1: Take square root of the number; if square root is an integer the number is not prime number, else follow the next step.

Step2: Say the largest integer closest to the square root of the number "n" is "m"; where $m < \sqrt{n}$;

Step3: Divide "n" by all the prime numbers from "2 to m," inclusive: if n is divisible by any number among them, the number "n" is not prime, else prime.

=> Let us take a number "$n = 101$."

Step1: $10 < \sqrt{101} < 11$; not an integer;

Step2: $\sqrt{101} =\approx 10 = m$;

Step3: Divide 101 by prime numbers from 2 to 10 i.e. 2, 3, 5, & 7. Since 101 is not divisible by any prime number among these, hence, 101 is a prime number.

=> Let us take a number "$n = 87$."

Step1: $9 < \sqrt{87}, 10$; not an integer;

Step2: $\sqrt{87} =\approx 9 = m$;

Step3: Divide 87 by prime numbers from 2 to 9 i.e. 2, 3, 5, & 7. Since 87 is divisible by 3, hence, 87 is not a prime number

3.1.8 Co-Prime numbers

Two numbers are co-prime to each other if they have no factor common between them except '1.'

Example: 14 & 15: The factors of "14" are {1, 2, 7, 14}, and the factors of "15" are {1, 3, 5, 15}. There is no factor common between "14 & 15" except '1,' so "14 & 15" are co-prime to each other.

It is important to know that the LCM of co-prime numbers is their product, and GCD or HCF is "1." Say two numbers "a" and "b" are co-prime to each other, then **LCM** = $a \times b$, and **GCD = 1**.

Note that any two consecutive numbers are always co-prime to each other. Also 1 is co-prime to any number.

3.1.9 Rounding off

Rounding off a number means to get a significant and practically usable number. Say, the distance between Manhattan to JFK airport is 15.24 miles; we can truncate the hundredth place digit to get a number which is rounded off to tenth place equalling 15.2 miles. However, by doing this, we do get a less accurate number. The degree of rounding off a number depends on how much accurate number you want. In this case, even the additional 0.2 miles may seem insignificant to a regular commuter. So, 15.2 can further be rounded off to the nearest tens as 15 miles which is further less accurate.

CASE I:

If the dropped digit is less than '5' (i.e. 0, 1, 2, 3, 4), the digit to the left of the dropped digit should be unchanged. For example, 5.312 rounded to the tenths digit is 5.3.

CASE II:

If the dropped digit is greater than '5' (i.e. 6, 7, 8 or 9), the digit to the left of the dropped digit should be increased by '1.' For example, 5.326 rounded to the hundredths digit is 5.33, but, rounded to the tenths digit is 5.3; similarly, 4.961 becomes 5.00, if rounded to the nearest integer.

CASE III:

If the dropped digit is '5,' and all the figures following the '5' are '0,' or if there is no digit after the '5':

- The digit to the left of the dropped digit should be unchanged, if that last figure is even, and
- The digit to the left of the dropped digit should be increased by 1 if that last figure is odd. For example, 3.195 becomes 3.20, whereas, 2.365 becomes 2.36, both when rounded to the hundredths digit.

CASE IV:

If the dropped digit is '5,' and all the figures following the '5' are not all '0,' the digit to the left of the dropped digit should be increased by '1.' For example, 4.7501 rounded to the hundredths digit is 4.8.

The same rules for rounding can be applied in the following cases as well:

- 2,134 rounded to the nearest tens is 2,130
- 2,139 rounded to the nearest tens is 2,140
- 2,135 rounded to the nearest tens is 2,140
- 2,145 rounded to the nearest tens is 2,140
- 2,145.1 rounded to the nearest tens is 2,150
- 2,145 rounded to the nearest hundreds is 2,100
- 2,155 rounded to the nearest hundreds is 2,200

- 2,168 rounded to the nearest hundreds is 2,200
- 2,134 rounded to the nearest hundreds is 2,100

Note that the number, rounded off, can be more than or less than the original number. 15.8454 rounded off to $15.8 < 15.8454$, whereas, another number 15.8554 rounded off to $15.9 > 15.8554$.

3.1.10 Order of operations—PEMDAS

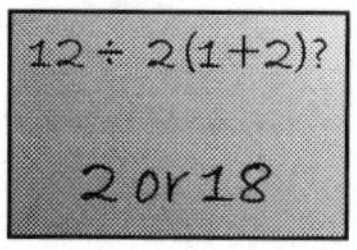

What is the value of $2\{32 - 2(6 \times 5^2 - 50 + 25 \div 5)\}$?

Follow the order of operation given as below:

P: Parentheses first

E: Exponents (i.e., Powers and Square Roots, etc.)

MD: Multiplication and Division (left-to-right)

AS: Addition and Subtraction (left-to-right)

Expression	**Comments**
$2\{10^2 - 2(6 \times 5^2 - 50 + 25 \div 5)\}$	Solve the parenthesis first. If there are more than two parentheses, follow the order: [ ], { }, & ( ). Here, we will solve { } first, and then ( ). **(P)**
$= 2\{\mathbf{100} - 2(6 \times \mathbf{25} - 50 \div 25)\}$	Among the operators exponent, multiplication, division, addition, subtraction, solve the exponent first. **(E)**
$= 2\{100 - 2(\mathbf{150} - 50 + 5)\}$	Multiplication and division can be done simultaneously. **(MD)**
$= 2\{100 - 2\mathbf{(105)}\}$ $= 2\{100 - \mathbf{210}\}$ $= 2\{-110\}$ $= -220$	Addition and subtraction can be done simultaneously. **(AS)**

3.1.11 Divisibility

A number is said to be divisible by another number if the division does not leave any remainder. Following are quick reckoner rules for divisibility test.

Divisor	Condition: A number is divisible	Examples
2	if the unit digit is Even: {0, 2, 4, 6, 8}	26**8**; 126,59**6**
3	if the Sum of all the digits is divisible by "3"	21,681; Sum of digits = 2 + 1 + 6 + 8 + 1 = 18; 18 div. by 3 implies that 21,681 is divisible by 3. 89,896; Sum of digits = 8 + 9 + 8 + 9 + 6 = 40; 40 not div. by 3 implies that 89,896 is not divisible by 3.
4	if the number formed out of last two digits is divisible by "4"	1,22**4** and 2,3**56** are div. by 4 as 24 and 56 are divisible by 4. 1,2**54** and 2,569,8**66** are not div. by 4 as 54 and 66 are not divisible by 4.
5	if the unit digit is "0" or "5."	1,22**0** and 2,35**5** are div. by 5. 1,25**4** and 2,569,86**6** are not div. by 5.
6	if the number is divisible by both "2" and "3"; follow the divisibility rules of "2" and "3."	1,254, and 89,892 are div. by 6.
8	if the number formed out of the last three digits is divisible by "8"	891,**224** and 2,542,**352** are div. by 8. 561,**354** and 2,569,**866** are not div. by 8.
9	if the Sum of all digits of the number is divisible by "9"	21,681; Sum of digits = 2 + 1 + 6 + 8 + 1 = 18; 18 div. by 9 implies that 21,681 is divisible by 9. 89,896; Sum of digits = 8 + 9 + 8 + 9 + 6 = 40; 40 not div. by 9 implies that 89,896 is not divisible by 9.

10	if the unit digit is "0."	12,340 and 5,245,870.
11	if the difference of the sum of digits at odd places, and that at even places is either "0" or divisible by "11."	251,384; S.O. = Sum of digits at odd places = $4 + 3 + 5 = 11$; S.E. = Sum of digits at even places = $8 + 1 + 2 = 11$; S.O. − S.E. = 11 − 11 = **0** implies that 251384 is div. by 11. 21,927,092; S.O. − S.E. = 5 − 27 = **- 22** div. by 11 implies that 21,927,092 is div. by 11.

3.1.12 Factors

All the possible divisors of an integer which leave no reminder are called factors.
Example: The factors of '6' are 1, 2, 3, & 6. We can write 6 as 1×6 or 2×3.

Factors of 60:

$$\begin{aligned} 60 &= 1 \times 60; \\ &= 2 \times 30; \\ &= 3 \times 20; \\ &= 4 \times 15; \\ &= 5 \times 12; \\ &= 6 \times 10. \end{aligned}$$

So, the factors of '60' are $\{1, 2, 3, 4, 5, 6, 10, 12, 15, 20, 30, 60\}$. Remember that '1' and the number itself are also the factors of a number.

Number of factors

Say, a number N can be written as $\boldsymbol{N = a^x b^y c^z}$, where a, b, and c are prime factors of the number N, and x, y, and z are positive integers, then

=> **Number of factors** (including '1' and the number itself) = $(x + 1)(y + 1)(z + 1)$;

=> **Number of factors** (excluding '1' and the number itself) = $(x + 1)(y + 1)(z + 1) - 2$.

Number of factors of 24:

$24 = 2 \times 2 \times 2 \times 3 = 2^3 \times 3^1$; here $a = 2$ & $b = 3$ are prime numbers, and $x = 2$, & $y = 1$.

Hence, the number of factors of 24 including '1' and the number itself
$= (3 + 1)(1 + 1) = 4 \times 2 = 8$.

Let us verify this.

Factors of 24 are: $\{1, 2, 3, 4, 6, 8, 12, 24\}$; total number of factors equals to 8.

Let us take one more example.

Number of factors of 2,520:

$2,520 = 2 \times 2 \times 2 \times 3 \times 3 \times 5 \times 7 = 2^3 \times 3^2 \times 5 \times 7$; here $2, 3, 5,$ & 7 are prime numbers.

Hence, the number of factors of 2,520 including '1' and the number itself
$= (3 + 1) \times (2 + 1) \times (1 + 1) \times (1 + 1) = 4 \times 3 \times 2 \times 2 = 48$.

Sum of all the factors:

Say, a number N can be written as $N = a^x b^y c^z$, where a, b, and c are prime factors of number N, x, y, and z are positive integers, then

Sum of all the factors (including 1 and the number itself) = $\left[\frac{a^{x+1} - 1}{a - 1}\right]\left[\frac{b^{y+1} - 1}{b - 1}\right]\left[\frac{c^{z+1} - 1}{c - 1}\right]$

Sum of all the factors of 60:

$60 = 2^2 \times 3^1 \times 5^1$;

Sum of all the factors (including 1 and the number itself) = $\left[\frac{2^{2+1} - 1}{2 - 1}\right]\left[\frac{3^{1+1} - 1}{3 - 1}\right]\left[\frac{5^{1+1} - 1}{5 - 1}\right]$

$= 7 \times (8/2) \times (24/4) = 168$.

Let us cross check this. Sum of all the factors of 60 $= 1 + 2 + 3 + 4 + 5 + 6 + 10 + 12 + 15 + 20 + 30 + 60 = 168$.

3.1.13 Perfect Square numbers

A perfect square number is a non-negative integer that can be expressed as the product of an integer with itself. In other words, a number which is a square of any integer is called Perfect square number. Perfect square numbers are always non-negative integers.

Example: $\{0, 1, 4, 9, 16, 25, 36, 49, ...\}$. Some mathematicians do not consider that "0" is a perfect square number. Moreover, the GMAT does not assume that you know the meaning of "perfect square number." The phrase used typically is "square of an integer."

We can see that $1 = 1 \times 1 = -1 \times -1$; and $4 = 2 \times 2 = -2 \times -2$.

Some properties of perfect square:

- Perfect square number ALWAYS has even number of powers of prime factors.
- The number of distinct factors of a perfect square number is ALWAYS Odd. (By factors, we refer to only the 'positive' numbers that divide the number without remainder.)
- The sum of distinct factors of a perfect square number is ALWAYS Odd.
- A perfect square number ALWAYS has an Odd number of Odd-factors, and Even number of Even-factors.

3.1.14 Multiples

While the factors divide the number completely, multiples are completely divisible by the number itself.

Example: The factors of '6' are 1, 2, 3 & 6; and the multiples of '6' are 6, 12, 18, 24, 30,...

=> A number is both a factor and a multiple of itself.

=> There are a finite number of factors of a number, but there are an infinite number of multiples.

In other words, we can write that the **multiples of a number $N = k \times N$,** where k is a positive integer. So the multiples of a number, say, xy^2 would be xy^2, $2xy^2$, $3xy^2$, $4xy^2$,... and the factors of xy^2 would be $\{1, x, xy, y^2, xy^2\}$.

3.1.15 Lowest Common Multiple—LCM

We all know what the LCM is; but occasionally we may get confused with HCF.

The LCM has three properties:

(1) It is a **multiple;** for numbers 6 & 10, the multiples of 6 are $\{6, 12, 18, 24, 30, 36, 42, 48, 54, 60, ...\}$, and the multiples of 10 are $\{10, 20, 30, 40, 50, 60, 70, 80, 90, ...\}$, so LCM is one among these multiples.

(2) It is a **common multiple** among all the multiples of 6, & 10; for numbers 6 & 10, the common multiples are $\{30, 60, 90, ...\}$, so it is one among these multiples.

(3) It is the **least common multiple (LCM)** among all the common multiples; for numbers 6 & 10, the least common multiple is **30: LCM.**

So, in other words, LCM of numbers is the smallest possible number which gets divided by each number without leaving any reminder.

Did you notice that in math, we never come across a term—Highest Common Multiple? It is because it is insignificant to discuss this as the Highest Common Multiple for any set of numbers would always be infinite; multiples of numbers are never ending.

Computation of LCM by factor method:

Example: Computation of LCM of 24, 54, and 70:

Step1: Factorise the numbers into prime factors

=> $24 = \mathbf{2^3} \times 3$;
=> $54 = 2 \times \mathbf{3^3}$;
=> $70 = 2 \times \mathbf{5 \times 7}$

Step2: Select the term with the highest exponent for each prime factor

=> The prime factors with their highest exponents are $2^3, 3^3, 5,$ & 7.

Step3: Multiply the numbers selected in step 2; it is the LCM of numbers.

=> **LCM** = $2^3 \times 3^3 \times 5 \times 7 =$ **7,560.**

Computation of LCM of algebraic expressions:

Example: Computation of LCM of $10xy^2z^5, 15x^2y^3z^4$, & $24x^3y^3z$:

=> $10xy^2z^5 = 2 \times \mathbf{5} \times x \times y^2 \times \mathbf{z^5}$;
=> $15x^2y^3z^4 = \mathbf{3} \times 5 \times x^2 \times \mathbf{y^3} \times x^4$;
=> $24x^3y^3z = \mathbf{2^3} \times 3 \times \mathbf{x^3} \times y^3 \times z$

LCM of $10xy^2z^5, 15x^2y^3z^4$, & $24x^3y^3z = 2^3 \times 3 \times 5 \times x^3 \times y^3 \times z^5 = \mathbf{120x^3y^3z^5}$

3.1.16 Highest Common Factor—HCF OR Greatest Common Divisor—GCD

Like LCM, the HCF/GCD also has three properties:

(1) It is a **factor;** for numbers 18 & 24, the factors of 6 are $\{1, 2, 3, 6, 9, 18\}$, and the factors of 24 are $\{1, 2, 3, 4, 6, 8, 12, 24\}$, so HCF is one among these factors.

(2) It is a **common factor** among these factors; for numbers 18 & 24, the common factors are $\{1, 2, 3, 6\}$, so it is one among these factors.

(3) It is the **highest common factor (HCF)** among all the factors; for numbers 18 & 24, the highest common factor is **6: HCF**.

So, in other words, HCF of numbers is the largest possible number which divides each number without leaving any reminder.

Did you notice that in math, we never come across a term—Lowest Common Factor? It is because it is insignificant to discuss this as the Lowest Common Factor for any set of numbers would always be '1.'

Computation of HCF by factor method:

Example: Computation of HCF of 24, 54, and 210.

Step1: Factorize the numbers into prime factors

=> $24 = 2^3 \times 3$;
=> $54 = 2 \times 3^3$;
=> $210 = 2 \times 3 \times 5 \times 7$.

Step2: Select the term with the lowest common exponent for each **common** prime factor

=> All the common prime factors with their lowest exponents are 2, & 3.

Step3: Multiply the numbers selected in step 2. It is the HCF/GCD of numbers.

=> HCF = $2 \times 3 = \mathbf{6}$.

Computation of HCF for algebraic expressions:

Example: Computation of HCF of $10xy^2z^5$, $15x^2y^3z^4$, & $24x^3y^3z$:

=> $10xy^2z^5 = 2 \times 5 \times \boldsymbol{x} \times \boldsymbol{y^2} \times z^5 = 2 \times 5 \times (\boldsymbol{x}) \times (\boldsymbol{y^2}) \times (\boldsymbol{z}) \times z^4$;
=> $15x^2y^3z^4 = 3 \times 5 \times x^2 \times y^3 \times z^4 = 3 \times 5 \times (\boldsymbol{x}) \times x \times (\boldsymbol{y^2}) \times y \times (\boldsymbol{z}) \times z^3$;
=> $24x^3y^3z = 2^3 \times 3 \times x^3 \times y^3 \times \boldsymbol{z} = 2^3 \times 3 \times (\boldsymbol{x}) \times x^2 \times (\boldsymbol{y^2}) \times y \times (\boldsymbol{z})$

HCF of $10xy^2z^5$, $15x^2y^3z^4$, & $24x^3y^3z = \boldsymbol{xy^2z}$.

LCM and HCF revisited:

Numbers: N_1, N_2, N_3	**LCM of $N_1, N_2, N_3 = 120x^3y^3z^5$** LCM is completely divisible by the numbers	**HCF of $N_1, N_2, N_3 = xy^2z$** Numbers are completely divisible by the HCF
$N_1 = 10xy^2z^5$	$\frac{LCM}{N_1} = \frac{120x^3y^3z^5}{10xy^2z^5} = 12x^2y$	$\frac{N_1}{HCF} = \frac{10xy^2z^5}{xy^2z} = 10z^4$
$N_2 = 15x^2y^3z^4$	$\frac{LCM}{N_2} = \frac{120x^3y^3z^5}{15x^2y^3z^4} = 8xz$	$\frac{N_2}{HCF} = \frac{15x^2y^3z^4}{xy^2z} = 15xyz^3$
$N_3 = 24x^3y^3z$	$\frac{LCM}{N_3} = \frac{120x^3y^3z^5}{24x^3y^3z} = 5z^4$	$\frac{N_3}{HCF} = \frac{24x^3y^3z}{xy^2z} = 24x^2y^2$

3.2 Exponents

Exponents, also called indices, are powers given to bases. Exponents tell the number of times a number is multiplied by itself. **Example:** $2^4 = 2 \times 2 \times 2 \times 2 = 16$. Here '2' is the base and '4' is the exponent or index.

Here are some basic operations of indices:

Operations	Examples
$a^m \times a^n = a^{m+n}$	$2^3 \times 2^5 = 2^{3+5} = 2^8$
$(ab)^m = a^m \times b^m$	$(2 \times 3)^4 = 2^4 \times 3^4$
$a^m \div a^n = a^{m-n}$	$2^3 \div 2^5 = 2^{3-5} = 2^{-2}$
$(a^m)^n = a^{mn}$	$(2^2)^3 = 2^{2.3} = 2^6$
$\frac{1}{a^m} = a^{-m}$	$\frac{1}{2^3} = 2^{-3}$
$\sqrt[n]{a^m} = (a^m)^{1/n} = a^{m/n}$	$\sqrt[3]{2^6} = (2^6)^{1/3} = 2^{6/3} = 2^2 = 4$
$a^0 = 1$, if $a \neq 0$	$2^0 = 1$
$(-a)^{\text{Even}} = +(a)^{\text{Even}}$; a = positive number	$(-2)^4 = -2 \times -2 \times -2 \times -2 = 16$
$(-a)^{\text{Odd}} = -(a)^{\text{Odd}}$; a = positive number	$(-2)^3 = -2 \times -2 \times -2 = -8$ or $-(2)^3 = -(2 \times 2 \times 2) = -8$

Note that $(a^m)^n \neq a^{m^n}$

- $(2^2)^3 = 2^{2.3} = 2^6 = 64$;
- $2^{3^2} = 2^9 = 512 \neq 64$

3.2.1 Unit digits of numbers with exponents

You may come across a question in which you are asked to deduce the unit digit of a number with a positive integer exponent.

Example: What is the unit digit of 2^4?

We know that $2^4 = 16$, so the unit digit is 6, however, if the question asks: What is the unit digit of 2^{17}?, then it would be a time-consuming exercise to calculate the value of 2^{17}.

Let us see the power cycle of '2.' It is $2, 4, 8, 16, 32, 64, 128, 256, 512, 1024, 2048, \ldots$ You may observe that the unit digit is the same for every 5th indexed number in the series. For example, the unit digit is '2' for 1st, 5th, 9th, 13th numbers in the series; while the unit digit is '8' for 3rd, 7th, 11th numbers in the series. So, for Base, '2,' the unit digit repeats after every 4th position, i.e. there exists a cycle of '4' positions, so the power cycle of base '2' is '4.'

Table showing the power cycles of the 10 digits:

Base	Series	Digits of cycle	Cycle	Unit digit
0	0, 0, 0, 0, ...	0	1	0 for any exponent
1	1, 1, 1, 1, ...	1	1	1 for any exponent
2	**2**, 4, 8, 16, 3**2**, 64, 128, 256, 51**2**, ...	2, 4, 8, 6	4	repeats after every 4th number
3	**3**, 9, 27, 81, 24**3**, 729, ...	3, 9, 7, 1	4	repeats after every 4th number
4	**4**, 16, 6**4**, 216....	4, 6	2	repeats after every 2nd number
5	**5**, 2**5**, 12**5**, 52**5**, ...	5	1	5 for any exponent
6	**6**, 3**6**, 21**6**, 1,29**6**, ...	6	6	6 for any exponent
7	**7**, 49, 343, 2,401, 16,80**7**, ...	7, 9, 3, 1	4	repeats after every 4th number
8	**8**, 64, 512, 4,096, 32,76**8**, ...	8, 4, 2, 6	4	repeats after every 4th number
9	**9**, 81, 72**9**, ...	9, 1	4	repeats after every 2nd number

Summary:

(1) The unit digits for the bases 0, 1, 5, & 6 are the bases itself.

(2) The unit digits for the bases 2, 3, 7, & 8 follow the cycle of '4.'

(3) The unit digits for the bases 4, & 9 follow the cycle of '2.'

3.2.2 Deduction of unit digit

Example: What is the unit digit of 7^{34}?

Step1: Identify the base, its power cycle, and the exponent.
=> We know that the power cycle of the base, 7, is 4 (7, 9, 3, 1), and the exponent is 34.

Step2: Divide the exponent by the power cycle, and get the remainder.
=> The remainder of 34/4 is 2.

Step3: Identify the digit of the power cycle corresponding to the remainder.
=> We know that the remainder is 2, and the power cycle of 7 is (7, 9, 3, 1), hence, the 2nd digit of the power cycle would be the unit digit of 7^{34}, which is 9: (7, **9**, 3, 1).

What if the questions had been: What is the unit digit of 7^{36}?

Again, we know that the power cycle of the base, 7, is 4 (7, 9, 3, 1), and the exponent is 36 and dividing 36 by 4 gives a remainder "0." If the remainder is "0," we must take the last number in the power cycles which is "1." So the unit digit of 7^{36} is 1.

Example: What is the unit digit of $7^2 \times 9^3 \times 3^2$?

Step1: Deduce the unit digit of each number with exponent
=> The unit digit of 7^2 is "9"; similarly, the unit digit of 9^3 is "9"; similarly, the unit digit of 3^2 is "9."

Step2: Multiply all the unit digits deduced
=> The unit digit of $7^2 \times 9^3 \times 3^2$ => $9 \times 9 \times 9 = 9^3$ is "9."

Note: For solving questions asking for the unit digit, keep in mind that '5' multiplied with any odd digit always results in a number ending with '5,' while '5' multiplied with any even digit always results in a number ending with '0.'

3.2.3 Roots and Surds

Root is a familiar word; you have heard this before—square root, cube root etc.

Square root:

It is opposite of square. While square of a number means multiplying a number with itself to get a new number, "Square root" means to get a number such that multiplying it with itself would give the original number.

Example: Say a number is $x = 4$, then the 'square = $x^2 = 4^2 = 16$,' and 'square root = $\sqrt[2]{x} = \sqrt[2]{4} = 2; 2 \times 2 = 4$.' Some consider that $\sqrt[2]{4} = 2$ or -2. However, it is not so. Its value is +2 only.

The confusion starts from here:

If $x^2 = 16$, what is x?
While taking the square root of both the sides, we get $\sqrt[2]{x^2} = \sqrt[2]{16} \Rightarrow \pm x = 4 \Rightarrow x = \pm 4$. Note that $\sqrt[2]{x} = 2$, not ± 2.

Remember that **square** of a number is NOT always greater than its **square root.** If a number is $x = 4$, then the '$x^2 = 16$,' and $\sqrt{x} = 2; 16 > 2$; however, if a number is $x = 1/4$, then the '$x^2 = 1/16$,' and '$\sqrt{x} = 1/2$'; $1/16 < 1/2$.

Square of a negative number is possible, but the **square root** is not possible, it would be an imaginary number. If a number is $x = -4$, then '$x^2 = 16$,' and '$\sqrt{x} = \sqrt[2]{-2}$'; an imaginary number.

Cube root:

Like square root, which is opposite of square, cube root is opposite of cube. Cube of a number means multiplying a number with itself thrice to get a new number, "Cube root" means to get a number such that multiplying it with itself thrice would give the original number.

Example: Say a number is $x = 8$, then the 'cube $= x^3 = 8^3 = 512$,' and 'cube root $= \sqrt[3]{x} = \sqrt[3]{8} = 2$'; $2 \times 2 \times 2 = 8$.

Like square, and its square root, **cube** of a number is NOT always greater than its **cube root.** If a number is $x = 8$, then the '$x^3 = 512$,' and '$\sqrt[3]{x} = 2$'; $512 > 2$; however, if a number is $x = 1/8$, then the '$x^3 = 1/512$,' and '$\sqrt[3]{x} = 1/2$'; $1/512 > 1/2$.

Cube root gives unique value of x:

We have seen that if $x^2 = 4, \Rightarrow x = \pm 2$.; no unique value of x; it may be either 2 or -2. However if $x^3 = 8, \Rightarrow x = 2$; a unique value of x; because if $x = -2, x^3 = -8$. So cube root gives the unique value.

n^{th} root:

Like square root, and cube root, n^{th} root means multiplying a number with itself n times to get the original number. Say a number is x, then its nth root would be $\sqrt[n]{x}$.

Example: Say a number is $x = 32$, $\sqrt[5]{32} = 2; 2 \times 2 \times 2 \times 2 \times 2 = 32$.

3.2.3.1 Surds

If the root of a number renders an irrational number, its exponent is called **Surd**.

The square root of 2, $\sqrt[2]{2} = 1.414$; an irrational number, hence, a surd.

You must know a few values: $\mathbf{\sqrt{2} = 1.414, \sqrt{3} = 1.732, \sqrt{5} = 2.236}$.

3.2.3.2 Rationalization of surds

Keeping a surd in the denominator is not a standard way of writing a number or expression in mathematics. Surds should be placed in the numerator. If a number is $\frac{3}{\sqrt{2}}$, we can get rid of $\sqrt{2}$ placed in the denominator by multiplying and dividing the fraction by $\sqrt{2}$.

$$=> \frac{3}{\sqrt{2}} = \frac{3}{\sqrt{2}} \times \left(\frac{\sqrt{2}}{\sqrt{2}}\right) = \frac{3\sqrt{2}}{(\sqrt{2})^2} = \frac{3\sqrt{2}}{2}.$$

Example 1: Rationalize $\frac{\sqrt{2}+1}{\sqrt{2}-1}$.

We cannot multiply the numerator and the denominator by $(\sqrt{2}-1)$ to remove the surd in the denominator; the denominator will result into $(\sqrt{2}-1)^2 = (\sqrt{2})^2 - 2.\sqrt{2}.1 + 1^2 = 2 - 2\sqrt{2} + 1 = 3 - 2\sqrt{2}$; again we would have a surd.

The optimum approach to do this is to make use of the formula $(a-b)(a+b) = a^2 - b^2$. Assume $a = \sqrt{2}$, and $b = 1$. By multiplying $\frac{\sqrt{2}+1}{\sqrt{2}+1}$ to the number, we will have a pair of $(\sqrt{2}-1)(\sqrt{2}+1)$ in the denominator, which will give $[(\sqrt{2})^2 - 1^2] = 2 - 1 = 1$; thus the surd is removed.

The term $(\sqrt{2}+1)$ is called the conjugate of $(\sqrt{2}-1)$.

$$=> \frac{\sqrt{2}+1}{\sqrt{2}-1} \times \left(\frac{\sqrt{2}+1}{\sqrt{2}+1}\right) = \frac{(\sqrt{2}+1)^2}{(\sqrt{2})^2 - 1^2} = \frac{(\sqrt{2}+1)^2}{1} = (\sqrt{2}+1)^2.$$

Example 2: Rationalize $\frac{\sqrt{x}-\sqrt{y}}{\sqrt{x}+\sqrt{y}}$.

$=> \frac{\sqrt{x}-\sqrt{y}}{\sqrt{x}+\sqrt{y}} \times \left(\frac{\sqrt{x}-\sqrt{y}}{\sqrt{x}-\sqrt{y}}\right) = \frac{(\sqrt{x}-\sqrt{y})^2}{(\sqrt{x})^2 + (\sqrt{y})^2} = \frac{(\sqrt{x}-\sqrt{y})^2}{x-y}$; $(\sqrt{x}-\sqrt{y})$ is the conjugate of $(\sqrt{x}+\sqrt{y})$.

Key: If the expression in the denominator has '+' sign, then multiply and divide the number with its conjugate, i.e. an expression with '–' sign and vice-versa. $(x-y)$ and $(x+y)$ are conjugates of each other, provided x is a rational number and y is a surd.

3.3 Multiplication and Division of decimals with 10^x

We are often faced with multiplications or division of decimal numbers with exponents such as 20.0987×10^3 or $\frac{5482.75}{10^{-4}}$.
Follow the following approach to do such operations in less time.

Scenario 1: Multiplication

Here, we either multiply a number by a number having a positive exponent of 10 or divide it with a negative exponent of 10.

Step 1: Move the decimal to its right for the same number of times as the exponent value

Step 2: If number of digits to the right of the decimal is less than the value of the exponent, add the requisite number of '0s' after the last digit of the number

Example 1: 20.0987×10^3?

The value of the exponent is 3. By moving the decimal to the right 3 times in the number 20.0987, we will get the number 20,098.7.

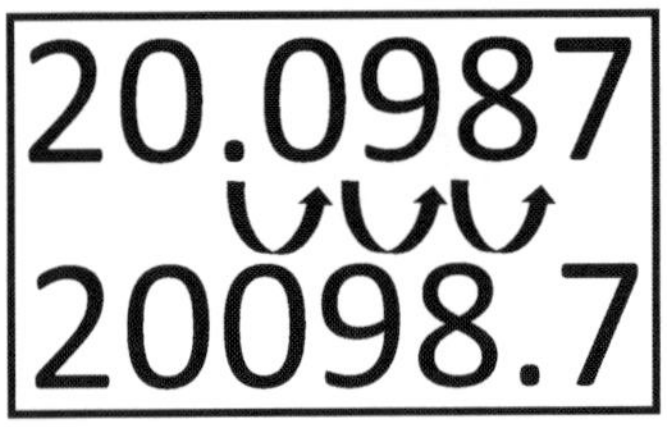

Example 2: 20.09×10^4?

As per the process described in the example above, we move the decimal to the right 4 times in the number 20.09. Since the number has only two digits after the decimal, we will add two 0s after '9.' Thus, $20.09 \times 10^4 = 200,900$.

Example 3: $20.0987 \div 10^{-3}$?

The value of the exponent is -3. Let us first make it positive. We can bring 10^{-3} in the numerator, thus, $20.0987 \div 10^{-3} = 20.0987 \times 10^3$. Now the situation is the same as was in the example 1. By moving the decimal to the right 3 times in the number 20.0987, we will get the number 20,098.7.

Scenario 2: Division

Here, we either divide a number by a number having a positive exponent of 10 or multiply with a negative exponent of 10.

Step 1: Move decimal to its left for the same number of times as the exponent value

Step 2: If number of digits to the left of the decimal is less than the value of the exponent, add requisite number of '0s' before the first digit of the number

Example 4: 20.0987×10^{-3}?

The value of the exponent is -3. By moving decimal to the left 3 times in the number 20.0987, we will get number the 0.0200987.

Example 5: $20.0987 \div 10^{-3}$?

First, convert division into multiplication. So, $20.0987 \div 10^{-3} = \frac{20.0987}{10^{-3}} = 20.0987 \times 10^3$. Now the question fits into scenario 1. Here, the value of the exponent is 3. By moving decimal to the right 3 times in the number 20.0987, we will get the number 20,098.7.

In a nut shell, following numbers are equal; $20.0987 \times 10^3 = 2.00987 \times 10^4 = 0.200987 \times 10^5 = 200.987 \times 10^2 = 20,098.7 = 20,0987 \times 10^{-1}$. Similarly, $\frac{20.0987}{100} = \frac{200.987}{1,000} = \frac{2009.87}{10^4}$.

Example 6: $3,000,000 \times 0.0003$?

To transform the second number, 0.0003, equals to '3 times the exponent of 10,' it needs an exponent of '4' to the base 10 or the multiplication of 10,000; we can borrow four '0s' from the first number, 3,000,000, so $3,000,000 \times 0.0003$ becomes 300~~0000~~ $\times$ 3 equalling 900.

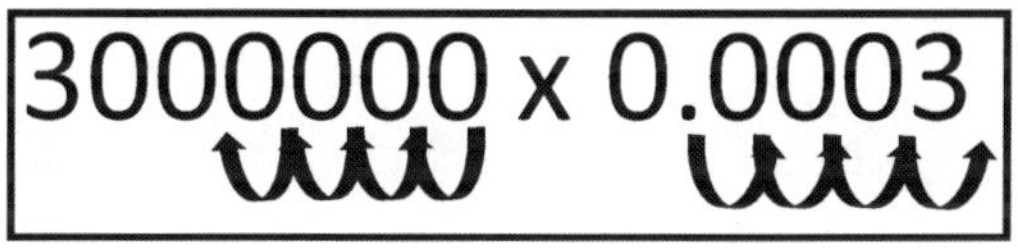

Example 7: $\frac{1.96[10^3(6.03 - 3 \times 10^{-2})]}{(1.4)^2} = ?$

$$=> \frac{1.96[10^3(6.03 - 3 \times 10^{-2})]}{(1.4)^2} = \frac{\cancel{1.96}[10^3(6.03 - 3 \div 100)]}{\cancel{1.96}} = 10^3(6.03 - 3 \div 100)$$

$=> 10^3(6.03 - 0.03) = 1,000 \times 6 = \mathbf{6,000}$.

3.4 Common Binomial expressions

(1) $a(b+c) = ab + ac$

$150 \times 51 = 150(50+1) = 150 \times 50 + 150 \times 1 = 7,500 + 150 = 7,650$
$38 \times 19 = 38(20-1) = 38 \times 20 - 38 \times 1 = 760 - 38 = 722$

(2) $(a+b)^2 = a^2 + 2ab + b^2$

$102^2 = (100+2)^2 = 100^2 + 2 \times 100 \times 2 + 2^2 = 10,000 + 400 + 4 = 10,404$

(3) $(a-b)^2 = a^2 - 2ab + b^2$

$98^2 = (100-2)^2 = 100^2 - 2 \times 100 \times 2 + 2^2 = 10,000 - 400 + 4 = 9,604$

(4) $(a^2 - b^2) = (a+b)(a-b)$

$98^2 - 2^2 = (98+2)(98-2) = 100 \times 96 = 9,600$

(5) $(a+b+c)^2 = a^2 + b^2 + c^2 + 2(ab + bc + ca)$

(6) $(a+b)^3 = a^3 + 3a^2b + 3ab^2 + b^3$

$102^3 = (100+2)^3 = 100^3+3\times100^2\times2+3\times100\times2^2+2^3 = 1,000,000+60,000+1,200+8 = 1,061,208$

or $(a+b)^3 = a^3 + 3ab(a+b) + b^3$

$51^3 = (50+1)^3 = 50^3 + 3 \times 50 \times 1 \times (50+1) + 1^3 = 125,000 + 150 \times 51 + 1 = 132,651$

(7) $(a-b)^3 = a^3 - 3a^2b + 3ab^2 - b^3$

$98^3 = (100-2)^3 = 100^3-3\times100^2\times2+3\times100\times2^2-2^3 = 1,000,000-60,000+1,200-8 = 941,192$

or $(a-b)^3 = a^3 - 3ab(a-b) - b^3$

$49^3 = (50-1)^3 = 50^3 - 3 \times 50 \times 1 \times (50-1) - 1^3 = 125,000 - 150 \times 49 - 1 = 117,649$

(8) $(a^3 + b^3) = (a+b)(a^2 - ab + b^2)$

$11^3 + 9^3 = (11+9) \times (11^2 - 11 \times 9 + 9^2) = 20 \times (121 - 99 + 81) = 20 \times 103 = 2,060$

(9) $(a^3 - b^3) = (a-b)(a^2 + ab + b^2)$

$11^3 - 9^3 = (11-9) \times (11^2 + 11 \times 9 + 9^2) = 2 \times (121 + 99 + 81) = 2 \times 301 = 602$

3.5 Some important summation of series

(1) Sum of first positive integers = $1 + 2 + 3 + 4 + 5 + ...n = \frac{n(n+1)}{2}$

Example: $12 + 13 + 14 + 15 + 16 + 17 + 18 =?$

$12 + 13 + 14 + 15 + 16 + 17 + 18 = (1 + 2 + 3 + 4 + ...18) - (1 + 2 + 3 + 4 + ...11)$

$$\frac{18(18+1)}{2} - \frac{11(11+1)}{2} = \frac{18 \times 19}{2} - \frac{11 \times 12}{2} = 171 - 66 = 105$$

Alternate approach:

=> $12 + 13 + 14 + 15 + 16 + 17 + 18 = (11 + 1) + (11 + 2) + (11 + 3) + (11 + 4) + (11 + 5) + (11 + 6) + (11 + 7)$

=> $(11+11+11+11+11+11+11)+(1+2+3+4+5+6+7) = 77+\left[\frac{7 \times 8}{2}\right] = 77+28 = 105$

(2) Sum of squares of first positive integers = $1^2+2^2+3^2+4^2+5^2+............n^2 = \frac{n(n+1)(2n+1)}{6}$

(3) Sum of cubes of first positive integers = $1^3 + 2^3 + 3^3 + 4^3 + 5^3 + ...n^3 = \left[\frac{n(n+1)}{2}\right]^2$

(4) Sum of odd integers = $1 + 3 + 5 + ...(2n - 1) = n^2$; Note that the nth term is $(2n - 1)$.

(5) Sum of even integers = $2 + 4 + 6 + ...2n = n(n + 1)$

3.6 Fractions

Fraction is a number which is formed out of numerator and denominator. It is represented as $\frac{\text{Numerator}}{\text{Denominator}}$. Both numerator and denominator are integers, but the denominator should not be '0.'

Example: $\frac{3}{4}, \frac{7}{8}, -\frac{9}{5}$ etc.

Proper fraction: For a fraction, if the absolute value of the numerator is less than the absolute value of the denominator, it is called a proper fraction. **Example:** $\frac{3}{4}, \frac{7}{8}, -\frac{9}{10}$ etc.

Improper fraction: If the absolute value of the numerator of a fraction is greater than the absolute value of the denominator, it is called improper fraction. **Example:** $\frac{4}{3}, \frac{11}{8}, -\frac{9}{5}$ etc.

Mixed fraction: If the fraction is a mix of both an integer and proper fraction, it is called mixed fraction. **Example:** $2\frac{3}{4}, -5\frac{2}{3}$ etc.

Converting mixed fraction to improper fraction: Say, a mixed fraction is $5\frac{2}{3}$.

Step 1: Multiply the integer part with the denominator. $5 \times 3 = 15$.
Step 2: Add the numerator to the number attained in step 1. So, $15 + 2 = 17$. It is the numerator of improper fraction.
Step 3: The equivalent improper fraction is $\frac{\text{new numerator}}{\text{denominator}} = \frac{5 \times 3 + 2}{3} = \frac{17}{3}$

Equivalent Fraction: Fractions that are equal in values are called equivalent fractions, although, the numerator and denominator of the equivalent fractions may not be the same.

Example: $1/2 = 2/4 = 3/6 = 4/8 = 5/10$.

3.6.1 Concept of fraction

Let us understand the concept of fraction through graphical representation. Fraction is a part of a whole. For example, the fraction 3/8 means that out of 8 parts, only 3 parts are taken; it can also be thought of as eating 3 slices of pizza of an 8-slice pizza.

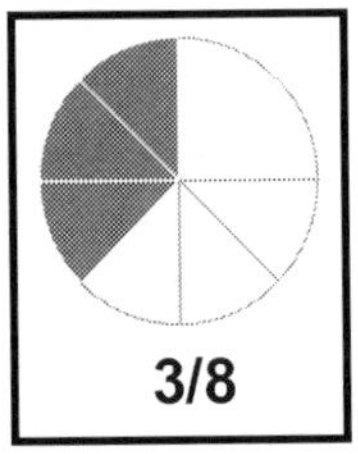

Reciprocal of fraction: You get reciprocal of a fraction when you interchange numerator and denominator. **Example:** Reciprocal of $\frac{3}{8}$ is $\frac{8}{3}$.

What is the significance of $\frac{8}{3}$?

To understand this, first convert $\frac{8}{3}$ into mixed fraction, which is $2\frac{2}{3}$. It means eating 2 full pizzas and 2 slices of pizza of a 3-slice pizza.

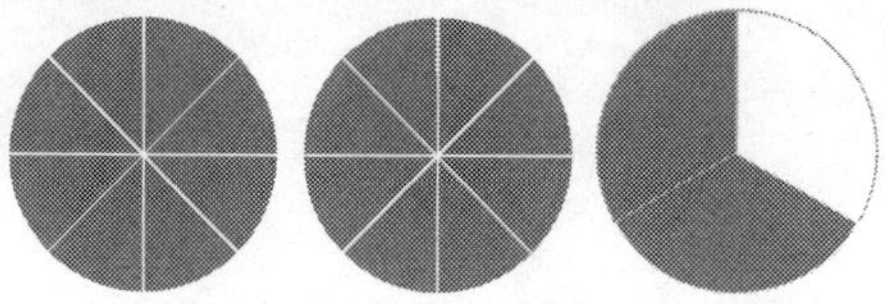

8/3 = Reciprocal of 3/8

Decimal Representation:

A fraction can be represented as a decimal number by dividing the numerator by the denominator. For example, the fraction 3/8 represents the numerical value 0.375; it is a decimal number.

Percent Representation:

A fraction can be represented as a percent by dividing the numerator by the denominator and then multiplying it by 100. For example, $3/4 = 0.75 => 0.75 \times 100\% = 75\%$ and similarly, $1/2 = 50\%$.

3.6.2 Mathematical operations on fractions

Mathematical operation	Example
Addition/ Subtraction	$\frac{1}{2} + \frac{1}{2} = 1$; two halves makes 1; $\frac{1}{2} - \frac{1}{4} =? => \frac{1\times2}{2\times2} - \frac{1}{4} = \frac{2}{4} - \frac{1}{4} = \frac{2-1}{4} = \frac{1}{4}$; make denominators equal
Multiplication	$\frac{1}{2} \times \frac{1}{2} = \frac{1}{4}$; half of a half makes it $\frac{1}{4}$; $\frac{3}{5} \times \frac{15}{6} = \frac{3}{2}$; any numerator can cancel any denominator; two numerators will be multiplied, and similarly, two denominators will be multiplied
Division	$\frac{1}{2} \div \frac{1}{2} = \frac{1}{2} \times \frac{2}{1} = 1$; '÷' will change to '×' and the fraction to the right will become its reciprocal; $\frac{3}{5} \div \frac{3}{5} = \frac{3}{5} \times \frac{5}{3} = 1$; a fraction divided by itself is always '1'; product of a fraction with its reciprocal is always 1.

Some confusing divisions:

- $\frac{2/3}{3/2} = \frac{2}{3} \div \frac{3}{2} = \frac{2}{3} \times \frac{2}{3} = \frac{4}{9}$;
- $\frac{2}{3/2} = 2 \div \frac{3}{2} = 2 \times \frac{2}{3} = \frac{4}{3}$;
- whereas, $\frac{2/3}{3} = \frac{2}{3} \div 3 = \frac{2}{3} \times \frac{1}{3} = \frac{2}{9}$

Example: $\frac{2}{2 - \frac{2}{1 - \frac{4}{2 - \frac{5}{2 + \frac{6}{7}}}}} = ?$

=> Start solving from the lowermost fraction

=> $\frac{2}{2 - \frac{2}{1 - \frac{4}{2 - \frac{5}{2 + \frac{6}{7}}}}} = \frac{2}{2 - \frac{2}{1 - \frac{4}{2 - \frac{5}{\frac{\mathbf{2 \times 7 + 6}}{7}}}}} = \frac{2}{2 - \frac{2}{1 - \frac{4}{2 - \frac{5}{\frac{\mathbf{20}}{7}}}}}$;

Remember that the primary division operator is between "5" and "20."

So $\frac{5}{\frac{20}{7}} = 5 \div \frac{20}{7} = 5 \times \frac{2}{3} = \frac{7}{20} = \frac{7}{4}$; Do not get it confused with $\frac{\frac{5}{20}}{7} \neq \frac{5}{\frac{20}{7}}$.

Coming back to the question:

=> $\frac{2}{2 - \frac{2}{1 - \frac{4}{2 - \frac{5}{\frac{\mathbf{20}}{7}}}}} = \frac{2}{2 - \frac{2}{1 - \frac{4}{2 - \frac{7 \times 5}{20}}}} = \frac{2}{2 - \frac{2}{1 - \frac{4}{\frac{1}{4}}}} = \frac{2}{2 - \frac{2}{1 - 4 \times 4}} = \frac{2}{2 - \frac{2}{1 - 16}} = \frac{2}{2 - \frac{2}{-15}}$

=> $\frac{2}{2 + \frac{2}{15}} = \frac{2}{\frac{32}{15}} = \frac{2 \times 15}{30} = \frac{15}{16}$

3.6.3 Comparing fractions

You may come across situation where you have to compare the two fractions—which fraction is greater or smaller?

A general rule: For positive factions, larger the numerator, greater is the fraction (with the same denominator), and smaller the denominator, greater is the fraction (with the same numerator) and vice-versa; opposite is true for negative fractions.

Example 1: Which is greater $\frac{1}{2}$ or $\frac{1}{3}$? Well, this is a simple one as the numerators of the fractions are equal. For positive factions, if the numerators of the fractions are equal, the fraction with smaller denominator is greater, so $\frac{1}{2} > \frac{1}{3}$; however, $-\frac{1}{2} < \frac{1}{3}; -\frac{1}{2} < -\frac{1}{3}; \frac{1}{2} > -\frac{1}{3}$.

Example 2: Which is greater $\frac{2}{5}$ or $\frac{3}{5}$? Again, this is a simple one as the denominator of the fractions are equal. For positive factions, if the denominators of the fractions are equal, the fraction with the larger numerator is greater, $\frac{2}{5} < \frac{3}{5}$; however, $\frac{2}{5} > -\frac{3}{5}; -\frac{2}{5} > -\frac{3}{5}; \frac{2}{5} > -\frac{3}{5}$.

Example 3: Which is greater $\frac{11}{15}$ or $\frac{9}{16}$? This is the easiest of all to deduce. As the numerator of the first fraction is greater than that of the other and its denominator is smaller than that of the other, so on both the counts, the first fraction scores over the other, hence, it is greater than the second fraction, so $\frac{11}{15} > \frac{9}{16}$; however, $-\frac{11}{15} < -\frac{9}{16}$.

Example 4: Which is greater $\frac{5}{7}$ or $\frac{7}{9}$? This is not an easier one to deduce. The numerator of the first fraction is greater than that of the other but its denominator is also larger than that of the other, so on one count, the first fraction scores over the second and on the other count, the second one scores over the first. So how to deduce which fraction is greater? There are three approaches to deduce this.

Approach 1:

Convert fractions to Decimals

Each fraction can be converted into a decimal by dividing the numerator by the denominator. In decimal format, the value of each fraction can be compared.

=> $\frac{5}{7} = 0.714$; $\frac{7}{9} = 0.777$, so $0.777 > 0.714$.

Since the decimal equivalent of 7/9 is larger than that of 5/7, so 7/9 > 5/7.

Approach 2:

Take 50%

The approach is to take 50% of the denominator and deduct it from the numerator, and then compare the remaining fractions.

Which is greater $\frac{5}{7}$ or $\frac{4}{5}$?

=> 50% of 7 (denominator of the first fraction) is 3.5, and 50% of 5 (denominator of the second fraction) is 2.5, so we can write $\left[\frac{5}{7} = \frac{3.5}{7} + \frac{1.5}{7}\right] <> \left[\frac{4}{5} = \frac{2.5}{5} + \frac{1.5}{5}\right]$

=> $(50\% + \frac{1.5}{7}) <> (50\% + \frac{1.5}{5})$ => It is obvious that $\frac{1.5}{7} < \frac{1.5}{5}$, so, $\frac{5}{7} < \frac{4}{5}$.

This approach may not be applicable in each case; after seeing all the approaches, you should decide which approach works best for a specific scenario.

Approach 3:

Cross Multiplication

Another approach to compare fractions is: Cross Multiplication. Numerator of the first fraction is multiplied to the denominator of the second fraction, and numerator of the second fraction is multiplied to the denominator of the first fraction. Now compare the results, and identify the greater of the two products; the fraction which has its numerator in the greater of the two products is a greater fraction than the other fraction.

=> Which is greater $\frac{5}{7}$ or $\frac{7}{9}$?

=> (Numerator of I × Denominator of II) = $5 \times 9 = 45$; (Numerator of II × Denominator of I) = $7 \times 7 = 49$, since $49 > 45$ and “49” includes the numerator of the second fraction, hence, the second fraction is greater than the first fraction.

3.6.4 More on fractions

(1) If $\frac{a}{b} > 1$, then $\frac{a+c}{b+c} < \frac{a}{b}$; where a, b & c are positive numbers

=> Say $\frac{a}{b} = \frac{3}{2}$, and $c = 1$, then $\frac{a+c}{b+c} = \frac{3+1}{2+1} = \frac{4}{3} = 1.33 < \left[\frac{a}{b} = \frac{3}{2} = 1.5\right]$

We can understand it this way: Since the fraction a/b > 1, it means that a > b. When we add ‘c’ to each ‘a’ and ‘b,’ larger proportion of ‘c’ is added to ‘b’ i.e. denominator and relatively a smaller proportion of ‘c’ is added to ‘a’ i.e. the numerator. Let us see how.

$$\frac{a+c}{b+c} = \frac{a\left(1+\frac{c}{a}\right)}{b\left(1+\frac{c}{b}\right)} = \frac{a}{b} \times \frac{\left(1 + \text{smaller fraction compared to } \frac{c}{b}\right)}{\left(1 + \text{larger fraction compared to } \frac{c}{a}\right)}$$

$$= \left[\frac{a}{b} \times \frac{\text{smaller number}}{\text{larger number}}\right] < \frac{a}{b}$$

(2) If $0 < \frac{a}{b} < 1$, then $\frac{a+c}{b+c} > \frac{a}{b}$; where a, b & c are positive numbers

=> Say $\frac{a}{b} = \frac{2}{3}$, and $c = 1$, then $\frac{a+c}{b+c} = \frac{2+1}{3+1} = \frac{3}{4} = 0.75 > \left[\frac{a}{b} = \frac{2}{3} = 0.667\right]$

(3) If $\frac{a}{b} > 1$, then $\frac{a-c}{b-c} > \frac{a}{b}$; where a, b & c are positive numbers

=> Say $\frac{a}{b} = \frac{3}{2}$, and $c = 1$, then $\frac{a-c}{b-c} = \frac{3-1}{2-1} = \frac{2}{1} = 2 > \left[\frac{a}{b} = \frac{3}{2} = 1.50\right]$

(4) If $0 < \frac{a}{b} < 1$, then $\frac{a-c}{b-c} < \frac{a}{b}$; where a, b & c are positive numbers

=> Say $\frac{a}{b} = \frac{2}{3}$, and $c = 1$, then $\frac{a-c}{b-c} = \frac{2-1}{3-1} = \frac{1}{2} = 0.50 < \left[\frac{a}{b} = \frac{2}{3} = 0.667\right]$

3.6.5 Fractions and their percent equivalents

Fraction	Percent
$\frac{1}{1}$	100%
$\frac{1}{2}$	50%
$\frac{1}{3}$	33.33%
$\frac{1}{4}$	25%
$\frac{1}{5}$	20%
$\frac{1}{6}$	16.67%
$\frac{1}{7}$	14.28%
$\frac{1}{8}$	12.50%

Fraction	Percent
$\frac{1}{9}$	11.11%
$\frac{1}{10}$	10%
$\frac{1}{11}$	9.09%
$\frac{1}{12}$	8.33%
$\frac{1}{25}$	4%
$\frac{1}{50}$	2%
$\frac{1}{100}$	1%
$\frac{1}{200}$	0.50%

Application: Sometimes you may come across calculating an odd-looking percent of an odd-looking number. The fraction to percent equivalent table may help you make the calculations easier.

Example 1: What is 14.25% of 705?

The traditional way of calculation is time-consuming: $\frac{14.25}{100} \times 705 = 100.47$. In the GMAT, seldom you need to calculate the exact value; the options would be wide apart. GMAT does not test your ability on how precisely you calculate the values, it rather tests your ability on how efficiently you attain the closest to the correct value.

Smarter approach:

From the table, we know that 14.25% is $\approx$ 14.28% which is equal to $\frac{1}{7}$. So, we can rather calculate $\frac{1}{7}$ of 705; again, we can approximate 705 to 707 as there is a difference of only '2' between 705 and 707. So, $\frac{1}{7}$ of 707 is easier to calculate which is equal to '101.' But '101' is an approximate value, the actual value would be less than '101' as we increased 14.25% to 14.28% and 705 to 707. The actual value already calculated as 100.47, which is quite close to the approximate value '101.'

Example 2: What is 12.61% of 802?

A. 95.68

B. 105.94

C. 101.13

Again, from the table, we know that 12.61% is $\approx$ 12.50% which is $\frac{1}{8}$. So, we can rather calculate $\frac{1}{8}$ of 802; again, we can approximate 802 as 800 as there is a difference of only '2' between them. So, 1/8 of 800 is easier to calculate which is equal to '100.' Again, '100' is an approximate value, the actual value must be little more than '100' as we had decreased 12.61% to 12.50% and 802 to 800. Looking at the options, we can safely lock '101.13' as the actual value since the other higher value is far greater than '100.'

3.6.6 See-Saw Approach

How do you solve a question which asks, "*If gasoline price increased by 10%, by what percent should one reduce its consumption so that the expense on gasoline remains the same?*"

Traditional Approach:

Say, before the increase, the price of gasoline was \$100/gallon, and the consumption was 100 gallons. So, the expense = $100 \times 100 = \$10,000$.

Now, after the increase, the price of gasoline is \$110/gallon (10% up), and say the consumption is reduced to x gallons. So, the expense = $110 \times x = \$110x$.

Since the expense needs to be the same as before, hence, $10,000 = 110x \Rightarrow x = 90.90$ gallons

$$\Rightarrow \text{Percent reduction in consumption} = \frac{100 - 90.90}{100} \times 100\% = 9.09\%.$$

See-Saw Approach:

Scenario 1: When $x \times y = C$; where x is one element, y is second element, and C is a constant.

If x goes up by certain percent and y must go down with some percent such that after the change $x \times y = x' \times y' = C$, follow the following short-cut steps to get the answer. Where after the change, x becomes x', and y becomes y'.

Step 1: Covert the given percent figure into fraction, say it is $\frac{1}{a}$; you should take the help of the conversion table.
Step 2: Add "1" to the denominator of the fraction attained; it would be $\frac{1}{a+1}$.
Step 3: Convert $\frac{1}{a+1}$ fraction into percent; it is the requisite answer; you should take the help of the table.

Let us get the answer to the above question with the See-Saw approach.

Step 1: Price goes up by 10%; equivalent fraction of 10% is $\frac{1}{10}$, so $\frac{1}{a} = \frac{1}{10}$.
Step 2: By adding "1" to the denominator of the fraction $\frac{1}{a}$, we get $\frac{1}{a+1} = \frac{1}{11}$.
Step 3: With the help of the table, we get $\frac{1}{11} = 9.09\%$; it is the requisite answer. The consumption must go down by 9.09%.

See the pictorial representation of the See-Saw approach.

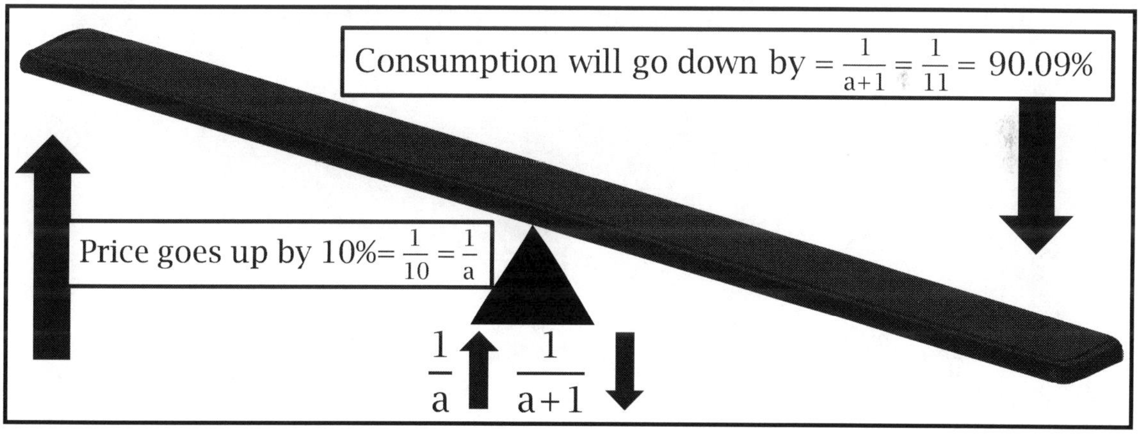

Scenario 2: When $x \times y = C$; where x is one element, y is second element, and C is constant.

If x goes down by certain percent and y must go up with some percent such that after the change $x \times y = x' \times y' = C$, follow the following short-cut steps to get the answer.

Step 1: Covert the given percent figure into fraction, say it is $\frac{1}{a}$; take the help of the table.
Step 2: Subtract "1" from the denominator of the fraction attained; it would be $\frac{1}{a-1}$.
Step 3: Convert $\frac{1}{a-1}$ fraction into percent; it is the requisite answer; you should take the help of the table.

Let us apply the See-Saw approach in another question: "*If milk price decreased by 10%, by what percent should one increase its consumption so that the expense on milk remains the same?"*

Step 1: Price goes down by 10%; equivalent fraction of 10% is $\frac{1}{10}$, $\frac{1}{10}$, so $\frac{1}{a} = \frac{1}{10}$;
Step 2: By subtracting "1" from the denominator of the fraction $\frac{1}{a}$, we get $\frac{1}{a-1} = \frac{1}{9}$.
Step 3: With the help of the table, we get $\frac{1}{9}$ = 11.11%; it is the requisite answer. The consumption must go up by 11.11%.

See the pictorial representation of the same.

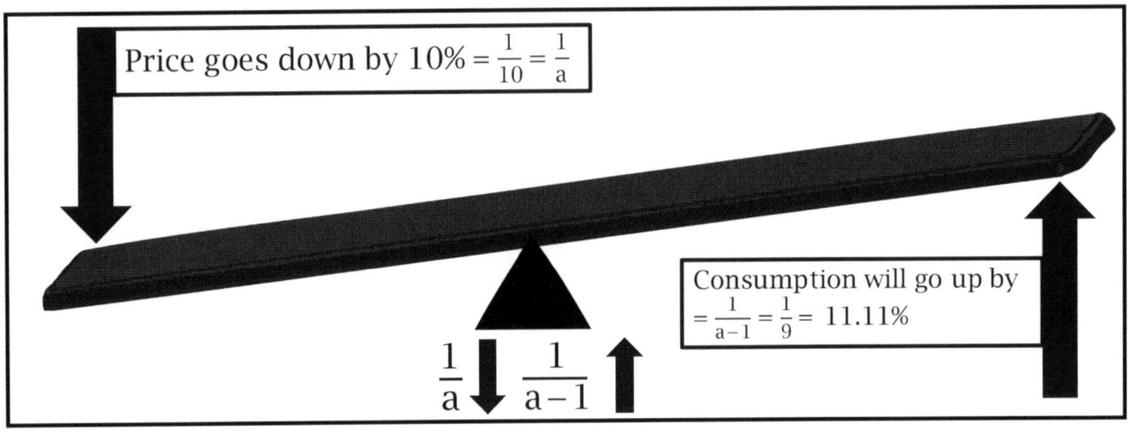

Example 1: If the sale of a brand is reduced by 17%, by what percent should the company increase its sales price such that the revenue earned remains the same as before?

A. 14.28%

B. 14.53%

C. 20%

D. 20.48%

First, identify whether the question fits into the category discussed above. Since "Quantity × Price = Revenue," and the condition is that the revenue must remain constant, hence, the question fits into the See-Saw approach model. Out of the two scenarios, the second scenario fits as an element goes down and the complementary element must go up.

When we look at the table, we do not find 17%, however, we can safely assume that it is closer to 16.67%; so the equivalent of 17% = ≈ 16.67% =≈ $\frac{1}{6} = \frac{1}{a}$ => $\frac{1}{a-1} = \frac{1}{5}$ = 20%. The sales price must be increased by 20%. The correct answer would be a little more than 20% as we approximated 17% to 16.67%. So, the correct answer must be option D—the only greater option than 20%. The company should increase the sales price of the brand by 20.48%.

Example 2: If the price of milk increased by 11%, by what percent should its consumption be reduced such that the expenses on milk remains the same as before?

We identified that the question fits into the category discussed above, and it fits into the first scenario. Now, 11% =≈ 11.11% = $\frac{1}{9}$ = $\frac{1}{a}$ => $\frac{1}{a+1}$ = $\frac{1}{10}$ =≈ 10%. The correct answer would be a little less than 10%.

3.7 Absolute numbers

We know that any real number can either be 0, positive or negative. When we ignore the negative sign of a number, we seem to be interested in its absolute value and not in its directed value.

In other words, absolute value of a number is its distance from '0.' If a number is presented between two-vertical bar (pipe ||) symbol, it means we are interested in its absolute value.

Say, a number is x, then its absolute value is $|x|$, as depicted in the figure below.

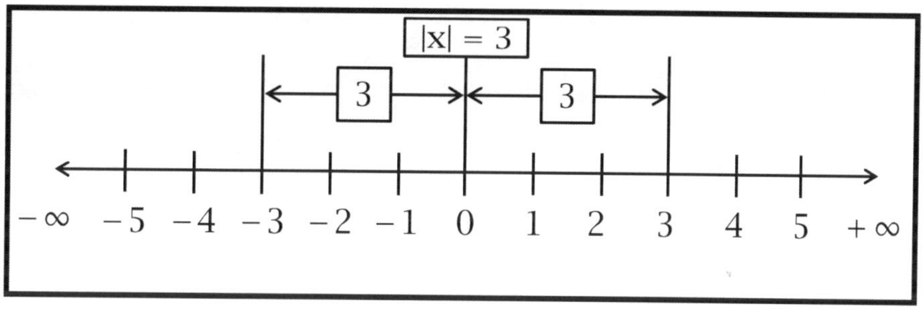

Key properties:

(1) $\sqrt{x^2} = |x|$

(2) $|\pm x| = |x|$

(3) $|x| = |-x|; -|x| \neq |-x|$

(4) $|x| \geq 0$

(5) $|x \pm y| = |y \pm x|$

Say $x = 10$ and $y = -10$, then
$|x - y| = |10 - (-10)| = |10 + 10| = 20$;
$|y - x| = |-10 - 10| = |-20| = 20$.

(6) $|x + y| \leq |x| + |y|$

Say $x = 10$ and $y = 10$, then
$|x + y| = |10 + 10| = 20$;
$|x| + |y| = |10| + |10| = 10 + 10 = 20$, hence, $|x + y| = |x| + |y|$

But, say $x = 10$ and $y = -10$, then
$|x + y| = |10 + (-10)| = |10 - 10| = 0$;
$|x| + |y| = |10| + |-10| = 10 + 10 = 20$, hence, $|x + y| < |x| + |y|$

This deduction is an important finding and is tested by test-makers.

Remember that the equality holds true only when both x and y are of the same sign and the inequality holds true when x and y are of opposite signs.

(7) $|x - y| \geq |x| - |y|$

Here too, the equality holds true only when both x and y are of the same sign and the inequality holds true when x and y are of opposite signs.

(8) $|x \times y| = |y \times x| = |x| \times |y| = |y| \times |x|$

(9) $|x \div y| = |x| \div |y|$

Example: If $7|2x - 3| = 21$, what is the value of x?

Simplify $7|2x - 3| = 21$ by dividing both the sides by '7'; we get $|2x - 3| = 3$.

=> $2x - 3 = 3$ or $-(2x - 3) = 3$.

=> $2x - 3 = 3$ => $2x = 6$ => $x = 3$; or

=> $-(2x - 3) = 3$ => $-2x + 3 = 3$ => $-2x = 0$ => $x = 0$.

So $x = 0$ or 3.

3.7.1 Inequality

To understand Inequality, first understand Equality.

$x = 5$ means that the value of the variable x equals 5.

The inequality could be anything $x \neq 4$, $x < 6$ or $x > 4$.

When the two sides of an expression are not equal, it is called Inequality. Inequalities use symbols $<, >, \leq$, or $\geq$ to describe the relationship between two expressions, for example: $10 < 12$, $12 > 8$, $x \leq 4$, $x + 2y \geq 4$, etc.

Understanding inequality:

Inequality	Meaning	Example
$x < 5$	x is less than 5	x can be any value less than 5 on the number line; it can be 0, 1, -155, or any other value up to negative infinity. Do not assume that x is an integer. x can even be 4.999. Always consider that x is a real number until stated that it is an integer.
$y > 10$	y is greater than 10	y can be any value greater than 10 on the number scale. It can be 10.0001, 10.56, 21, or any other value up to infinity.
$p \leq 10$	p is less than or equal to 10	p can be any value less than or equal to 10 on the number line. It can be 10, 4, -2.3, or any other value up to negative infinity.
$q \geq 10$	q is greater than or equal to 10	q can be any value greater than or equal to 10 on the number line. It can be 10, 14, 23.5, or any other value up to infinity.
$-2 \leq x < 4$; x is an integer	x is greater than or equal to -2 but less than 4	$x = \{-2, -1, 0, 1, 2, 3\}$
$-2 > x \geq -4$; x is an integer	x is less than -2 and greater than or equal to -4	$x = \{-3, -4\}$

Representation of inequalities on number line:

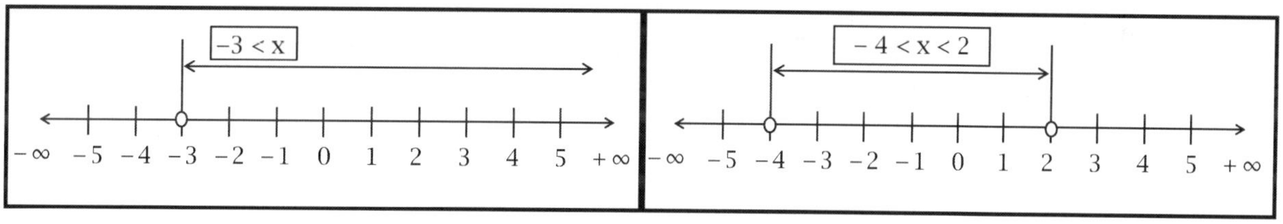

Key properties of Inequalities:

(1) You can **add or subtract anything from an inequality**.

Say, $x + 3 > 10$.

You can subtract 3 from both the sides, so the inequality becomes,
$x + 3 - 3 > 10 - 3 \Rightarrow x > 7$.

Take another example; say, $y - 3 > -10$.

You can add 3 on both the sides, so the inequality becomes,
$y - 3 + 3 > -10 + 3 \Rightarrow y > -7$.

(2) You can multiply or divide an inequality by a **positive** number.

Say, $2x > 10$.

You can divide the inequality by 2, so the inequality becomes,
$\frac{2x}{2} > \frac{10}{2} => x > 5$.

Say, $\frac{y}{3} > -10$. You can multiply the inequality by 3, so the inequality becomes,
$\frac{y}{3} \times 3 > -10 \times 3 => y > -30$.

(3) You can multiply or divide an inequality by a **negative** number, but doing so will ask for **reversing** the sign of inequality.

We know that $7 > 5$;

if we multiply an inequality by -1, it will become $-7 > -5$ which is wrong to say because $-7 < -5$; so when we multiply or divide an inequality with a negative number, we have to flip the sign.

Say, $-2x > 10$.

You can divide the inequality by -2, so the inequality becomes,

$\frac{-2x}{-2} \not> < \frac{10}{-2} => x < -5$.

Say $\frac{-y}{3} > -10$. You can multiply the inequality by -3, so the inequality becomes,

$\frac{-y}{3} \times -3 \not> < -10 \times -3 => y < 30$.

3.7.2 Inequalities with absolute numbers

Few tough questions may be asked combining inequalities and absolute numbers.

Example 1: If $|x + 1| \leq 4$, what is x?

$|x + 1| \leq 4$

=> If $x + 1 \leq 4$ or $-(x + 1) \leq 4$

$x + 1 \leq 4 => x \leq 3$

Or, $-(x + 1) \leq 4 => x + 1 \geq -4 => x \geq -5$. (Pay attention to reversal of sign of inequality)

So, $\mathbf{3 \geq x \geq -5}$

It can be represented graphically as below.

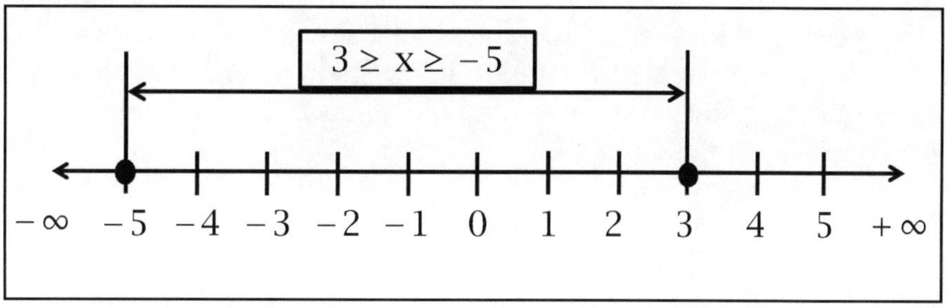

Example 2: If $|2x + 1| \geq 3$, what is x?

$|2x + 1| \geq 3 \Rightarrow 2x + 1 \geq 3$ or $-(2x + 1) \geq 3$

If $2x + 1 \geq 3 \Rightarrow 2x \geq 2 \Rightarrow x \geq 1$

Or, $-(2x + 1) \geq 3 \Rightarrow 2x + 1 \leq -3 \Rightarrow 2x \leq -4 \Rightarrow x \leq -2$. (Pay attention to reversal of sign of inequality)

So, $\mathbf{-2 \geq x}$ or $\mathbf{x \geq 1}$

It can be represented graphically as below.

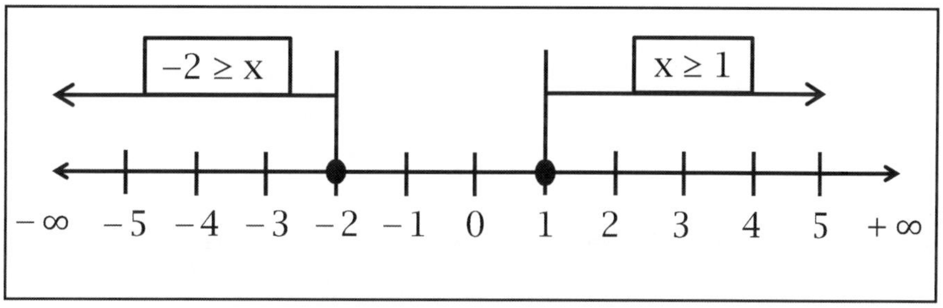

Example 3: If $8 + x^2 \geq 12$, what is the range of x?

$8 + x^2 \geq 12$

$\Rightarrow x^2 \geq 4;$

Taking square root of both the sides, we get,

$|x| \geq 2 \Rightarrow x \geq 2$ or $x \leq -2$.

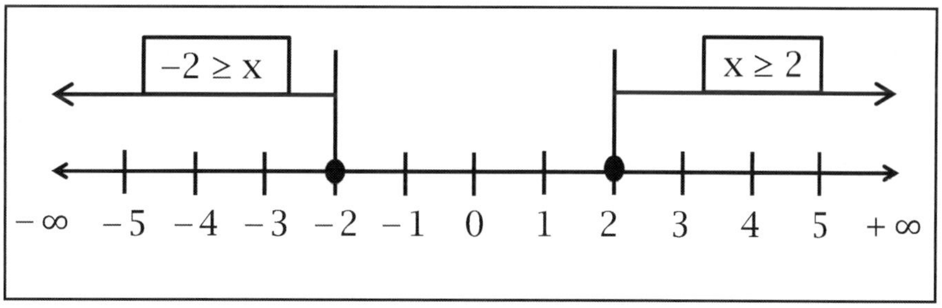

3.7.3 Compound Inequality

An inequality that has more than one inequality symbol in an expression is called compound inequality. $2 > x \geq 4$ is an example of compound inequality. GMAT inequality questions may involve more than one inequality. To solve such questions, you may have to transform several inequalities to a compound inequality.

Example 4: If $x > 4$, $x \leq 10$, and $x < 12$, what is the range of x?

Step 1: Fix 'more than x' or 'more than or equal to x' inequality.

There is only one 'more than x' inequality, so we fix $x > 4$.

Step 2: Fix 'less than x' or 'less than or equal to x' inequality.

There is one 'less than x' and one 'less than or equal to x' inequalities, but between the two inequalities $x \leq 10$, and $x < 12$, we choose the first one i.e. $x \leq 10$ since $x \leq 10$ is a subset of $x < 12$, not vice-versa; so we fix $x \leq 10$.

Step 3: Combine both the inequalities derived in steps 1 and 2.

Finally, we get $\mathbf{4 < x \leq 10}$.

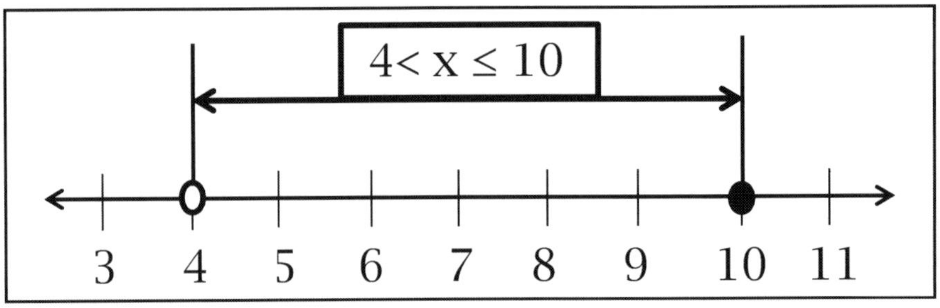

Note:

If $x > y$, and $y > z$, then we can combine the inequalities as $x > y > z$ or $x > z$.

However,

If $x > y$, and $y < z$, then though we can combine the inequalities as $x > y < z$, but we **cannot** deduce that $x > z$.

3.7.4 Mathematical operations on inequalities

(1) **Addition and subtraction:**

If $x > y$ and $a < b$, we cannot deduce that $x + a < y + b$; it is wrong.

We must make both the inequalities in the same direction. So, we can rewrite the inequalities as $x > y$ and $b > a$, and now we can add these.

So, $\boldsymbol{x + b > y + a}$.

Let's take another example.

If $x - 3 < y \le x + 6$, then we **cannot** write the inequality as $x < y \le x + 6 + 3 \Rightarrow x < y \le x + 9$. It is wrong!

If you subtract a number from one part of a compound inequality, you must subtract the same number from all the parts.

The simplified and corrected inequality is: $\boldsymbol{x < y + 3 \le x + 9}$.

(2) **Multiplication and division:**

If $x < -\frac{y}{3} \le 3x$, then we can simplify the inequality by multiplying it by -3.

So, $-3\left(x < -\frac{y}{3} \le 3x\right) \Rightarrow -3x > (-\frac{y}{3} \times -3) \ge -3 \times 3x \Rightarrow \boldsymbol{-3x > y \ge -9x}$.

Always be wary of changing the sign of inequality when multiplying or dividing it by a negative number.

The inequality can also be written as $\boldsymbol{3x < -y \le 9x}$.

Similarly, if $3x > 6y - 3 > 6z$, we can divide the inequality by 3, so we get $\boldsymbol{x > 2y - 1 > 2z}$. Always remember that each part of the compound inequality should be divided.

Chapter 4

Practice Questions: Number Properties

1. If x and y are integers, and $5x + 3y = 15$, which of the following could NOT be the value of y?

(A) -5

(B) 0

(C) 3

(D) 5

(E) 1,005

2. If $\frac{m}{n}$ is an integer, which of the following statements can be true EXCEPT?

(A) Both m and n are integers

(B) m is an integer

(C) Either m or n is negative

(D) $\frac{n}{m}$ is an integer

(E) $m = kn$; where k is non-integer

3. If $m \# n = \frac{nm}{2}$, then what is the value of $(4 \# 3) \# 1$?

(A) 1

(B) $\sqrt{6}$

(C) 3

(D) 6

(E) 9

4. There are 10 integers; out of these, at least two integers are negative. If the product of these integers is negative, at most how many of the integers can be negative?

(A) Two

(B) Three

(C) Five

(D) Nine

(E) Ten

5. If $abc = 1$, $cde = 0$, and $ebc = 0$, which of the following must be 0?

(A) a

(B) b

(C) c

(D) d

(E) e

6. If $(5x + 11)$ is a positive integer, for how many integer values of x will the number $(5x + 11)$ be less than 300?

 (A) 56
 (B) 57
 (C) 58
 (D) 59
 (E) 60

7. Which of the following numbers does not have a factor (other than 1) which is a perfect square?

 (A) 48
 (B) 117
 (C) 245
 (D) 297
 (E) 345

8. How many numbers between 10 and 10,000 have the digit '4' in the units place and the digit '2' in the tens place?

 (A) 90
 (B) 99
 (C) 100
 (D) 190
 (E) 1,000

9. A student's scrapbook had a large collection of stamps which he wanted to number starting from '1.' If he used up a total of 855 digits in the process, how many stamps did he have in his scrapbook?

 (A) 285
 (B) 321
 (C) 420
 (D) 823
 (E) 855

10. How many five-digit numbers divisible by 3 can be formed using the digits 0, 1, 2, 3, 4 and 5 without repeating any digit?

 (A) 96
 (B) 120
 (C) 192
 (D) 216

(E) 240

11. $a12a$ is a four-digit number with the digit a in both the thousands and units places. What is the value of a, if the number is divisible by 6?

(A) 2
(B) 3
(C) 4
(D) 6
(E) 8

12. What is the units digit of $\left(13^{22} \times 12^{22}\right) + \left(17^{27} \times 25^{25}\right)$?

(A) 1
(B) 3
(C) 5
(D) 7
(E) 9

13. What is the value of $\sqrt{7}$ percent of $7\sqrt{7}$?

(A) 0.049
(B) 0.14
(C) 0.49
(D) 0.50
(E) 4.90

14. What is the value of $(5 \times 10^6) + (9 \times 10^4) + (2 \times 10^2)$?

(A) $500,000$
(B) $509,200$
(C) $590,200$
(D) $5,090,200$
(E) $5,900,200$

15. If $x^2 - yz = -5, y^2 - xz = 1$ and $z^2 - xy = 7$, what is the value of $(x - y)^2 + (y - z)^2 + (z - x)^2$.

(A) 3
(B) 6
(C) 8
(D) 12
(E) 24

16. What is the largest possible positive integer value of a that satisfies $a^2 + 5b = 315$; where b is a prime number?

(A) 4

(B) 5

(C) 10

(D) 15

(E) 17

17. If m and n are positive integers, and n is odd, which of the following must be even?

(A) $mn + 2n$

(B) $mn + 2m$

(C) $(n + 3)(m - 3)$

(D) $(n + 2)(m - 3)$

(E) $(n + 2)(m - 2)$

18. If n is even, which of the following cannot be odd?

I. $2n + 7$

II. $3n - 2$

III. $5n^2 + 1$

(A) I only

(B) II only

(C) III only

(D) I and II only

(E) I, II and III

19. What is the value of p, if the sum of consecutive odd numbers from 1 to p equals 225 (p being a term of the series)?

(A) 15

(B) 21

(C) 29

(D) 30

(E) 31

20. How many even numbers between 1 and 300 are divisible by 13 but not by 7?

(A) 9

(B) 10

(C) 11

(D) 21

(E) 22

21. How many numbers of the form $a \times b$, where a and b are prime numbers, have sum of factors, excluding the factor as the number itself, equal to 31?

(A) One

(B) Two

(C) Three

(D) Four

(E) Six

22. How many numbers, which are squares of prime numbers have sum of factors equal to 57?

(A) One

(B) Two

(C) Three

(D) Seven

(E) Nine

23. A perfect number is one whose sum of factors other than the number itself is equal to the number. Which of the following numbers is a perfect number?

(A) 12

(B) 18

(C) 20

(D) 21

(E) 28

24. How many even numbers, from 10 to 100 (both inclusive), are not multiples of 7?

(A) 32

(B) 38

(C) 39

(D) 45

(E) 46

25. If p^2, a perfect square number, is divisible by 5, and q, an integer, is not divisible by 5, which of the following must be divisible by 10?

(A) $pq + 5$

(B) $p + q$

(C) $2(p + 5q)$

(D) $2\,(5p + q)$

(E) $p^2 + 5q$

26. How many integers between 150 and 250, both inclusive, are divisible neither by 6 nor by 5?

(A) 34

(B) 60

(C) 67

(D) 68

(E) 71

27. A positive integer X, divided by 7 has a remainder 2. Which of the following must be true?

I. X is odd

II. X cannot be a prime number

III. $(X + 5)$ divided by 7 has remainder 0.

(A) I only

(B) I and II only

(C) III only

(D) I and III only

(E) I, II and III

28. If $N = 1! + 2! + 3! + 4! + \cdots + 8!$, what is the remainder when N is divided by 10?
$p! = 1 \times 2 \times 3 \times \cdots \times (p - 1) \times p$.

(A) 0

(B) 1

(C) 2

(D) 3

(E) 4

29. How many numbers between 1 and 250, excluding, are divisible by either 5 or 7 but not both?

(A) 69

(B) 70

(C) 71

(D) 77

(E) 78

30. What is the highest exponent of '2' that can divide $(10! + 11! + 12!)$ without leaving any remainder?

(A) 8

(B) 9

(C) 12

(D) 16

(E) 25

31. For how many positive integer values of p other than '1,' is $(p + 4)(p + 6)$ divisible by p?

(A) Four

(B) Seven

(C) Eight

(D) Twelve

(E) Twenty four

32. If $2^{20} = 2^{15}x + y$, where x and y are non-negative integers, what is the minimum possible value of $|x - y|$?

(A) 0

(B) 1

(C) 2^5

(D) 2^{10}

(E) 2^{15}

33. What is the digit in the units place of $7^{123} + 3^{42}$?

(A) 0

(B) 2

(C) 4

(D) 6

(E) 8

34. A perfect cube is a number whose cube root is an integer. For example, 8, 27, and 64 are perfect cubes. If m and n are perfect cubes, which of the following is NOT necessarily a perfect cube?

(A) m^6n^3

(B) mn

(C) $mn + 125$
(D) $-m$
(E) $(m - n)^9$

35. If $a \neq 0$, which of the following must be greater than a?

I. $10a$
II. $10a^2$
III. $10 - a$

(A) I only
(B) II only
(C) III only
(D) II and III only
(E) None

36. What is the rightmost non-zero digit in the product $6^{26} \times 14^{43} \times 15^{69}$?

(A) 2
(B) 4
(C) 5
(D) 6
(E) 8

37. How many trailing zeros does 30! have? Given that $n! = 1 \times 2 \times 3 \ldots (n - 1) \times n$.

(A) 3
(B) 6
(C) 7
(D) 15
(E) 26

38. Which of the following expressions has the highest value?

(A) 15^{50}
(B) 33^{40}
(C) 2^{200}
(D) 3^{100}
(E) 11^{50}

39. A number of coins are to be distributed among a number of children. If the coins were distributed equally among three, four or five children, there were one, two or three coins left over, respectively. What is the least number of coins?

(A) 55

(B) 56

(C) 57

(D) 58

(E) 62

40. Two numbers are such that their highest common factor is 18 and the sum is 216. How many such pairs of numbers exist?

(A) Two

(B) Four

(C) Six

(D) Ten

(E) Eleven

41. Two numbers are such that their highest common factor is 12 and the least common multiple is 360. How many such pairs of the numbers exist?

(A) One

(B) Two

(C) Four

(D) Seven

(E) Nine

42. If $A = 12! + 3 \times 13! + 24$, which of the following is a factor of A? Given that $n! = 1 \times 2 \times \ldots (n-1) \times n$.

(A) 5

(B) 10

(C) 15

(D) 20

(E) 24

43. A store has a stock of 18 pens, 45 pencils and 63 erasers. The owner wants to put them in boxes so that each box has an equal number of items of the same type. What is the minimum number of boxes required?

(A) 5

(B) 6

(C) 10

(D) 11

(E) 14

44. Which of the following must be true, if p and q are positive integers and p is a factor of q?

(A) $q = p \times k$, where k is an integer and $k > 1$

(B) Number of factors of p is less than that of q

(C) The highest common factor of p and q is more than 1

(D) The least common multiple of p and q is equal to q

(E) p^2 is greater than q

45. If $0 < mn < 1$, which of the following can be true?

(A) $m < -1$ and $n < -1$

(B) $m < -1$ and $n > 0$

(C) $m > -1$ and $n < -1$

(D) $m > 1$ and $n < -1$

(E) $m > 1$ and $n > 1$

46. If p and q are non-negative integers, $p + q < 11$ and $p - q > 8$, which of the following must be true for all the qualified values of p?

(A) $q < 3$

(B) $q > 2$

(C) $2 < q < 10$

(D) $q = 0$ or 1

(E) $q < 1$

47. If $\left|\frac{x}{y}\right| < 1$ and $y \neq 0$, which of the following must be greater than 1?

(A) $\frac{y}{x}$

(B) $\left(\frac{x}{y}\right)^2$

(C) $\frac{x}{y}$

(D) $\frac{|y|}{x}$

(E) $\sqrt{\left|\frac{y}{x}\right|}$

48. If $x \leq 2$ and $5x + 3y \geq 13$, which of the following could be the value of y?

(A) -2

(B) -1

(C) 0

(D) $\frac{1}{5}$

(E) 2

49. If $m > 0$, which of the following must be true?

I. $m - m^2 < 0$

II. $9m - 1 > 0$

III. $m^3 > 0$

(A) I only

(B) II only

(C) III only

(D) II and III only

(E) I, II, and III

50. How many integer values of a are possible, if a is an integer such that $a^2 - 4a < 221$?

(A) 27

(B) 28

(C) 29

(D) 30

(E) 31

Chapter 5

Answer Key: Number Properties

(1) C

(2) E

(3) C

(4) D

(5) E

(6) E

(7) E

(8) C

(9) B

(10) D

(11) D

(12) A

(13) C

(14) D

(15) B

(16) C

(17) C

(18) B

(19) C

(20) B

(21) C

(22) A

(23) E

(24) C

(25) C

(26) C

(27) C

(28) D

(29) B

(30) C

(31) B

(32) C

(33) B

(34) C

(35) E

(36) D

(37) C

(38) B

(39) D

(40) A

(41) C

(42) E

(43) E

(44) D

(45) C

(46) E

(47) E

(48) E

(49) C

(50) C

Chapter 6

Solution: Number Properties Practice Questions

1. Let us reduce the equation $5x + 3y = 15$ to $x = \frac{15-3y}{5} = 3 - \frac{3y}{5}$.

   $\frac{3y}{5}$ must be an integer so that x must be an integer. To get $\frac{3y}{5}$ as an integer, y must be multiplicand of 5. So set values of y would be $\{0, \pm 5, \pm 10, \pm 15, ...\}$.

   All the options except option C qualify the criterion.

   The correct answer is option C.

2. Say $\frac{m}{n} = k$; where k in an integer. It means that $m = kn$. This negates option E, so option E is the correct answer.

   Alternate approach:

   We apply process of elimination. Since this question is 'CAN BE TRUE' type of question, we must plug in at least one value for m and n in the options that satisfy the question.

   (A) Both m and n are integers - Say $m = 2$, and $n = 1$. It gives $\frac{m}{n} = \frac{2}{1} = 1$, an integer.

   (B) m is an integer - Again, say $m = 2$, and $n = 1$. It gives $\frac{m}{n} = \frac{2}{1} = 1$, an integer.

   (C) Either m or n is negative - Say $m = -2$, and $n = 1$. It gives $\frac{m}{n} = -\frac{2}{1} = -1$, an integer.

   (D) $\frac{n}{m}$ is an integer - Say $m = 1$, and $n = 1$. It gives $\frac{n}{m} = \frac{1}{1} = 1$, an integer, while $\frac{m}{n} = \frac{1}{1} = 1$ is also an integer.

   (E) $m = kn$; where k is non-integer - If $m = kn$, we can rewrite it as $\frac{m}{n} = k$. The question already states that $\frac{m}{n}$ is an integer, but the option says k or $\frac{m}{n}$ is not an integer. This option cannot be true.

   The correct answer is option E.

3. With the given condition, we get $(4 \# 3) = \frac{3 \times 4}{2} = 6$.

   Again $(4 \# 3) \# 1 = 6 \# 1 = \frac{1 \times 6}{2} = 3$.

   The correct answer is option C.

4. Given that the product of integers is negative means that there must be an odd number of negative integers in the list, because even number of integers will make product positive. So possible numbers of negative integers are 1, 3, 5, 7, and 9. The maximum number of possible negative integers would be 9.

   The correct answer is option C.

   The information *at least two integers are negative* is redundant here.

5. $abc = 1$ means that none of a, b, and c is 0. Similarly, if $cde = 0$, it means that either d or e or both are 0, because we already deduced that c cannot be 0.

Again, $ebc = 0$, means that e is 0, because we already deduced that b and c cannot be 0.

The correct answer is option E.

You might have observed that condition $cde = 0$ is redundant here. We inferred $e = 0$, by only two conditions: $abc = 1$, and $ebc = 0$.

6. Given: $0 < 5x + 11 < 300$.

We split this inequality into two.

(1) $0 < 5x + 11$
$=> x > -\frac{11}{5} = -2.20$
$=> x$ will have set of integers values as $\{-2, -1, 0, \ldots\}$

(2) $300 > 5x + 11$
$=> x < \frac{289}{5}$ or $x < 57.80$
$=>$ It means that x will have set of integers values as $\{57, 56, 55, 54 \ldots, -2\}$

So, the total number of integer values of x would be $3 + 57 = 60$.

The correct answer is option E.

7. We need to find a number which does not have a perfect square as a factor.

A perfect square is one which has an even exponent i.e. the number is of the form x^{2k}; where x and k are integers.

Working with the options:

Option A: $48 = 16 \times 3 = 4^2 \times 3$.

Thus, 4^2, a perfect square is a factor of 48.

Option B: $117 = 9 \times 13 = 3^2 \times 13$.

Thus, 3^2, a perfect square is a factor of 117.

Option C: $245 = 49 \times 5 = 7^2 \times 5$.

Thus, 7^2, a perfect square is a factor of 245.

Option A: $297 = 9 \times 33 = 3^2 \times 33$.

Thus, 3^2, a perfect square is a factor of 297.

Hence, the correct option must be option E.

Let us verify: $345 = 3 \times 5 \times 23$. Thus, there are no perfect squares as a factor of 345.

The correct answer is option E.

8. There is only 'one' two-digit number having the digit '4' in the units place and the digit '2' in the tens place i.e. 24.

Among three-digit numbers, the units and tens digits are 4 and 2, respectively.

The hundreds digit may be chosen from '1' to '9' in nine ways.

Thus, there are 'nine' such numbers.

Among four-digit numbers, the units and tens digits are 4 and 2, respectively.

The thousands digit may be chosen from '1' to '9' in nine ways and the hundreds digit may be chosen from '0' to '9' in 10 ways.

Total number of ways of choosing the two digits $= 9 \times 10 = 90$.

Thus, there are 90 such numbers.

There are no five-digit numbers having the digit '4' in the units place and the digit '2' in the tens place since the last number is 10,000 which is the least possible five-digit number.

Thus, we have $1 + 9 + 90 = 100$ such numbers.

Alternate approach:

There is only 'one' two-digit number having the digit '4' in the units place and the digit '2' in the tens place i.e. 24.

Thus, the least such number is 24.

Since the last two digits should always be '2' and '4,' successive numbers are obtained by adding 100 to the least number.

Thus, the largest such number is 9,924.

Thus, the numbers are 24, 124, 224, ... 9,924.

Thus, number of terms in the above series

$$= \frac{\text{Last term} - 1^{\text{st}} \text{ term}}{100} + 1 = \frac{9,924 - 24}{100} + 1 = 100.$$

The correct answer is option C.

9. The first nine stamps are numbered 1, 2 . . . 9.

Thus, the number of digits used up to number the first nine stamps = 9.

From numbers '10' to '99,' each number has two digits.

There are $99 - 10 + 1 = 90$ numbers from '10' to '99.'

Thus, number of digits used $= 90 \times 2 = 180$.

From 100 to 999, each number has three digits.

There are $999 - 100 + 1 = 900$ numbers.

Thus, number of digits required $= 900 \times 3 = 2,700$.

However, then the total number of digits would be exceeded i.e. the last number is much less than 999.

We have already used up $9 + 180 = 189$ digits.

Number of digits left to be used $= 855 - 189 = 666$.

Since each number has three digits, # of numbers used $= \frac{666}{3} = 222$.

Thus, 222 numbers starting from 100 would end in $100 + 222 - 1 = 321$.

Thus, the student wrote all numbers from 1 to 321 to use 855 digits.

Thus, he had 321 stamps.

The correct answer is option B.

10. There are six digits available, of which we need to select five to make a number.

A number is divisible by '3,' if the sum of its digits is divisible by '3.'

Hence, we need to choose five of the six digits in such a way that their sum is divisible by '3.'

The sum of all the six digits $= 0 + 1 + 2 + 3 + 4 + 5 = 15$, which is a multiple of '3.'

Since we need to remove a digit such that the sum remains divisible by '3,' the digit to be removed should also be divisible by '3' i.e. digits '0' or '3.'

Thus, we have the following cases:

(1) Digits used: 1, 2, 3, 4, 5: Number of five-digit numbers $= 5! = 120$.

(2) Digits used: 0, 1, 2, 4, 5: Number of five-digit numbers $= 4 \times 4 \times 3 \times 2 \times 1 = 96$ (since '0' cannot be used in the ten-thousands position, it can be filled in four ways and not five).

Hence, total number of five-digit numbers divisible by three = 120 + 96 = 216.

The correct answer is option D.

11. A number is divisible by '6' implies that it is divisible by both '2' and '3.'

Since $a12a$ is divisible by '2,' the digit in the units place i.e. a must be even.

Thus, possible values of a are: 2, 4, 6 or 8 (a cannot be '0' since the digit in the thousands place of a four-digit number cannot be '0').

Since the number is divisible by '3,' the sum of digits must be divisible by '3.'

Sum of the digits of the number $a12a = a + 1 + 2 + a = 2a + 3$.

Thus, $(2a + 3)$ must be divisible by '3.'

Since in the number $(2a + 3)$, 3 is divisible by 3, $2a$ must be divisible by 3 i.e. a is divisible by 3.

Thus, checking with the possible values of a i.e. 2, 4, 6 or 8, we see than only $a = 6$ satisfies the above condition.

Hence, we have $a = 6$.

The correct answer is option D.

Alternatively, you can plug in the values of a from the options, and check which number thus formed is divisible by 6.

12. The units digit of 13^{22} is the same as that of 3^{22}; it is governed by the digit at the unit place.

The same logic works for the other numbers too.

Thus, the question is equivalent to finding the units digit of $\left(13^{22} \times 12^{22}\right) + \left(17^{27} \times 25^{25}\right)$.

Since the exponents of 3 and 2 are the same, we can write $3^{22} \times 2^{22} = (3 \times 2)^{22} = 6^{22}$.

We know that the last digit of 6 raised to any positive integer exponent is always 6.

Similarly, for $7^{27} \times 5^{25}$, we observe that 7^{27} is odd and 5^{25} has units digit of 5 (since the last digit of 5 raised to any positive integer exponent is always 5).

Thus, the last digit of $7^{27} \times 5^{25}$ is essentially an odd number multiplied with 5 which will always be 5 irrespective of the odd number being multiplied to.

Hence, the last digit of $\left(13^{22} \times 12^{22}\right) + \left(17^{27} \times 25^{25}\right)$ is the last digit of $6 + 5 = 1$.

The correct answer is option A.

13. $\sqrt{7}$ percent of $7\sqrt{7} = \sqrt{7}\,\%$ of $7\sqrt{7} = \frac{\sqrt{7} \times 7\sqrt{7}}{100} = \frac{49}{100} = 0.49$.

The correct answer is option C.

14. $(5 \times 10^6) + (9 \times 10^4) + (2 \times 10^2) = 5,000,000 + 90,000 + 200 = 5,090,200$.

The correct answer is option D.

Alternate approach:

The value can be inferred mentally. Since the maximum power of the number $\left(5 \times 10^6\right) + \left(9 \times 10^4\right) + \left(2 \times 10^2\right)$ is 10^6, which will have six 0s, the number would be a seven digit number; thus, the correct answers could be either D or E. The second term (9×10^4) must have four digits after 9, so the correct answer is 5,090,200.

15. $(x - y)^2 + (y - z)^2 + (z - x)^2$

=> $(x^2 - 2xy + y^2) + (y^2 - 2yz + z^2) + (z^2 - 2zx + x^2)$

=> $2\,(x^2 + y^2 + z^2 - xy - yz - zx)$

=> $2\,\{(x^2 - yz) + (y^2 - xz) + (z^2 - xy)\}$

=> $2\,\{(-5) + 1 + 7\}$

=> 6

The correct answer is option B.

16. In this problem, both a and b are positive integers.

In the equation, we have two terms which are multiples of 5 viz. $5b$ and 315.

Hence, a^2 must be a multiple of 5.

However, since a^2 is a perfect square and a multiple of 5, a^2 must be a multiple of 25 implying that a is a multiple of 5.

Since we need to find the largest value of a, we choose the largest multiple of 5 such that its square is less than 315.

Thus, we choose $a = 15$ (since $15^2 = 225 < 315$)

Hence, we have $15^2 + 5b = 315 \Rightarrow 5b = 315 - 225 = 90 \Rightarrow b = 18$.

However, in this case $b = 18$ is not a prime number; the question states that b is a prime number, so this is not a qualified value.

Taking $a = 10$ (the next largest possible multiple of 5)

We have $10^2 + 5b = 315 \Rightarrow 5b = 215 \Rightarrow b = 43$: a prime number (a qualified value).

Thus, the largest possible value of $a = 10$.

The correct answer is option C.

17. Results of multiplication of Even and Odd numbers

- Even $\times$ Even = Even
- Even $\times$ Odd = Even
- Odd $\times$ Odd = Odd

It must be noted that if only one even number is present, the product is even; whereas, if product of two numbers is Odd, then both the numbers must be Odd.

Let us look at the options one by one.

(A) $mn + 2n$: We can write this expression as
[(odd or even) $\times$ odd + even $\times$ odd] = [(odd or even) $\times$ odd + even]. Since we do not know the nature of m, we cannot predict whether $(mn + 2n)$ is odd or even.

(B) $mn + 2m$: We can write this expression as
[(odd or even) $\times$ odd + even $\times$ (odd or even)] = [(odd or even) $\times$ odd + even].

Since product of even or odd with any even number is even, we can write even. [(odd or even)] = even, but we cannot predict whether [(odd or even) $\times$ odd] is odd or even.

(C) $(n + 3)(m - 3)$: We can write this expression as
[(odd + odd) $\times$ (even or odd $-$ odd)] = [even $\times$ (even or odd $-$ odd)] . We need not bother about the fate of (even or odd $-$ odd), because whether it is odd or even, it is multiplied by an even number. Further, we already know that any number multiplied by an even number will result in an even number. So $(n+3)(m-3)$ must be an even number.

The correct answer is option C.

18. The question is if n is even, which of the following CANNOT be odd?

Results of Even and Odd numbers

- Even $\times$ Even = Even
- Even $\times$ Odd = Even
- Odd $\times$ Odd = ODD

It must be noted that if there is only one even number is present, the product is even, whereas, if product of two numbers is Odd, then both the numbers must be Odd.

Let us look at the options one by one.

I. $2n + 7$: In this option, 2 and n are even, whereas, 7 is odd.
So, $2n + 7$ = even × even + odd = even + odd = odd. Thus, $2n + 7$ is an odd number.

II. $3n - 2$: In this option, 2 and n are even, whereas, 3 is odd.
So, $3n - 2$ = odd × even − even = even − even = even. Thus, $3n - 2$ is an even number. Correct answer.

III. $5n^2 + 1$: In this option n is even, whereas, 5 and 1 are odd.
So, $5n^2 + 1$ = odd × even × even + odd = even + odd = odd. Thus, $5n^2 + 1$ is an odd number.

The correct answer is option B.

19. Consecutive odd integers from one are: 1, 3, 5, 7, 9, ...

We see that there is a constant gap of '2' between consecutive terms of the above series and hence, it constitutes an arithmetic progression.

For an arithmetic progression, the sum of terms is given as:

$$\left(\frac{\text{first term} + \text{last term}}{2}\right) \times (\text{number of terms})$$

In the above arithmetic progression, the 1st term is '1' and the last term is 'p'.

Since we start with '1' (an odd number) and end with 'p,' another odd number, the number of terms in the series $= \left(\frac{p-1}{2} + 1\right) = \left(\frac{p+1}{2}\right)$ (for example: For the series 1, 3, 5, 7, 9; the number of terms $= \frac{9-1}{2} + 1 = 5\Big)$.

Thus, we have

$$\left(\frac{1+p}{2}\right) \times \left(\frac{p+1}{2}\right) = 225$$

$$=> \left(\frac{p+1}{2}\right)^2 = 225$$

$$=> \frac{p+1}{2} = 15$$

$$=> p = 29$$

Alternate approach:

Let us take a few terms of the series and see if there is any pattern apparent in the value of the sum of the terms:

- One term: 1: Sum = 1 = 1^2
- Two terms: 1, 3: Sum = 4 = 2^2
- Three terms: 1, 3, 5: Sum = 9 = 3^2

Thus, we observe that the sum is actually the square of the number of terms added.

Since $225 = 15^2$, it implies that we need to add 15 terms to get the sum as 225.
Thus, we need to find the 15th term of the series: 1, 3, 5, 7 ...
The nth term of an arithmetic progression is given by: $a + (n - 1)d$; where a is the 1st term, n is the number of terms and d is the common difference.

Here: $a = 1, n = 15$ and $d = 2$.

Thus, the 15th term $= 1 + (15 - 1) \times 2 = 29$.

The correct answer is option C.

20. Since we need even numbers divisible by '13,' we should look at numbers $13 \times 2, 13 \times 4, 13 \times 6 \ldots$

The last number less than 300 divisible by '13' is $299 = 13 \times 23$.

Since we need even numbers, the last multiple of '13' would be 13×22.

Thus, the required set of numbers is: $13 \times 2, 13 \times 4, 13 \times 6, \ldots 13 \times 20, 13 \times 22$, i.e., 11 numbers.

However, we do not want numbers divisible by '7.'

In the above set, there is only one multiple of '7' i.e. 13×14.
Thus, we need to exclude it.

Hence, the # of such numbers $= 11 - 1 = 10$.

The correct answer is option B.

21. The factors of $a \times b$, where a and b are prime, are $1, a, b$ and $a \times b$.

Thus, we have

$1 + a + b = 31$; the question asks to exclude $a \times b$ as a factor.

$=> a + b = 30$

Since a and b are prime, we can have the following cases for $a + b = 30$:

a	***b***
7	23
11	19
13	17

Thus, there are three possible numbers: $7 \times 23 = 161, 11 \times 19 = 209$, and $13 \times 17 = 221$.

The correct answer is option C.

22. Let the number be a^2, where a is a prime number.

The factors of a^2 are 1, a and a^2.

Thus, we have

$1 + a + a^2 = 57$

$=> a + a^2 = 56$

$=> a^2 + a - 56 = 0$

$=> a^2 + 8a - 7a - 56 = 0$

$=> a(a + 8) - 7(a + 8) = 0$

$=> (a - 7)(a + 8) = 0$

$=> a = 7$ or -8

Since a is a prime number, we have $a = 7$.

Hence, we have only one such number.

The correct answer is option A.

You may cross check this. For $a^2 = 49$, we have factors $1, 7$, & 49, and their sum $= 1 + 7 + 49 = 57$.

23. We need to work with the options:

Option A: Factors of 12 (other than 12) are: 1, 2, 3, 4, and 6.
Sum of the above factors = $1 + 2 + 3 + 4 + 6 = 16 \neq 12$.

Hence, option A is incorrect.

Option B: Factors of 18 (other than 18) are: 1, 2, 3, 6, and 9.
Sum of the above factors = $1 + 2 + 3 + 6 + 9 = 21 \neq 18$.

Hence, option B is incorrect.

Option C: Factors of 20 (other than 20) are: 1, 2, 4, 5, and 10.
Sum of the above factors = $1 + 2 + 4 + 5 + 10 = 22 \neq 20$.

Hence, option C is incorrect.

Option D: Factors of 21 (other than 21) are: 1, 3, and 7.

Sum of the above factors = $1 + 3 + 7 = 11 \neq 21$.

Hence, option D is incorrect.

Hence, option E must be correct. Let us verify:

Option E: Factors of 28 (other than 28) are: 1, 2, 4, 7, and 14.

Sum of the above factors = $1 + 2 + 4 + 7 + 14 = 28$.

The correct answer is option E.

A few perfect numbers are 6, 28, and 496.

24. Since we need even numbers from 10 to 100, we express them as multiples of 2:

- $10 = 2 \times 5$ (first one)
- $100 = 2 \times 50$ (last one)

Thus, there are $50 - 5 + 1 = 46$ even integers from 10 to 100.

From 2×5 to 2×50, the multiples of 7 are 2×7; 2×14; 2×21; 2×28; 2×35; 2×42 and 2×49; there are seven numbers.

Hence, the number of even numbers not a multiple of 7 are: $46 - 7 = 39$.

The correct answer is option C.

25. Since p^2 is a perfect square number and is divisible by 5, which is not a perfect square, p^2 must be a multiple of 25 or p itself must be a multiple of 5.

Now, we do not know whether p is an even multiple of 5 (i.e. divisible by 10) or an odd multiple of 5 (i.e. not divisible by 10).

Since the question asks us to find a term divisible by 10, among the options, we must have $2p$ as one of the term.

Thus, the correct answer can only be options C or D.

Again, since q is not divisible by 5, option D, $2(5p + q) = (10p + 2q)$ will not be divisible by 5 (though $10p$ is divisible by 5, $2q$ is not divisible by 5) or will not be divisible by 10 either.

Hence, the answer must be option C.

Let us see how.

Option C: $2(p + 5q) = 2p + 10q$

It has two terms: $2p$ (which is divisible by 10 as discussed earlier) and $10q$ (also divisible by 10). Hence, $(2p + 10q)$ is divisible by 10.

The correct answer is option C.

Alternate approach:

This question can solved with a plug-in value approach.

Let us assume that $p^2 = 25$ (perfect square and divisible by 5) => $p = 5$ and $q = 3$ (not divisible by 5). By plugging in the values in options, we find that only option C works.

26. Let us calculate the number of integers divisible by 6.

The first integer divisible by 6 within the given range of 150 to 250 is 150.

Every 6th integer after this will be divisible by 6.

Thus, the set of integers divisible by 6 is $\{150, 156, 162, \ldots, 246\}$

So, number of integers in the series $= \dfrac{246 - 150}{6} + 1 = 17.$

We can also calculate the same in the following manner:

The first integer divisible by 6 within the given range of 150 to 250 is $150 = 6 \times 25$.

The last integer in the series is $246 = 6 \times 41$.

So, the number of integers in the series $= 41 - 25 + 1 = 17$.

Similarly, let us calculate the number of integers divisible by 5.

The first integer divisible by 5 within the given range of 150 to 250 is $150 = 5 \times 30$.

The last integer in the series is $250 = 5 \times 50$.

So, the number of integers in the series $= 50 - 30 + 1 = 21$.

Now, few integers would be common in both the above series. These integers would be a multiple of the LCM of 6 and 5 i.e. 30; we must exclude them once.

The first integer divisible by 30 within the given range of 150 to 250 is $150 = 30 \times 5$.

The last integer in the series is $240 = 30 \times 8$.

So, the number of integers in the series $= 8 - 5 + 1 = 4$.

So, the number of integers which are divisible by 6 or 5 or both are: $17 + 21 - 4 = 34$.

Total number of integers in the range from 150 to 250 $= 250 - 150 + 1 = 101$.

Thus, the number of integers neither divisible by 6 nor by 5 is $101 - 34 = 67$.

The correct answer is option C.

27. Say the number X is such that $X = 7n + 2$; where n is a non-negative integer; $n = \{0, 1, 2, 3 \ldots\}$

Option analysis:

I. X is odd: It is not must be true. If $n = 2, X = 7 \times 2 + 2 = 16$, which is even.

II. X cannot be a prime number: It is not must be true. If $n = 0, X = 7 \times 0 + 2 = 2$, which is prime.

III. $(X + 5)$ divided by 7 has remainder 0: This condition must be true.

$X + 5 = 7n + 2 + 5 = 7n + 7 = 7(n + 1)$, which is a multiplicand of 7 and hence, completely divisible by 7 or leaves a remainder 0.

The correct answer is option C.

Alternate approach:

If X, divided by 7 leaves a remainder 2, X would have either value from set {2, 9, 16, 23, ...}.

Option analysis

I. X is odd: It is not must be true. Have a look at the set above. 16 is even.

II. X cannot be a prime number: It is not must be true. Have a look at the set above. 2 and 23 are prime.

III. $(X + 5)$ divided by 7 has remainder 0: This condition must be true. For any value from the set, if you add 5, the number would be completely divisible.

Another way of understanding this is: Since X, divisible by 7 has 2 as remainder. This means that X is short by 5 to make it completely divisible. Hence $X + 5$ would be completely divisible or leaving a reminder of 0.

The correct answer is option C.

28. We know that $p! = 1 \times 2 \times 3 \times \cdots \times (p - 1) \times p$.

Thus, we have

- $1! = 1$
- $2! = 1 \times 2 = 2$
- $3! = 1 \times 2 \times 3 = 6$
- $4! = 1 \times 2 \times 3 \times 4 = 24$
- $5! = 1 \times 2 \times 3 \times 4 \times 5 = 120$
- $6! = 1 \times 2 \times 3 \times 4 \times 5 \times 6 = 5! \times 6$
- $7! = 1 \times 2 \times 3 \times 4 \times 5 \times 6 \times 7 = 5! \times 42$, etc.

Thus, we observe that all factorials from $5!$ onwards are multiples of 10 and hence, would not leave any remainder when divided by 10.

The remainder would only come from the first four terms which are not a multiple of 10.
Thus, we have $1! + 2! + 3! + 4! = 1 + 2 + 6 + 24 = 33$.
Thus, the remainder when N is divided by 10 is the same as when 33 is divided by 10, which is 3.

Let's test this.

$1! + 2! + 3! + 4! + 5! = 1 + 2 + 6 + 24 + 120 = 153.$

Similarly, for others...

Hence, the remainder is 3.

The correct answer is option D.

29. Number of multiples of '5' between 1 and 250 = $\frac{250}{5} - 1 = 50 - 1 = 49$ (since we want the numbers to be between 1 and 250, we cannot include 250 as a possible multiple of '5').

Number of multiples of '7' between 1 and 250 is the quotient on dividing 250 by 7 (we do not subtract '1' since 250 is not a multiple of '7') i.e. quotient of $\frac{250}{7}$ i.e. 35.

However, there are some numbers which are multiples of both 5 and 7 i.e. 35 which have been counted in both the above cases. We need to remove them since we do not want numbers divisible by both 5 and 7.

Number of multiples of '35' between 1 and 250 is the quotient on dividing 250 by 35 i.e. quotient of $\frac{250}{35}$ i.e. 7.

Thus, the total # of such numbers = $(49 - 7) + (35 - 7) = 42 + 28 = 70$.

The correct answer is option B.

It is to be noted that $49 + 35 - 7 = 77$ is a wrong answer as '77' includes those numbers too that are divisible by both 5 and 7.

30. Given that, $10! + 11! + 12!$

$= 10! + 11 \times 10! + 12 \times 11 \times 10!$

$= 10!\,(1 + 11 + 12 \times 11)$

$= 10!\,(12 + 12 \times 11)$

$= 10! \times 12 \times (1 + 11)$

$= 10! \times 12 \times 12$

The highest exponent of 2 in 10! is calculated as shown below. We have seen this process in detail in earlier questions.

- $\frac{10}{2} = \mathbf{5}$

- $\frac{5}{2} = 2.5 = \mathbf{2}$ (Integer part)
- $\frac{2}{2} = \mathbf{1}$

Thus, the exponent of 2 in 10! is **5 + 2 + 1 = 8.**

Again, $12 \times 12 = 12^2 = (2^2 \times 3)^2 = 2^4 \times 3^2$.

Thus, the exponent of 2 in 12×12 is **4.**

Since the terms are multiplied, the exponents of 2 will be added.

Thus, the exponent of 2 in 10! + 11! + 12! is **8 + 4 = 12.**

Hence, the highest exponent 2 in (10! + 11! + 12!) is 12.

The correct answer is option C.

31. $(p + 4)\,(p + 6)$
$=> p^2 + 10p + 24$
$=> p\,(p + 10) + 24.$

Thus, $p\,(p + 10) + 24$ must be divisible by p.

This is possible when 24 is divisible by p.

Thus, p should be the factors of 24 (other than 1) i.e. 2, 3, 4, 6, 8, 12 and 24.

Hence, there are seven possible values of p.

The correct answer is option B.

32. $2^{20} = 2^{15}x + y$
$=> x = \frac{2^{20} - y}{2^{15}} = \frac{2^{20}}{2^{15}} - \frac{y}{2^{15}}.$

Since x is an integer, y must be divisible by 2^{15}.
Thus, $y = 2^{15}k; k$ is an integer.

Hence, $2^{20} = 2^{15}x + 2^{15}k => 2^{20} = 2^{15}(x + k) => x + k = \frac{2^{20}}{2^{15}} = 2^5.$

We need to find such values of x and k such that $|x - y|$ is minimum or the difference between x and y is minimum.

Putting $k = 0$, we have $x = 2^5 = 32$.
Thus, $y = 2^{15}k = 0$.

Hence, the difference between x and y is $32 - 0 = 32$.

Putting $k = 1$, we get $x = 2^5 - 1 = 31$.
Thus, $y = 2^{15}k = 2^{15}$ = a higher value.

Hence, the difference between x and $y = |x - y|$ = a higher value.

Similarly, putting $k = 2$, we get $x = 2^5 - 2 = 30$.
Thus, $y = 2^{15} \times 2 = 2^{16}$ = a higher value.

Hence, the difference between x and $y = |x - y|$ = a higher value.

In this case, the difference between x and y is much greater than 32.
As k increases, the difference between x and y would increase.

Hence, the minimum difference is 32 or 2^5.

The correct answer is option C.

33. The units digit of exponents of 7 and 3 follow a cycle of 4 places as shown below:

Number with exponent	Units digit
7^1	7
7^2	9
7^3	3
7^4	1

Number with exponent	Units digit
3^1	3
3^2	9
3^3	7
3^4	1

Thus, $7^{123} = (7^4)^{30} \times 7^3$=> Units digit is 3.

$3^{42} = (3^4)^{10} \times 3^2$=> Units digit is 9.

Hence, the units digit of $7^{123} + 3^{42}$ is the units digit of $3 + 9 = 12$ i.e. 2.

The correct answer is option B.

34. Find out the cube root of options and figure out whether the value is an integer value or not? The option does not return an integer is not a perfect cube.

(A) Cube root of $m^6n^3 = [(m^6\ n^3)]^{\frac{1}{3}} = m^2n$. This is a integer, hence, option A is a perfect cube.

(B) Cube root of mn: Since m and n are perfect cube, their products will also be perfect cube.

(C) Cube root of $mn + 125$: This is the correct answer. Perfect cube + Perfect cube will not necessarily be a perfect cube. For example, if $m = n = 1$, then $mn + 125 = 1 + 125 = 126$; not a perfect cube.

(D) Cube root of $-m$: Since m is perfect cube, $-m$ will also be perfect cube.

(E) Cube root of $(m-n)^9 = \left[(m-n)^9\right]^{\frac{1}{3}} = (m-n)^3$. This is an integer; hence, option E is a perfect cube.

The correct answer is option C.

Alternate approach:

Pick up some value. Say we take $m = 1$, and $n = 8$.

(A) $m^6n^3 = 1^6.8^3 = 1.8^3 = 8^3$. Cube root of 8^3 is 8, which is an integer; hence, option A is a perfect cube.

(B) $mn = 1.8 = 8$. Cube root of 8 is 2, which is an integer; hence, option B is a perfect cube.

(C) $mn + 125 = 1.8 + 125 = 133$. Cube root of 133 is not an integer; hence, option C is not a perfect cube. Correct answer.

(D) $-m = -1$. Cube root of -1 is -1, which is an integer; hence, option D is a perfect cube.

(E) $(m-n)^9 = (1-8)^9 = (-7)^9$. Cube root of $(-7)^9$ is $(-7)^3$, which is an integer; hence, option E is a perfect cube.

35. This is a MUST BE TRUE type of question, we must pick at least one value for each option such that the option breaks. If the value does holds true for all cases, the option is the correct answer.

It is to be noted that $a \neq 0$ means that a can take any values–negative, positive or fractional, except '0.'

I. $10a$: Say $a = -1$; thus, $10a = -10$. Since $-10 \ngtr a(-1)$, this option is incorrect.

II. $10a^2$: Say $a = \frac{1}{10}$; thus, $10a^2 = 10 \times \left(\frac{1}{10}\right) \times \left(\frac{1}{10}\right) = \frac{1}{10}$. Since $\frac{1}{10} \ngtr a(= \frac{1}{10})$, this option is also incorrect.

III. $10 - a$: Say $a = 10$; thus, $10 - a = 10 - 10 = 0$. Since $0 \ngtr a(= 10)$, this option is also incorrect.

The correct answer is option E.

36. $6^{26} \times 14^{43} \times 15^{61}$

$=> (2 \times 3)^{26} \times (2 \times 7)^{43} \times (3 \times 5)^{61}$

$=> 2^{26+43} \times 3^{26+61} \times 5^{61} \times 7^{43}$

$=> 2^{69} \times 3^{87} \times 5^{61} \times 7^{43}$

Every '2' would combine with a '5' to result in '10' i.e. a '0' at the right in the product.

Since there are sixty nine 2s and sixty one 5s, there would be 69 – 61 = eight 2s left over.

Thus: $2^{69} \times 3^{87} \times 5^{61} \times 7^{43} = (2^8 \times 3^{87} \times 7^{43}) \times 10^{61}$

Since we need the rightmost non-zero digit, we can ignore 10^{61}.

The units digit of exponents of 7, 3 and 2 follow a cycle of 4 places as shown below:

Number with exponent	Units digit
7^1	7
7^2	9
7^3	3
7^4	1

Number with exponent	Units digit
3^1	3
3^2	9
3^3	7
3^4	1

Number with exponent	Units digit
2^1	2
2^2	4
2^3	8
2^4	6

$2^8 => (2^4)^2 =>$ Units digit = 6.

$3^{87} => (3^4)^{21} \times 3^3 =>$ Units digit = $1 \times 7 = 7$.

$7^{43} => (7^4)^{10} \times 7^3 =>$ Units digit = $1 \times 3 = 3$.

Thus, the units digit of $2^8 \times 3^{87} \times 7^{43}$ is the units digit of $6 \times 7 \times 3 = 6$.

The correct answer is option D.

37. Trailing zeros refers to the zeroes at the end of a number before the appearance of any non-zero digit. For example, the number of trailing zeroes in 10200 is two and not three.

The trailing zero appears due to 10 appearing as a multiplicand i.e. if there are k trailing zeroes, then the number has 10^k as a factor.

'10' itself has two factors '2' and '5.'

Thus, if a number has p as the exponent of 2 and q as the exponent of 5, then the exponent of 10 would be the lesser of p and q.

In order to find the exponent of 2 in the factorial of a number n, i.e. $n!$, we need to keep dividing n successively by 2 and then adding the integer parts of the quotients attained in each step.

Similarly, to find the exponent of 5 in $n!$, we need to keep dividing n successively by 5 and then adding the integer parts of the quotients attained in each step.

It is obvious that the sum of the quotients for 5 would be less than that for 2.

Hence, the exponent of 10 would be the same as the exponent for 5, which would also be the number of trailing zeroes.

Here, we have to find the number of trailing zeroes in 30!.

Hence, according to the argument above, we find the exponent of 5 in 30!, the process of which is shown below:

- $\frac{30}{5} = \mathbf{6}$
- $\frac{6}{5} = 1.2 \Rightarrow$ **1 (Integer part)**

Hence, the exponent of 5 in 30! = **6** + **1** = **7**.

Hence, the number of trailing zeroes in 30! is 7.

The correct answer is option C.

Alternate approach:

As discussed above that the exponents of 10 will bring in 0s in a number, and we know that 2×5 makes 10, thus between a 2 and a 5, whichever has fewer 0s will make that many 10s.

Since $5 > 2$, it is obvious that there would be less number of 5s than 2s.

So let us count 5s in 30. We see that multiples of 5 are: 5, 10, 15, 20, 25(two 5s), and 30. Each multiplicand except 25 would yield one 5; thus, total number of 5s are 7.

38. We need to have either a common base or a common exponent in order to compare the values. However, either method is not possible in the above example.

Let us approach in a different way:

Let us work with the options:

Option A: $15^{50} < 16^{50} = (2^4)^{50} = 2^{200} \ldots$(i)

Option B: $33^{40} > 32^{40} = (2^5)^{40} = 2^{200} \ldots$(ii)

Option D: $3^{100} < 4^{100} = (2^2)^{100} = 2^{200} \ldots$(iii)

Option E: $11^{50} < 16^{50} = (2^4)^{50} = 2^{200} \ldots$(iv)

Thus, we observe that only $33^{40} > 2^{200}$ and the others are less than 2^{200}.

Hence, 33^{40} is the greatest of all the options.

The correct answer is option B.

39. Let the number of coins be n.

We observe that:

- n when divided by 3 leaves a remainder 1 i.e. $(n + 3 - 1) = (n + 2)$ would be divisible by 3
- n when divided by 4 leaves a remainder 2 i.e. $(n + 4 - 2) = (n + 2)$ would be divisible by 4
- n when divided by 5 leaves a remainder 3 i.e. $(n + 5 - 3) = (n + 2)$ would be divisible by 5

Thus, we have $(n + 2)$ is divisible by the LCM of 3, 4 and 5 i.e. 60.

Thus, the minimum value of $n = 60 - 2 = 58$. (LCM is the least value)

Let us verify: 58 divided by 3 leaves a remainder 1, divided by 4 leaves a remainder 2, and divided by 5 leaves a remainder 3.

The correct answer is option D.

40. Since the numbers have 18 as their HCF, they must be multiples of 18.

Let the numbers be $p = 18a$ and $q = 18b$; where a and b are numbers having no common factor except '1,' we can also call them co-primes.

Thus, we have

$18a + 18b = 216$

$=> a + b = \frac{216}{18} = 12.$

Since a and b have no common factors (except '1'), the possible cases are:

a	b	$p = 18a$	$q = 18b$
1	11	18	198
5	7	90	126
7	5	126	90
11	1	198	18

Thus, there are only two pairs of such numbers: (18, 198) and (90, 126).

The correct answer is option A.

41. Since the numbers have 12 as their HCF, they must be multiples of 12.

Let the numbers be $p = 12a$ and $q = 12b$; where a and b are numbers having no common factor except '1.'

Thus, their least common multiple = $12ab$.

Thus, we have

$12ab = 360$

$=> ab = \frac{360}{12} = 30.$

Since a and b have no common factors (except '1'), the possible cases are:

a	b	$p = 12a$	$q = 12b$
1	30	12	360
2	15	24	180
3	10	36	120
5	6	60	72
6	5	72	60
10	3	120	36
15	2	180	24
30	1	360	12

Thus, there are four pairs of such numbers: (12, 360), (24, 180), (36, 120) and (60, 72).

The correct answer is option C.

42. Given that, $A = 12! + 3 \times 13! + 24$

$=> 12! + 3 \times 13 \times 12! + 24$

$=> 12!\,(1 + 39) + 24$

$=> 12! \times 40 + 24$

$=> 8\,(12! \times 5 + 3)$

Thus, A is definitely a multiple of '8'; however, there is no option as 8, so we analyze further.

Again, 12! is a multiple of 3.

Thus, $(12! \times 5 + 3)$ is also a multiple of 3.

Hence, A is a multiple of both 3 and 8 i.e. 24.

We can verify that A is not a multiple of 5.

Referring to the expression $A = 12! + 3 \times 13! + \mathbf{24}$, we see that both 12! and 13! are multiples of 5; however, 24 is not.

Hence, A is not a multiple of 5, thus, options A, B, C and D are incorrect.

The correct answer is option E.

Alternate approach:

Given that, $A = 12! + 3 \times 13! + 24$, the third term of A is 24, and we know that $12! = 1 \times 2 \times \mathbf{3} \cdots \times \mathbf{8} \cdots \times 11 \times 12$ and $13! = 1 \times 2 \times \mathbf{3} \cdots \times \mathbf{8} \cdots \times 11 \times 12 \times 13$ are multiples of 24, thus A must be a multiple of 24.

43. The number of boxes required would be a minimum if each box contains the maximum number of items of a single type.

Since we need to have the same number of items in each box, the number of items should be the highest common factor of the number of items of each category

Highest common factor of 18, 45 and 63 = 9.

Thus, number of boxes required for pens = $\frac{18}{9} = 2$.

Similarly, number of boxes required for pencils = $\frac{45}{9} = 5$.

Similarly, number of boxes required for erasers = $\frac{63}{9} = 7$.

Hence, the number of boxes = 2 + 5 + 7 = 14.

The correct answer is option E.

44. We need to work with the options:

Option A: Since p is a factor of q, we can have a situation such that $p = q$ (since q is also a factor of itself).

Thus, $q = p \times 1$ is a possible scenario; here $k = 1 \ngtr 1$.

Hence, option A is incorrect.

Option B: Going by the above logic, we can have $p = q$.

Thus, number of factors of p and q can be equal.

Hence, option B is incorrect.

Option C: Since p is a factor of q, we can have $p = 1$.

Then, the highest common factor of p and q can be 1.

Hence, option C is incorrect.

Option D: Since p is a factor of q, p is less than or equal to q and we can say that $q = p \times k$, where $k \geq 1$.

Thus, we can say that q is a multiple of p.

Thus, the least common multiple of p and q must be q.

Hence, option D is correct.

Option E: Since p is a factor of q, p may be at least 1 or at the most q.

If $p = 1$, then obviously $p^2 \leq q$.

Hence, option E is incorrect.

The correct answer is option D.

45. This is a **'can be true'** type of question.

Let us see the difference between **'must be true'** and **'can be true'** types of questions.

For **'must be true'** types of questions, under all circumstances, all the possible values of variable(s) must satisfy the equality or the inequality.

For **'can be true'** types of questions, at least one, not necessarily all, values of variable(s) should satisfy the equality or the inequality.

Let's draw the range of $0 < mn < 1$ on number line.

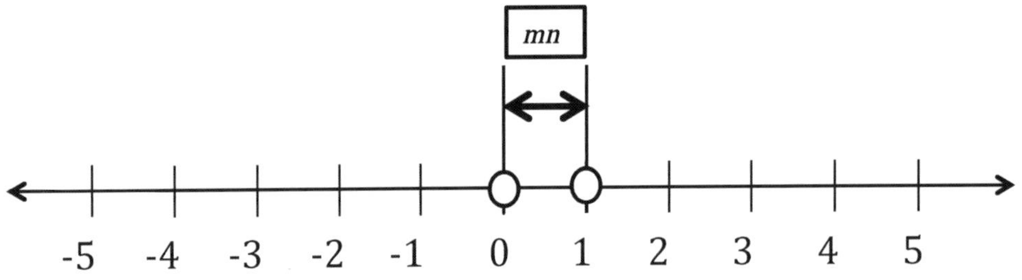

It is evident that $\boldsymbol{mn}$ is a positive number.

Since the product of two variables is positive, both are either positive or negative.

Let us analyze the options one by one.

(A) **$m < -1$ and $n < -1$**

$m < -1$ & $n < -1$ mean that both m & n are negative numbers. m and n both being negative satisfies above criterion. But this is not the correct answer, since product of m & n, if each equals -1, makes mn exactly equal to 1. But their actual values are less than -1.

On a number line, when a negative number decreases in value, its absolute value increases.

For the given condition, m & n have values whose absolute values are greater than 1. This makes the product of m & n greater than 1.

(B) **$m < -1$ and $n > 0$**

Here, $m < -1$ and $n > 0$ means that m is negative, and n is positive. It does not satisfy our criterion of either both the variables being positive or both the variables being negative, so this is not the correct answer.

(C) $m > -1$ **and** $n < -1$

Here, $n < -1$ means that n is negative, whereas, $m > -1$ means that m can be negative, positive, or even 0. This may or may not satisfy our criterion. It will depend on what value m assumes. Since the question is not a must be true kind, we must analyze further.

As stated earlier that for **'can be true'** type of questions, get at least one value that satisfies the criterion. Our criterion is that the product of m & n should lie between 0 & 1. Can you pick any suitable values for m and n?

These may be,

Say, $m = -3/4, -1/2, -1/4, 0, 1/4, 1/2, 1, 2, 3 \ldots\ldots$ and $n = -\frac{3}{2}, -2, -3, -4\ldots$

If we pick $m = -1/4$ and $n = -2$, $mn = \frac{1}{2}$, lying in the range of mn.

So option C is the correct answer.

There is no need to analyze other options as we are satisfied with option C, however, for the sake of your understanding, we discuss other options too.

(D) $m > 1$ **and** $n < -1$

Here, $m > 1$ means that m is positive, and n is negative. It does not satisfy our criterion of either both being positive or both being negative. It is not the correct answer.

(E) $m > 1$ **and** $n > 1$

Here, $m > 1$ & $n > 1$ mean that both m & n are positive. It satisfies our criterion, but this is not the correct answer. Since both m & n will have values greater than 1, which makes their product greater than 1. So mn would lie on the right hand side of the actual range of $0 < mn < 1$. It is not the correct answer.

The correct answer is option C.

46. $p + q < 11$, and $p - q > 8$ are two linear inequalities. Looking at the options, we find that we have to find the range of q. This calls for solving the inequalities. However, the challenge with the given inequalities is that their directions are opposite. We must make their directions same and then add them to get the range of q.

Also, given that p and q are non-negative integers.

=> p and q can be 0 or positive integers.

Approach 1:

Given that, $p + q < 11$, and $p - q > 8$.

Say, we decide to change the direction of the second inequality. To get that, we multiply it with -1.

=> $-1 \times (p - q > 8)$

=> $-p + q < -8$. (Note that the sign of inequality is reversed)

Now we can add both the inequalities.

$p + q < 11$ - - - - -(1)

$-p + q < -8$ - - - - -(2)

We get, $2q < 3$

=> $q < \frac{3}{2}$ => 0 or 1

=> Correct option seems to be D, but we must check the corresponding values of p for q.

If $q = 0$, x is either 9 or 10: both are qualified values of p.

But for $q = 1$, from the first inequality $p + q < 11$, we get $p < 10$. And from the second inequality $p - q > 8$, we get $p > 9$. This is not possible, it vanishes p altogether as there is no integer lying between 9 and 10; hence, this option is incorrect. So only possible value left to be considered is $q = 0$.

Option E suggests $q < 1$, which means that $q = 0$, and gives $p = 9$ or 10.

The correct answer is option E.

Approach 2:

Another approach could be "**Plug in value**" approach.

Let us analyze the options one by one.

(A) $q < 3$

This gives 0, 1, and 2 as three possible values of q. Let us try $q = 2$ in the inequalities. $p + q < 11$ gives $p < 9$, but $p - q > 8$ gives $p > 10$. This is not possible. This option is incorrect.

(B) $q > 2$

When these kinds of options are there, we must plug in an unusual yet qualified values such as $q = 100$. Note that $q = 100$ renders negative value for p, which is not possible as p is a non-negative integer. This option is incorrect.

(C) $2 < q < 10$

From option A, we already know that q cannot be greater than 2, hence, this option is also incorrect.

(D) $q = 0$ or 1

$q = 0$ gives $p = 9$ or 10: qualified values for x. But for $q = 1$, from the first inequality $p + q < 11$, we get $p < 10$. And from the second inequality $p - q > 8$, we get $p > 9$. As discussed in approach 1, it is not possible; hence, this option is also incorrect. So only possible value left is $q = 0$.

(E) This is the correct answer. $q < 1$ means that $q = 0$ and $p = 9$ or 10.

47. We are iven that $\left|\frac{x}{y}\right| < 1$ and $y \neq 0$. Thus,

=> $|x| < |y|$

The above result is possible in three scenarios:

(1) Say both x & y are +ive; say $x = 10$ then $y > 10$;

(2) Say both x & y are −ive; say $x = -10$ then $y < -10$;

(3) If any one of x & y is +ive and the other is −ive, then

 (a) say x = +ive = 10, then $y < -10$;

 (b) say x = −ive = −10, then $y > 10$

Based on these findings, we analyze each option one by one. Note that this is a "**must be true**" kind of question, which means that in each circumstance, the correct option should satisfy the constraints.

An optimum approach with "must be true" kind of questions is that you must try to prove an option WRONG. If you are satisfied that in each circumstance, the option satisfies the constraints, it is the correct answer.

Let us analyze the options one by one.

(A) $\frac{y}{x}$

It would fail if the scenario is # 3 i.e. any one of x & y is +ive and other is −ive.

$\frac{y}{x}$(Negative) < 1.

(B) $\left(\frac{x}{y}\right)^2$

It would fail in all the scenarios. Given that $\left|\frac{x}{y}\right| < 1 \Rightarrow \left(\frac{x}{y}\right)^2 < 1$.

(C) $\frac{x}{y}$

It would fail if the scenario is # 3 i.e. any one of x & y is +ive and the other is −ive.

$\frac{x}{y}$(Negative) < 1.

(D) $\frac{|y|}{x}$

It would fail if the scenario is # 3b i.e. x is –ive. $\frac{|y|}{x}$ (Negative) < 1.

(E) $\sqrt{\left|\frac{y}{x}\right|}$

Since $|x| < |y|$

=> $1 < \left|\frac{y}{x}\right|$; note that we can multiply or divide an inequality by a number provided we are sure of its sign. $|x|$ being +ive will not affect the sign of the inequality.

=> $\sqrt{1} < \sqrt{\left|\frac{y}{x}\right|}$; taking square root of both the sides.

=> $1 < \sqrt{\left|\frac{y}{x}\right|}$; greater than 1.

The correct answer is option E.

48. The inequality $5x + 3y \geq 13$ can be reduced to $x \geq \frac{13 - 3y}{5}$.

Since this is a 'Could be true' type of question, any option that is true for even a single value for y would be the answer.

Since $x \leq 2$, we can deduce that $2 \geq \frac{13 - 3y}{5}$

=> $10 \geq 13 - 3y$

=> $-3 \geq -3y$

=> $1 \leq y$; notice the sign change because of multiplication of a negative number.

Only one option qualifies and that is option E.

The correct answer is option E.

Alternatively, you can plug in the value of y in the inequality, and see whether $x \leq 2$. Only option E, i.e., $y = 2$ would qualify.

49. If $m > 0$, the set of possible values of m would be $\left\{\ldots, \frac{1}{3}, \frac{1}{2}, 1, 2, \ldots\right\}$.

This is a 'MUST BE TRUE' type of question. It means that for all possible values of m, the option values must be correct. We must approach the problem such that we find out at least one value that is incorrect.

I. It can be reduced to $m < m^2$. Let us try a random value of m in 0 to 1 interval. Always remember that for $1 > m > 0$, the indexed values of m would always be less than m. Say $m = \frac{1}{2}$, then $m^2 = \frac{1}{4}$, which is less than m. This option is not must true.

II. $9m - 1 > 0$ can be reduced to $m > \frac{1}{9}$. Incorrect. m may also lie between 0 and $\frac{1}{9}$.

III. $m^3 > 0$. Correct. If $m > 0, m^3$ would always be greater than 0.

The correct answer is option C.

50. $a^2 - 4a < 221$

$=> a^2 - 4a + 4 < 225$; adding '4' on each side.

$=> (a - 2)^2 < 225$

$=> (a - 2)^2 < 15^2$

$=> a - 2 < |15|$

$=> -15 < (a - 2) < 15$

$=> -15 + 2 < a < 15 + 2$

$=> -13 < a < 17.$

Thus, the possible integer values that a can assume are: $-12, -11, \ldots, -1, 0, 1, 2, \ldots 15, 16.$

Thus, there are 12 integers from -12 to -1 (both inclusive), '0,' and 16 integers from 1 to 16 (both inclusive); i.e. a total of $12 + 1 + 16 = 29$ integers.

The correct answer is option C.

Alternate approach:

$a^2 - 4a < 221$

$a^2 - 4a - 221 < 0$

$a^2 - 17a + 13a - 221 < 0$

$a(a - 17) + 13(a - 17) < 0$

$(a - 17)(a + 13) < 0$

$=> -13 < a < 17$

As explained above, there are 29 integers for a within $-13 < a < 17$.

Chapter 7

Algebra

7.1 Identities

Let us first look at a few basic terms.

Polynomial: A polynomial is an expression of variables that can only have rational coefficients.

For example, $x^3 + 2xy^2 + 6z$ is a polynomial.

Order and Degree: The Order of a polynomial refers to the number of variables involved in the polynomial.

For example, $x^2 + 2xy$ is a polynomial of order '2' since there are two variables, x and y.

The degree of a polynomial refers to the highest value of the exponent of any variable in the polynomial.

For example, $x^2 + 2xy$ has degree '2' since the highest exponent is 2.

Identities: An identity is a statement equating two expressions which are equal for all values of the variables involved.

For example, $5x = 2x + 3x$ is an identity, as $5x$ and $2x + 3x$ are equal for all values for x.

Some important identities are as follows:

- $(a \pm b)^2 = a^2 \pm 2ab + b^2$
- $\left(x \pm \frac{1}{x}\right)^2 = x^2 + \frac{1}{x^2} \pm 2$
- $a^2 - b^2 = (a + b)(a - b)$
- $(a \pm b)^3 = a^3 \pm 3a^2b + 3ab^2 \pm b^3 = a^3 \pm b^3 \pm 3ab(a \pm b)$

Let us take an example

If $x + \frac{1}{x} = 2$, what is the value of $x^4 + \frac{1}{x^4}$?

We have $\left(x + \frac{1}{x}\right)^2 = 2^2 => x^2 + \frac{1}{x^2} + 2 = 4$

$=> x^2 + \frac{1}{x^2} = 4 - 2 = 2$

$=> \left(x^2 + \frac{1}{x^2}\right)^2 = 2^2 => x^4 + \frac{1}{x^4} + 2 = 4$

$=> x^4 + \frac{1}{x^4} = 4 - 2 = 2.$

Let us take another example:

What is the value of $\left\{\dfrac{(0.21+0.17)^2+(0.21-0.17)^2}{\left(0.21^2+0.17^2\right)}\right\}$?

Let $a = 0.21$ and $b = 0.17$

Thus, we have

$$\frac{(0.21+0.17)^2+(0.21-0.17)^2}{\left(0.21^2+0.17^2\right)}$$

$$= \frac{(a+b)^2+(a-b)^2}{(a^2+b^2)}$$

$$= \frac{a^2+2ab+b^2+a^2-2ab+b^2}{a^2+b^2}$$

$$= \frac{2\,(a^2+b^2)}{a^2+b^2}$$

$$= 2$$

7.2 Linear Equations

Linear equations are equations in which the highest exponent/index of the variables is 1, i.e. the degree is 1.

For example, $2x + 5 = 11$ is a linear equation with a single variable, x, whose index is one.

Similarly, $3x + 5y = 9$ is a linear equation with two variables, x and y, and the indices of both are 1.

Solving linear equations with a single variable:

While solving any linear equation with a single term, we need to group all terms containing the variable on the left and all other constant terms on the right.

While shifting terms from left to right as addition or subtraction, the sign of the terms are negated. Similarly, while shifting terms from left to right as a multiplication or division, the processes are interchanged.

Let us take an example

$$\frac{2x+1}{3} - 4 = \frac{3}{2} - x$$

We group all terms with x to the left and the other constant terms to the right. Thus, the term $-x$, when taken to the left becomes $+x$, similarly, the term -4, when taken to the right, becomes $+4$:

$$\frac{2x+1}{3} + x = \frac{3}{2} + 4$$

Simplifying the left and right sides:

$$\frac{(2x+1)+3x}{3} = \frac{3+2\times 4}{2}$$

$$=> \frac{5x+1}{3} = \frac{11}{2}$$

The '3' on the left and the '2' on the right are both 'divisions', and hence, on changing sides, they become 'multiplication'

$$=> 2(5x+1) = 11 \times 3$$

$$=> 10x + 2 = 33$$

We take the constant '2' to the right:

$$=> 10x = 33 - 2 = 31$$

The '10' as a product on the left is taken to the right as a division:

$$=> x = \frac{31}{10}$$

Solving Linear Equations with two variables:

There are two possible ways of solving such equations:

- Method of substitution: From the first equation, we express one variable in terms of the other and substitute in the second equation.
- Solving simultaneously: Using both equations simultaneously we eliminate one of the two variables.

Let us take an example

$2x + 3y = 15 \ldots$(i)

$5x - 2y = 28 \ldots$(ii)

- Solving by the method of substitution:

  From (i): $x = \frac{15 - 3y}{2}$

  Substituting in (ii):

  $$\frac{5\,(15 - 3y)}{2} - 2y = 28$$

  The above can be solved as discussed above, for a single variable:

  $$=> 75 - 15y - 4y = 56 => -19y = -19$$

  $$=> y = 1$$

  Substituting $y = 1$ in the expression for x, we have

  $$x = \frac{15 - 3 \times 1}{2} = 6$$

- Solving simultaneously:

  We need to eliminate one of the variables by multiplying the equations with a suitable constant and then adding/subtracting them:

  Equation (i) $\times$ 2 + Equation (ii) $\times$ 3:

$4x + 6y = 30$

$15x - 6y = 84$

$\Rightarrow 19x = 114 \Rightarrow x = 6$

Substituting $x = 6$ in (i):

$2 \times 6 + 3y = 15$

$\Rightarrow y = \frac{15 - 12}{3} = 1$

Graphically, a pair of linear equations with two variables represents two straight lines, the solution of which represents their point of intersection.

Let us take two general equations:

$ax + by = c \ldots$ (i)

$Ax + By = C \ldots$ (ii)

Here, a, b, c, A, B & C are the constants and x & y are the variables.

There are three important results possible:

- $\frac{a}{A} = \frac{b}{B} = \frac{c}{C}$: This implies that the equations are identical equations, i.e. they are overlapping lines and hence, they have infinite solutions.

- $\frac{a}{A} = \frac{b}{B} \neq \frac{c}{C}$: This implies that the equations are equations of parallel lines and since parallel lines never intersect, there exists no possible solution.

- $\frac{a}{A} \neq \frac{b}{B}$: This refers to the case of intersecting lines, i.e. a unique solution can be obtained from the two equations.

    - $\frac{a}{A} \neq \frac{b}{B} = \frac{c}{C}$: Since the ratio of coefficients of y is the same as the ratio of the constant terms, the value of y in the equation is zero.

    - $\frac{b}{B} \neq \frac{a}{A} = \frac{c}{C}$: Since the ratio of coefficients of x is the same as the ratio of the constant terms, the value of x in the equation is zero.

Solving for x and y from a single equation given they are integers:

Let us take an example

Find all the possible values of x and y if $6x + 32y = 102$, where x & y are positive integers:

- First we need to reduce the coefficients to their lowest terms.

  Thus, we have $3x + 16y = 51$

- We find any possible integer solution of the above equation (it is not important to keep both x and y positive at this stage).

  We can see that $x = 17$ and $y = 0$ obviously satisfies.

- It is clear that we need to increase the value of y to make it positive. Thus, we add the coefficient of x to the value of y and simultaneously, we subtract the coefficient of y from the value of x to generate the next possible integer solution. Thus, we get

x	y
17	0
$17 - 16 = \mathbf{1}$	$0 + 3 = \mathbf{3}$
$1 - 16 = -15$	$3 + 3 = 6$

We see that on repeating the same process a second time, the value of x becomes negative and hence, inadmissible. Thus, there is only one possible solution: $x = 1,\ y = 3$.

Determining the value of an expression involving the variables without solving:

Let us take an example

$2x + 5y = 19 \ldots$(i)

$x + 2y = 8 \ldots$(ii)

We need to determine the values of:

(1) $3x + 7y$

(2) $x + 3y$

(3) $4x + 9y$

For (1): Simply adding (i) and (ii): $3x + 7y = 27$

For (2): Simply subtracting (ii) from (i): $x + 3y = 11$

For (3): (i) + 2 × (ii): $4x + 9y = 35$

Additional solved problems:

(1) In the GCAT test, for every correct answer, a student is awarded '+3' marks and for every wrong answer, '−1' mark. There are 40 questions and the student must answer all questions.

I. If the student gets 40 marks, how many questions did he get correct?

II. How many different scores can the student get, if he randomly marks the answers?

Explanation:

Let the student got x correct and y wrong answers.

I. Thus, we have
$x + y = 40$ (There are 40 questions and he answered all questions)

$3x - y = 40$ (Each correct answer is of 3 marks and each wrong is of −1 mark)

Adding the above two equations:

$4x = 80 \Rightarrow x = 20$

II. If the student answers all questions randomly, x changes from 0 to 40 and y, correspondingly changes from 40 to 0.

Thus, there are 41 possible values of the score: 120, 116, 112, ... −32, −36, −40.

(Note: the score forms a series with the difference between their consecutive terms equals to 4)

(2) A and B have a number of marbles. The total number of marbles is 24. If A gives B 6 marbles, the ratio of the number of marbles with A and B is reversed. How many marbles does A have?

Explanation:

Let the number of marbles with A be x.

Thus, the number of marbles with B = $(24 - x)$.

After A gives 6 marbles to B, we have

Number of marbles with A = $(x - 6)$.

Number of marbles with B = $(24 - x + 6) = (30 - x)$.

Since the ratio of the number of marbles with A and B is reversed, we have

$$\frac{x-6}{30-x} = \frac{24-x}{x}$$

=> $x^2 - 6x = 720 - 54x + x^2$

=> $48x = 720$

=> $x = 15$

Alternate approach

As we can see, the above approach is a bit complicated.

However, we can use a much simpler logic.

After A gives B 6 marbles, the total number of marbles with A and B remain unchanged.

Since the ratio of the number of marbles has reversed, we can simply say that the number of marbles with A and B have been interchanged.

Thus, we have

Number of marbles A has NOW = Number of marbles B had BEFORE

$x - 6 = 24 - x$

=> $x = 15$

(3) Abe has 20 coins with him. A few of the coins are dimes (a dime is equal to 10 cents) and the rest are quarters (a quarter is equal to 25 cents). If the total amount with him is \$4.25, how many dimes and how many quarters does he have? (Note: 100 cents make a dollar)

Explanation:

Let us solve this problem in three different ways.

Approach 1: Using two variables

Let the number of dimes be x and the number of quarters be y.

Thus, we have

$x + y = 20 \ldots$(i)

Total worth of dimes is $10x$ cents and the total worth of quarters is $25y$ cents.

Since the total amount with him is \$4.25, i.e. 425 cents, we have

$10x + 25y = 425$

$=> 2x + 5y = 85 \ldots$(ii)

(ii) $-$ (i) $\times$ 2:

$3y = 45$

$=> y = 15$

$=> x = 5$

Approach 2: Using one variable

Let the number of dimes be x.

Since there are 20 coins, the number of quarters = $(20 - x)$

Since the total amount is 425 cents, we have

$10x + 25(20 - x) = 425$

$=> 10x - 25x = 425 - 500$

$=> -15x = -75$

$=> x = 5$

=> Number of quarters = $y = 20 - 5 = 15$

Approach 3: Using no variables

Let us assume that all his coins are dimes.

Thus, Abe has 20 dimes, which accounts for $20 \times 10 = 200$ cents.

However, he actually has 425 cents.

Thus, there is a shortage of $425 - 200 = 225$ cents.

Thus, we can conclude that some of his dimes are actually quarters.

In that case, each such dime would increase in value from 10 cents to 25 cents, i.e. an increase of $25 - 10 = 15$ cents.

The number of times such a conversion from dime to quarter required to fulfill the shortage of 225 cents

$= \frac{225}{15} = 15$ times

Thus, of the 20 dimes, 15 need to be converted to quarters and the rest should remain as dimes.

Thus, number of quarters is 15 and the number of dimes is 5.

(4) Joe purchases 7 identical gift-packs from Shop A and 4 identical gift-packs from Shop B. Each gift pack has a number of pens. The total number of pens in the 11 gift-packs is 26. If he wants an additional 18 pens, what is the maximum number of gift-packs, similar to the ones he purchased initially, he needs to buy?

Explanation:

Let the number of pens present in gift-packs from Shop A = x

Let the number of pens present in gift-packs from Shop B = y

Thus, total number of pens = $(7x + 4y)$

Thus, we have

$7x + 4y = 26$

Since x and y must be positive integers, $4y$ is even. Thus, $7x$ must also be even since the sum of $7x$ and $4y$ is 26, an even number.

Since $7x$ is even, x must be even.

Thus, the only value of $x = 2$ (any value of x greater than 3 is not possible since $7x$ would exceed 26 and y would become negative).

Thus, we have

$$y = \frac{26 - 7 \times 2}{4} = 3$$

Thus, the gift-packs from Shop A have 2 pens each and those from Shop B have 3 pens each.

Thus, to maximize the number of gift-packs, he should purchase the ones from Shop A.

Thus, to buy 18 additional pens, the maximum number of gift-packs required

$$= \frac{18}{2} = 9$$

7.3 Quadratic Equations

An equation is said to be quadratic when the variable contains the highest exponent of '2', thus, $\boldsymbol{ax^2 + bx + c}$ is a quadratic expression in one variable i.e. of $\boldsymbol{x}$.

Solving a Quadratic Equation by Factorization:

A quadratic equation can be solved quickly if it can be written in the form of a product of two linear equations.

Let us take an example

$2x^2 - 7x + 6 = 0$

The given equation can be written as:

$2x^2 - 4x - 3x + 6 = 0$

While dividing the middle term in two parts, it should be seen that the product of the two terms equals the product of the first and last terms of the quadratic.

Thus, here we have $(-4x) \times (-3x) = (2x^2) \times (6) = 12x^2$

$=> 2x(x - 2) - 3(x - 2) = 0$

$=> (2x - 3)(x - 2) = 0$

This results in two separate linear equations:

$2x - 3 = 0$

OR

$x - 2 = 0$

$=> x = \frac{3}{2}$ OR $x = 2$

Thus, each quadratic equation can be written in the form $(x - p)(x - q) = 0$, where p and q are the roots of the quadratic equation.

Roots of a Quadratic Equation:

After suitable reduction, every quadratic equation can be written in the form:

$\boldsymbol{ax^2 + bx + c = 0}$

The solution or the roots of the above equation is given by:

$\boldsymbol{p} = \dfrac{-\boldsymbol{b} + \sqrt{\boldsymbol{b}^2 - 4\boldsymbol{ac}}}{2\boldsymbol{a}}$ and $\boldsymbol{q} = \dfrac{-\boldsymbol{b} - \sqrt{\boldsymbol{b}^2 - 4\boldsymbol{ac}}}{2\boldsymbol{a}}$, where $\boldsymbol{p}$ and $\boldsymbol{q}$ are the roots of the equation.

Let us see how

$ax^2 + bx + c = 0$

Dividing throughout by a:

$x^2 + \frac{b}{a}x + \frac{c}{a} = 0$

Let us combine the first two terms of the above equation to form a perfect square

$$=> x^2 + 2x\left(\frac{b}{2a}\right) + \left(\frac{b}{2a}\right)^2 - \left(\frac{b}{2a}\right)^2 + \frac{c}{a} = 0$$

$$=> \left(x + \frac{b}{2a}\right)^2 = \left(\frac{b}{2a}\right)^2 - \frac{c}{a} = \frac{b^2 - 4ac}{4a^2}$$

Taking square roots on both sides

$$=> x + \frac{b}{2a} = \pm\sqrt{\frac{b^2 - 4ac}{4a^2}} = \pm\frac{\sqrt{b^2 - 4ac}}{2a}$$

$$=> x = -\frac{b}{2a} \pm \frac{\sqrt{b^2 - 4ac}}{2a}$$

$$=> x = \frac{-b \pm \sqrt{b^2 - 4ac}}{2a}$$

Let us find the roots of the previous equation $2x^2 - 7x + 6 = 0$ using the above formula

We have $a = 2,\ b = -7\ \&\ c = 6$

Thus, the roots are:

$$\frac{-b \pm \sqrt{b^2 - 4ac}}{2a} = \frac{-(-7) \pm \sqrt{(-7)^2 - 4(2)(6)}}{2(2)} = \frac{7 \pm 1}{4} = \frac{7 - 1}{4} \text{ OR } \frac{7 + 1}{4} = \frac{3}{2} \text{ OR } 2$$

Discriminant:

The value given by $\boldsymbol{D}(\triangle) = \boldsymbol{b}^2 - 4\boldsymbol{ac}$, i.e. the quantity under the square root, is called the discriminant and depending upon its value we can determine the nature of the roots of the quadratic equations:

- If $D \geq 0$: The roots are real
- If $D = 0$: The roots are real and equal

- If $D > 0$: The roots are real and unequal
- If $D < 0$: The roots are imaginary
- If D is a perfect square: The roots are real and rational
- If D is not a perfect square: The roots are real and irrational (one root is conjugate of the other, i.e. the roots are of the form $(a \pm \sqrt{b})$

Let us take some example

- In the equation $x^2 + 2x + 1 = 0$, $D = b^2 - 4ac = 2^2 - 4 \times (1) \times (1) = 0$ => The roots are real and equal.
- In the equation $x^2 + 2x + 2 = 0$, $D = b^2 - 4ac = 2^2 - 4 \times (1) \times (2) = -4 < 0$ => The roots are complex.

Sum and Product of the roots:

If the roots of the quadratic equation $ax^2 + bx + c = 0$ are p and q, we have

$ax^2 + bx + c \equiv a(x - p)(x - q)$

$=> ax^2 + bx + c \equiv a(x^2 - x(p + q) + pq)$

$=> ax^2 + bx + c \equiv ax^2 - a(p + q)x + apq$

Comparing the coefficient of x and the constant term, we have

$b = -a(p + q)$ and $c = apq$

$=> p + q = -\frac{b}{a}$ and $pq = \frac{c}{a}$

Expressing a quadratic equation in terms of its roots:

Suppose we have to form the equation whose roots are p and q.

Thus, we have $(x - p) = 0$ and $(x - q) = 0$

$=> (x - p)(x - q) = 0$

$=> x^2 - x(p + q) + pq = 0$

$=> \boldsymbol{x^2}$ − (**Sum of roots**) $\boldsymbol{x}$ + (**Product of roots**) **= 0**

Let us take an example

The equation having roots -2 and 5 is

$x^2 - (-2 + 5)\,x + (-2)\,(5) = 0$

$=> x^2 - 3x - 10 = 0$

Additional solved problems

(1) A man can swim in still water (without any current) at a rate of 4 miles per hour. He undertakes to swim from point A to point B and back, in a river which has a current of its own. If the distance between the points A and B is 15 miles, and the total time he takes for the trip is 8 hours, what is the rate of flow of the river current in miles per hour?

Note: While swimming against the river, the man's normal swimming rate would be reduced by the amount of the rate of the current, similarly, while going along with the flow of the river, his speed would be increased by the rate of the current.

Explanation:

Let the rate of the river current be r miles per hour.

Since the man makes a round trip, one way he would be swimming against the river, while the other way, he would be swimming with the river.

Thus, while swimming against the river, the rate of the man = $(4 - r)$ miles per hour

Thus, time taken to cover 15 miles = $\left(\frac{15}{4 - r}\right)$ hours.

While swimming with the river, the rate of the man = $(4 + r)$ miles per hour

Thus, time taken to cover 15 miles = $\left(\frac{15}{4 + r}\right)$ hours.

Since the total time is 8 hours, we have

$$\frac{15}{4 - r} + \frac{15}{4 + r} = 8$$

$=> 15\,(4 + r + 4 - r) = 8(4^2 - r^2)$

$=> 8r^2 = 8 \times 4^2 - 15 \times 8 = 128 - 120$

$=> 8r^2 = 8 => r^2 = 1$

$=> r = 1$

(2) A man buys a number of pieces of chocolates for \$24. If the price of a piece of chocolate increases by \$2, he can buy 1 piece of chocolate less for the same amount. What is the price of 1 piece of chocolate?

Explanation:

Let the price of a piece of chocolate be $\$x$.

Thus, the number of chocolate pieces the man can buy for \$24 $= \left(\frac{24}{x}\right)$

New price of a piece of chocolate = $\$\ (x + 2)$.

Thus, the number of chocolate pieces the man can now buy for \$24 $= \left(\frac{24}{x+2}\right)$

Since the number of pieces of chocolate is 1 less than previously, we have

$$\frac{24}{x+2} = \frac{24}{x} - 1$$

$$=> \frac{24}{x} - \frac{24}{x+2} = 1$$

$$=> \frac{24\,(x + 2 - x)}{x\,(x+2)} = 1$$

$$=> x\,(x + 2) = 48$$

Since $48 = 6 \times 8$, we have

$$x = 6$$

Alternately, we can solve the quadratic:

$$x\,(x + 2) = 48 => x^2 + 2x - 48 = 0$$

$$=> (x + 8)\,(x - 6) = 0$$

$$=> x = 6$$

(3) The total cost, in dollars, of manufacturing n items of a product is given by $(2n^2 + 30)$. The selling price of each item is fixed at \$36. What is the number of items that must be sold to have maximum profit?

Explanation:

Total cost of manufacturing n items = \$ $(2n^2 + 30)$.

Selling price of each item = \$36.

Thus, total selling price of n items = \$ $(36n)$.

Thus, profit earned = Selling price - Cost price

$= 36n - (2n^2 + 30)$

$= -2\,(n^2 - 18n + 15)$

$= -2\,(n^2 - 18n + 81 - 66)$

$= -2(n - 9)^2 + 132$

The profit would be maximized if the negative square term, i.e. $-2(n - 9)^2$ becomes '0', which happens when $n = 9$.

The corresponding maximum value of the profit is \$132.

7.4 Inequalities

Inequalities: It deals with cases where variables or numbers are less than or more than other variables or numbers. The following symbols are used

- $>$: More than;
- $<$: Less than;
- $\leq$: Less than or equal to;
- $\geq$: More than or equal to

Let us look at some of the important rules:

(1) In any given inequality, we can add or subtract any quantity from both sides keeping the inequality the same.

For example, if we have $x > y$

Adding z to both sides: $x + z > y + z$

Subtracting z from both sides: $x - z > y - z$

(2) In any given inequality, we can multiply or divide any other 'positive' quantity on both sides keeping the inequality the same.

For example, if we have $x > y$

Multiplying a positive quantity z on both sides: $x \times z > y \times z$

Dividing by a positive quantity z on both sides: $\frac{x}{z} > \frac{y}{z}$

(3) In any given inequality, we can multiply or divide any other 'negative' quantity on both sides, thus, reversing the inequality.

For example, if we have $x > y$

Multiplying a negative quantity $(-z)$ on both sides: $-x \times z < -y \times z$

Dividing by a negative quantity $(-z)$ on both sides: $-\frac{x}{z} < -\frac{y}{z}$

(4) Any two inequalities having the same inequality can be added.

For example, if we have $x > y$ and $p > q$

Adding the inequalities, we have $x + p > y + q$

(5) In order to subtract one inequality from another, the former has to be negated, thus, reversing the inequality and then the inequalities can be added.

For example, if we have $x > y$ and $p > q$

Negating the former inequality, we have $-x < -y => -y > -x$

Thus, on adding the inequalities $p > q$ and $-y > -x$, we have

$p - y > q - x$

(6) Any term with an even exponent is always non-negative, its minimum value being '0'. Thus, we have

$x^2 \geq 0$, $y^4 \geq 0$, $z^6 \geq 0$, etc.

Thus, if we have $x^5y^2z > 0$, we have

$=> (x^4y^2)\,xz > 0$

Since x^4y^2 is always non-negative, we have

$=> xz > 0$

$=> x > 0$ and $y > 0$

OR

$x < 0$ and $y < 0$

(7) Inequality of ratio of two quantities and the inequality of the product of those quantities are similar.

For example, if we have $\frac{x}{y} > 0$

$=> \frac{x}{y} \times y^2 > 0$ (since $y^2 > 0$)

$=> xy > 0$

Note: However, there is a slightly different result for "greater than or equal to" or "less than or equal to" type of inequalities, as shown:

If we have $\frac{x}{y} \geq 0$

$=> \frac{x}{y} \times y^2 \geq 0$

$=> xy \geq 0$

On careful observation, it can be seen that $y = 0$ is a possible solution for $xy \geq 0$, however, not so in the case of $\frac{x}{y} \geq 0$.

(1) **Number line based inequalities**: There are four important regions on a number line, as shown below:

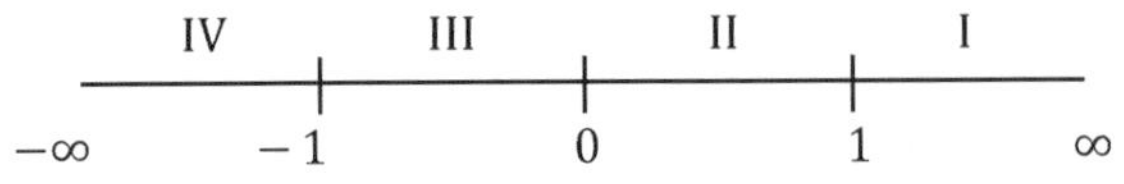

Some properties of the above four regions:

- **Region I**: A number x in the region satisfies $1 < x < \infty$

  Higher the exponent of x, higher is the value of the term and vice versa.

  Thus, we have
  $x < x^2 < x^3 \ldots$ For example $2 < 2^2 < 2^3$
  $x > \sqrt{x} > \sqrt[3]{x} \ldots$ For example $2 > \sqrt{2} > \sqrt[3]{2}$

- **Region II**: A number x in the region satisfies $0 < x < 1$

  Higher the exponent of x, smaller is the value of the term and vice-versa.

  Thus, we have
  $x > x^2 > x^3 \ldots$ For example $\frac{1}{2} > \left(\frac{1}{2}\right)^2 > \left(\frac{1}{2}\right)^3$
  $x < \sqrt{x} < \sqrt[3]{x} \ldots$ For example $\frac{1}{2} < \sqrt{\frac{1}{2}} < \sqrt[3]{\frac{1}{2}}$

- **Region III**: A number x in the region satisfies $-1 < x < 0$

  - For odd exponents (the values are always negative):

    Higher the exponent of x, higher is the value of the term and vice versa.

    Thus, we have
    $x < x^3 < x^5 \ldots$ For example $-\frac{1}{2} < \left(-\frac{1}{2}\right)^3 < \left(-\frac{1}{2}\right)^5$
    $x > \sqrt[3]{x} > \sqrt[5]{x} \ldots$ For example $-\frac{1}{2} > \sqrt[3]{-\frac{1}{2}} > \sqrt[5]{-\frac{1}{2}}$

  - For even exponents (the values are always positive):

    Higher the exponent of x, smaller is the value of the term and vice versa.

    Thus, we have
    $x^2 > x^4 > x^6 \ldots$ For example $\left(-\frac{1}{2}\right)^2 > \left(-\frac{1}{2}\right)^4 > \left(-\frac{1}{2}\right)^6$

    (Note: Square roots and fourth roots, etc. are not possible for negative numbers)

    An even exponent results in a positive value, which will always be greater than the value resulting from an odd exponent, which is always negative.

- **Region IV**: A number x in the region satisfies $-\infty < x < -1$

  - For odd exponents (the values are always negative):

    Higher the exponent of x, smaller is the value of the term and vice versa.

    Thus, we have
    $x > x^3 > x^5 \ldots$ For example $-2 > (-2)^3 > (-2)^5$

    $x < \sqrt[3]{x} < \sqrt[5]{x} \ldots$ For example $-2 < \sqrt[3]{-2} < \sqrt[5]{-2}$

  - For even exponents (the values are always positive):

    Higher the exponent of x, higher is the value of the term and vice versa.

    Thus, we have
    $x^2 < x^4 < x^6 \ldots$ For example $(-2)^2 < (-2)^4 < (-2)^6$

    (Note: Square roots and fourth roots, etc. are not possible for negative numbers)

    An even exponent results in a positive value, which will always be greater than the value resulting from an odd exponent, which is always negative.

(2) **Quadratic inequalities**: We have the following two rules:

- $x^2 < k^2 \Rightarrow -k < x < k$
- $x^2 > k^2 \Rightarrow x < -k$ OR $x > k$

Any quadratic can be converted to one of the above two forms.

Let us take an example

$x^2 - 8x + 12 < 0$

$\Rightarrow (x^2 - 2 \times x \times 4 + 4^2) - 4^2 + 12 < 0$

$\Rightarrow (x - 4)^2 < 4$

$\Rightarrow -2 < x - 4 < 2$

$\Rightarrow 2 < x < 6$

Alternately, we can use the rules below: If $(x - k)$ and $(x - m)$ are the factors of a quadratic, and $k > m$, we have

- $(x - k)(x - m) > 0 => x > k$ OR $x < m$

  Thus, x is greater than the greatest root OR smaller than the smallest root.

- $(x - k)(x - m) < 0 => m < x < k$

  Thus, x lies between the two roots.

  Let us take an example

  $x^2 - 8x + 12 < 0$

  $=> (x - 2)(x - 6) < 0$

  Since the roots are 2 and 6, with 6 being the greater root, we have

  $=> 2 < x < 6$

Modulus or Absolute value: Modulus of a number is a function that returns the magnitude of the number ignoring the sign.

Thus, we have $|2| = |-2| = 2$

Hence, we have $|x| = 2 => x = \pm 2$

Another way of interpreting the modulus of a number is distance of the number from a point on the number line.

For example, $|x - 3|$ denotes the distance of x from the point '3' on the number line.

Thus, $|x - 3| = 4$ implies the distance of the point x from the point '3' on the number line is 4 units. Since the distance can be measured on either side of '3', we attain the points $3 + 4 = 7$ and $3 - 4 = -1$ on the number line, as shown below.

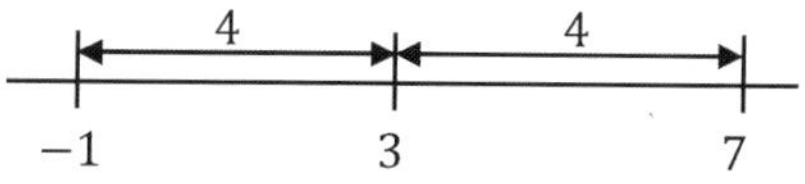

Thus, using the above concept, we have

$|x - a| = b => x - a = \pm b$

$=> x = a \pm b$

(1) **Modulus based inequality**: We have the following two rules:

- $|x - a| > b$
  $=> x - a > b$ OR $x - a < -b$
  $=> x > a + b$ OR $x < a - b$

  Alternately, the given inequality implies that the distance of x from a is greater than b units. If the distance were equal to b units, we would have got the points $(a + b)$ and $(a - b)$. Since the distance should be greater than b units, x should overshoot the above two points, i.e. x is greater than the greatest point or x is less than the smallest point.

  Thus, we have $x > a + b$ OR $x < a - b$

- $|x - a| < b$
  $=> -b < x - a < b$
  $=> a - b < x < a + b$

  Alternately, the given inequality implies that the distance of x from a is lesser than b units. If the distance were equal to b units, we would have got the points $(a + b)$ and $(a - b)$. Since the distance should be lesser than b units, x should remain within the above two points.

  Thus, we have $a - b < x < a + b$

- For any two positive numbers a and b, we always have:
  $\frac{(a+b)}{2} \geq \sqrt{ab}$

- For a given sum of two or more quantities, the product of the quantities is maximized if the quantities are made equal.

Similarly, for a given product of two or more quantities, the sum of the quantities is minimized if the quantities are made equal.

Let us take a few examples:

- What is the maximum value of $a \times b$ if $2a + 3b = 20$?

  Since we need to maximize $a \times b$, we need to maximize a and b.

  Since $2a + 3b = 20$, the product $2a \times 3b$ would be maximized if $2a = 3b$

  $=> 2a + 2a = 20 => a = 5$

  $=> 2a = 3b = 10$

Thus, the maximum value of $2a \times 3b = 10 \times 10 = 100$

=> The maximum value of $a \times b = \frac{(2a) \times (3b)}{6} = \frac{100}{6} = \frac{50}{3}$

- What is the minimum perimeter of a rectangle having an area of 100?

  Let the length and width of the rectangle be l and w, respectively.
  Thus, the area of a rectangle = $l \times w = 100$

  We need to minimize the perimeter, i.e. $2(l + w)$. Thus, we need to minimize the value of $(l + w)$.

  The value of $(l + w)$ will be minimized if $l = w$
  => $l \times l = 100$ => $l = 10$
  => $l = w = 10$

  Thus, minimum perimeter = $2(10 + 10) = 40$.

Additional solved problems:

(1) What are the integer values of x satisfying $\left(\frac{2x - 3}{x + 1}\right) \leq 0$?

Explanation:

We have $\left(\frac{2x - 3}{x + 1}\right) \leq 0$
Thus, we have

Case I:
$2x - 3 \leq 0$ AND $x + 1 > 0$
=> $x \leq \frac{3}{2}$ AND $x > -1$
=> $-1 < x \leq \frac{3}{2}$
=> The integer values of x are: 0 and 1.

Case II:
$2x - 3 \geq 0$ AND $x + 1 < 0$
=> $x \geq \frac{3}{2}$ AND $x < -1$, which is not a possible scenario
Thus, the only possible integer values of x are: 0 and 1.

(2) What are the integer values of x satisfying $|x^2 - 6| < 3$?

Explanation:

We have $|x^2 - 6| < 3$

$=> -3 < x^2 - 6 < 3$

$=> 3 < x^2 < 9$

Case I: $x^2 < 9$

$=> -3 < x < 3$

=> The possible integer values of x are: $-2, -1, 0, 1, 2$

Case II: $x^2 > 3$

$=> x > \sqrt{3}$ OR $x < -\sqrt{3}$

=> The possible integer values of x are: $\cdots - 4, -3, -2, 2, 3, 4, \ldots$

Thus, the common set of values of x are: -2 and 2.

(3) Which of the following is true about x if $x^2 > x > x^3$?

(A) $x > 1$

(B) $0 < x < 1$

(C) $-1 < x < 0$

(D) $x < -1$

Explanation:

We know that there are four major regions on the number line:

I. $x > 1$

II. $0 < x < 1$

III. $-1 < x < 0$

IV. $x < -1$

Let us pick 1 number from each region and check whether it satisfies the given inequality:

- $x = 2 : 2^2 > 2 \ngtr 2^3$

- $x = \frac{1}{2} : \left(\frac{1}{2}\right)^2 \ngtr \frac{1}{2} > \left(\frac{1}{2}\right)^3$
- $x = -\frac{1}{2} : \left(-\frac{1}{2}\right)^2 > -\frac{1}{2} \ngtr \left(-\frac{1}{2}\right)^3$
- $x = -2 : (-2)^2 > -2 > (-2)^3$ - Satisfies

Alternate approach:

Since x^2 is greater than both x, a term with a lower exponent and x^3, a term with a higher exponent, the value of x must be negative.

Since x, a term with a smaller odd exponent, is greater than x^3, a term with a higher odd exponent, the value of x must be less than -1.

Hence, correct answer is option D.

(4) If $-5 < x < 3$, $-9 < y < -5$, $-13 < p < 8$, $-4 < q < 6$ and $1 \leq r \leq 8$, where x, y, p, q and r are integers, what are the range of values of

I. $x + 2y$
II. $2x - y$
III. $p \times q$
IV. $\frac{q}{r}$

Explanation:

I. We have $-9 < y < -5$
$=> -18 < 2y < -10$
$=> -5 - 18 < x + 2y < 3 - 10$
$=> -23 < x + 2y < -7$

II. We have $-5 < x < 3$
$=> -10 < 2x < 6$
$=> -10 - 9 < 2x + y < -5 + 6$
$=> -19 < 2x + y < 1$

III. We have $-13 < p < 8$ and $-4 < q < 6$
$=> -12 \leq p \leq 7$ and $-3 \leq q \leq 5$
Thus, the maximum value of $p \times q$

= The greater among $\{(-12) \times (-3)\}$ and $\{(7) \times (5)\} = 36$
Also, the minimum value of $p \times q$

= The smaller among $\{(-12) \times (5)\}$ and $\{(7) \times (-3)\} = -60$
Thus, we have $-60 \leq pq \leq 36$

IV. We have $-4 < q < 6 => -3 \leq q \leq 5$ and $1 \leq r \leq 8$
To determine the maximum value of $\frac{q}{r}$, we must keep the ratio positive. Since r is entirely positive, we must take the positive value of $q = 5$ and the least positive value of $r = 1$.
Thus, the maximum value of $\frac{q}{r} = \frac{5}{1} = 5$
Again, to determine the minimum value of $\frac{q}{r}$, we must keep the ratio negative. Since r is entirely positive, we must take the negative value of $q = -3$ and the least positive value of $r - 1$.

(Note: the minimum value of $\frac{q}{r}$ is the negative number with the largest magnitude)

Thus, the minimum value of $\frac{q}{r} = \frac{-3}{1} = -3$
Thus, we have $-3 \leq \frac{q}{r} \leq 5$

7.5 Functions and graphs of functions

If y is expressed in terms of x, such that, for any x, a unique value of y is obtained, the expression of y in terms of x is called a function of x.

Functions of x are usually denoted by symbols of the form $f(x)$, $g(x)$, $h(x)$, etc.

Some examples of functions:

- $f(x) = 2x + 5$
- $f(x) = x^2 - 2x + 1$
- $f(x) = \frac{1}{x}$
- $f(x) = \sqrt{x} + 1$
- $g(x) = \frac{x - 2}{x^2 + 3}$
- $g(x) = 1$

The idea of a function can be represented by a simple diagram as shown below:

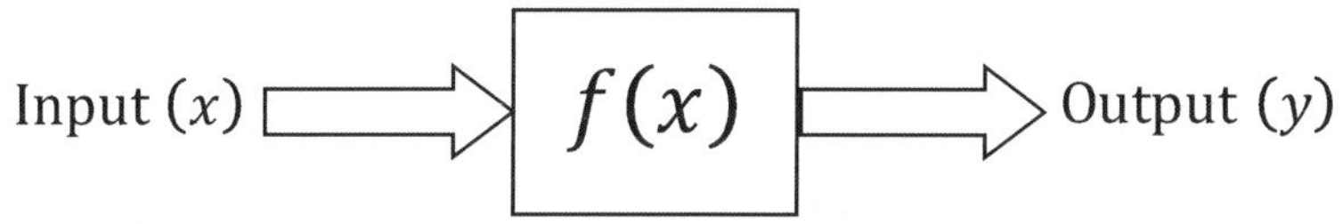

Method of substitution in functions:

For any given $f(x)$, the value of $f(a)$ is determined by substituting $x = a$ on the right side of the equality.

Let us take an example

$f(x) = 2x + 1$

Thus, we have

- $f(1) = 2(1) + 1$, i.e. we replace x with 1 on the right side of the equality
  $\Rightarrow f(1) = 3$
- $f(x^2) = 2(x^2) + 1 = 2x^2 + 1$
- $f(2x + 1) = 2(2x + 1) + 1 = 4x + 3$

Domain and Range:

The Domain refers to the set of values of x that can be used in the function.

The Range refers to the set of values of $f(x)$ obtained using the above values of x.

Let us take an example

$f(x) = x^2 - 1$

Let the set of values of x to be used be $\{-1, 0, 1\}$.

Thus, we have

- $f(-1) = (-1)^2 - 1 = 0$
- $f(0) = 0^2 - 1 = -1$
- $f(1) = 1^2 - 1 = 0$

Here, Domain = $\{-1, 0, 1\}$ and Range = $\{-1, 0\}$

To determine the Domain for a function, two things need to considered:

- For any term under a square-root or fourth-root, etc., the term should be non-negative.
  For example $f(x) = \sqrt{x-1}$
  We have $x - 1 \geq 0 => x \geq 1$
  Thus, the Domain is: $1 \leq x < \infty$

- For any term in the denominator, the term must be non-zero.
  For example $f(x) = \frac{2x}{x-3}$
  We have $x - 3 \neq 0 => x \neq 3$
  Thus, the Domain is: x is any real number except 3
  $=> -\infty < x < 3$ OR $3 < x < \infty$

Let us take an example

$$f(x) = \frac{x+3}{\sqrt{2x-4}-4}$$

Thus, we have

- $2x - 4 \geq 0 => x \geq 2 \ldots$(i)

- $\sqrt{2x-4} - 4 \neq 0 \Rightarrow \sqrt{2x-4} \neq 4 \Rightarrow 2x - 4 \neq 16 \Rightarrow x \neq 10 \ldots$(ii)

Thus, the domain: $2 \leq x < 10$ OR $10 < x < \infty$

Composite functions:

For any two functions $f(x)$ and $g(x)$, the functions defined as $f(f(x))$, $f(g(x))$, $g(g(x))$ and $g(f(x))$ are composite functions.

Let us take an example

$f(x) = 2x + 1$ and $g(x) = x^2 - 1$

Thus, we have

- $f(f(x)) = 2(f(x)) + 1 = 2(2x + 1) + 1 = 4x + 3$
- $f(g(x)) = 2(g(x)) + 1 = 2(x^2 - 1) + 1 = 2x^2 - 1$
- $g(g(x)) = (g(x))^2 + 1 = (x^2 + 1)^2 + 1 = x^4 + 2x^2 + 2$
- $g(f(x)) = (f(x))^2 + 1 = (2x + 1)^2 + 1 = 4x^2 + 4x + 2$

Special case in a composite function:

If $f(x)$ and $g(x)$ are two functions and it is observed that $f(g(x)) = g(f(x)) = x$, we have

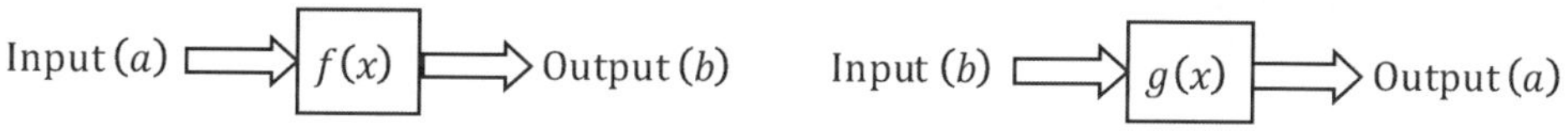

Thus, we have

If $f(a) = b \Rightarrow g(b) = a$

Note: Such functions $f(x)$ and $g(x)$ are inverse functions of one another.

Let us take an example

If $f(x) = (x + 1)^3 - 1$ and $g(x) = \sqrt[3]{x+1} - k$, such that $f(g(x)) = g(f(x))$, what is the value of k?

The normal way of solving, by evaluating the composite functions $f(g(x))$ and $g(f(x))$, is complicated. Instead, we use the above method:

$f(1) = (1 + 1)^3 - 1 = 7$

$=> g(7) = 1$

$=> \sqrt[3]{7+1} - k = 1 => 2 - k = 1$

$=> k = 1$

Periodic function:

A function $f(x)$ is periodic if there exists a number n so that $f(x+n) = f(x)$ for all x. Here, n is the period of the function.

Let us take an example

If $f(x+3) = f(x+2) - f(x+1)$, what is the value of n if $f(1) = -f(1+n)$?

We have $f(x+3) = f(x+2) - f(x+1)$

Substituting different values of x:

- $x = 0 : f(3) = f(2) - f(1) \ldots$(i)
- $x = 1 : f(4) = f(3) - f(2) \ldots$(ii)

Adding (i) and (ii):

$f(3) + f(4) = f(3) - f(1)$

$=> f(1) = -f(4)$

$=> f(1) = -f(1+3)$

$=> n = 3$

Piece-wise functions:

Functions which have different expressions over different values of x are piece-wise functions. Some examples are shown below:

- **Modulus function**: $f(x) = |x|$:
    - $f(x) = x$, if $x \geq 0$,
    - $f(x) = -x$, if $x < 0$

  The graph of $f(x) = |x|$ is shown:

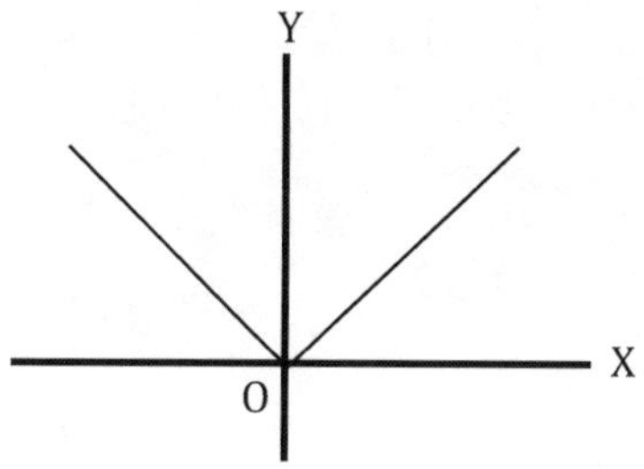

- **Greatest Integer Function**: $f(x) = [x]$: It is a function that returns the greatest integer less than or equal to x. Thus, we have

  - $[1.23]$: The greatest integer less than or equal to 1.23, i.e. the greatest integer among $1, 0, -1, -2, \cdots = 1$
  - $[1]$: The greatest integer less than or equal to 1, i.e. the greatest integer among $1, 0, -1, -2, \ldots$ is 1
  - $[-1.23]$: The greatest integer less than or equal to -1.23, i.e. the greatest integer among $-2, -3, -4, \cdots = -2$
  - $[-1]$: The greatest integer less than or equal to -1, i.e. the greatest integer among $-1, -2, -3, \ldots$ is -1

- **Least Integer Function**: $f(x) = \{x\}$: It is a function that returns the least integer greater than or equal to x. Thus, we have

  - $\{1.23\}$: The least integer greater than or equal to 1.23, i.e. the least integer among $2, 3, 4, \cdots = 2$
  - $\{1\}$: The least integer greater than or equal to 1, i.e. the least integer among $1, 2, 3, 4, \cdots = 1$
  - $\{-1.23\}$: The least integer greater than or equal to -1.23, i.e. the least integer among $-1, 0, 1, 2, \cdots = -1$
  - $\{-1\}$: The least integer greater than or equal to -1, i.e. the least integer among $-1, 0, 1, 2, \cdots = -1$

- **Max-Min function**:

  - $f(x) = \max(a, b)$ implies that $f(x) = a$ if $a > b$ OR $f(x) = b$ if $b > a$
  - $f(x) = \min(a, b)$ implies that $f(x) = a$ if $a < b$ OR $f(x) = b$ if $b < a$

Let us take an example

If $f(x) = \min(6x - 8, x^2)$, for what integer values of x is $f(x) = x^2$?

Since $f(x) = \min(6x - 8, x^2) = x^2$, we have

$x^2 < 6x - 8 \Rightarrow x^2 - 6x + 8 < 0$

$\Rightarrow (x-2)(x-4) < 0 \Rightarrow 2 < x < 4$

Thus, the only integer value of $x = 3$

Properties of graphs of functions:

- The graph of $f(x+p)$ is obtained by shifting the graph of $f(x)$ by p units left
- The graph of $f(x-p)$ is obtained by shifting the graph of $f(x)$ by p units right
- The graph of $f(x)+p$ is obtained by shifting the graph of $f(x)$ by p units up
- The graph of $f(x)-p$ is obtained by shifting the graph of $f(x)$ by p units down
- The graph of $f(-x)$ is obtained by reflecting the graph of $f(x)$ about the Y-axis
- The graph of $-f(x)$ is obtained by reflecting the graph of $f(x)$ about the X-axis

Graphs of some quadratic functions:

- $f(x) = x^2$:

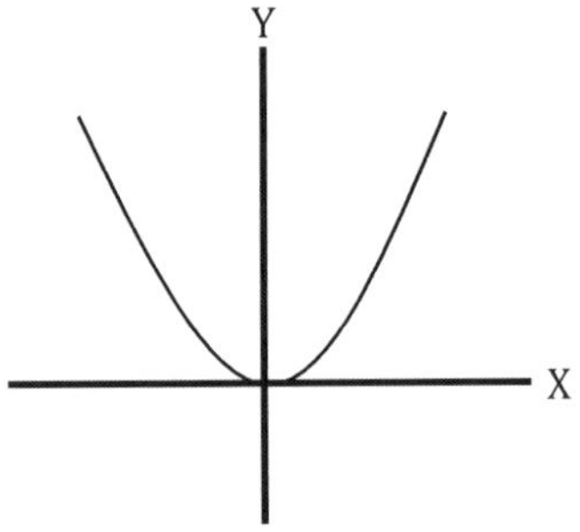

- $f(x) = x^2 + 1$:

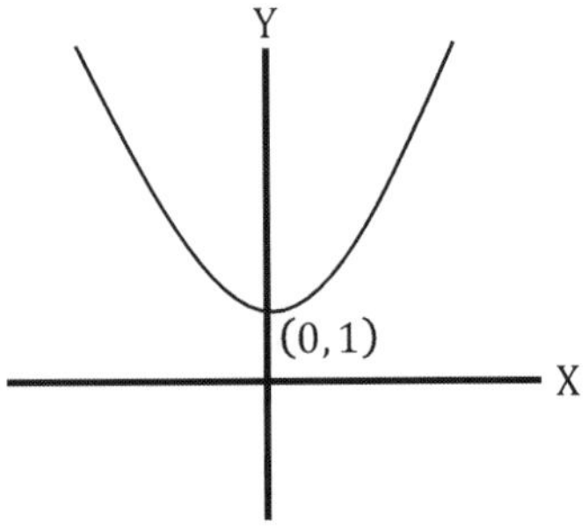

- $f(x) = (x-1)^2$:

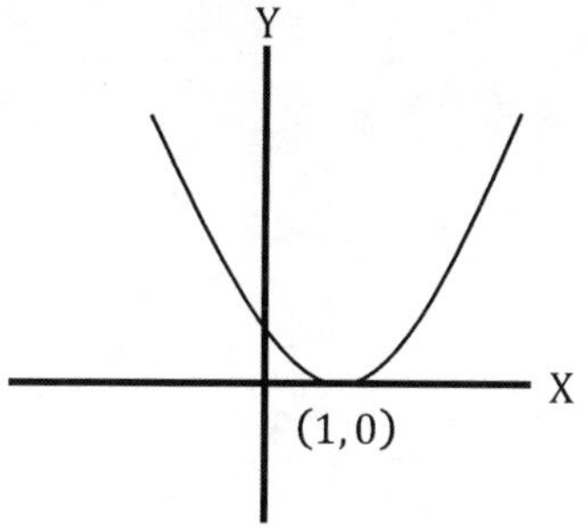

Additional solved problems:

(1) If $f(x) = \sqrt[4]{x} + \sqrt[3]{x} + \sqrt{x} + 1 + \frac{1}{\sqrt{x}} + \frac{1}{\sqrt[3]{x}} + \frac{1}{\sqrt[4]{x}}$, what is the value of $\left\{\frac{f(4)}{f\left(\frac{1}{4}\right)} + \frac{f(3)}{f\left(\frac{1}{3}\right)} + \frac{f(2)}{f\left(\frac{1}{2}\right)}\right\}$?

Explanation:

We have

$$f(x) = \sqrt[4]{x} + \sqrt[3]{x} + \sqrt{x} + 1 + \frac{1}{\sqrt{x}} + \frac{1}{\sqrt[3]{x}} + \frac{1}{\sqrt[4]{x}}$$

$$=> f\left(\frac{1}{x}\right) = \sqrt[4]{\frac{1}{x}} + \sqrt[3]{\frac{1}{x}} + \sqrt{\frac{1}{x}} + 1 + \frac{1}{\sqrt{\frac{1}{x}}} + \frac{1}{\sqrt[3]{\frac{1}{x}}} + \frac{1}{\sqrt[4]{\frac{1}{x}}}$$

$$=> f\left(\frac{1}{x}\right) = \frac{1}{\sqrt[4]{x}} + \frac{1}{\sqrt[3]{x}} + \frac{1}{\sqrt{x}} + 1 + \sqrt{x} + \sqrt[3]{x} + \sqrt[4]{x}$$

$$=> f\left(\frac{1}{x}\right) = f(x) => \frac{f(x)}{f\left(\frac{1}{x}\right)} = 1$$

$$=> \frac{f(2)}{f\left(\frac{1}{2}\right)} = \frac{f(3)}{f\left(\frac{1}{3}\right)} = \frac{f(4)}{f\left(\frac{1}{4}\right)} = 1$$

Thus, we have

$$\frac{f(4)}{f\left(\frac{1}{4}\right)} + \frac{f(3)}{f\left(\frac{1}{3}\right)} + \frac{f(2)}{f\left(\frac{1}{2}\right)}$$

$$= 1 + 1 + 1 = 3$$

(2) If $f(x) = \frac{4^x}{4^x + 2}$, what is the value of $f(a) + f(1-a)$?

(A) $\frac{1}{2}$

(B) 1

(C) 2

Explanation:

We have

$$f(x) = \frac{4^x}{4^x + 2}$$

$$\Rightarrow f(a) = \frac{4^a}{4^a + 2}$$

Also, we have

$$f(1-a) = \frac{4^{(1-a)}}{4^{(1-a)} + 2}$$

$$= \frac{\frac{4}{4^a}}{\frac{4}{4^a} + 2} = \frac{4}{4^a}\left(\frac{1}{\frac{4 + 2 \times 4^a}{4^a}}\right)$$

$$= \frac{4}{4^a} \times \frac{4^a}{2(2 + 4^a)}$$

$$= \frac{2}{2 + 4^a}$$

Thus, we have

$$f(a) + f(1-a)$$

$$= \frac{4^a}{4^a + 2} + \frac{2}{2 + 4^a} = \frac{4^a + 2}{4^a + 2}$$

$$= 1$$

Alternate approach

Let $a = 1 \Rightarrow 1 - a = 0$:

$$f(1) = \frac{4^1}{4^1 + 2} = \frac{4}{6} = \frac{2}{3}$$

$$f(0) = \frac{4^0}{4^0 + 2} = \frac{1}{3}$$

$$\Rightarrow f(1) + f(0) = \frac{2}{3} + \frac{1}{3} = 1$$

Since the answer options are constant values, the answer must be Option B.

(3) Are the following functions the same?

(A) $f(x) = x^x$

(B) $g(x) = x \times x \times x \times x \times \cdots \times x$ (x times)

Explanation:

(A) We have $f(x) = x^x$

For example

If $x = 3 \Rightarrow f(x) = 3^3 = 27$

If $x = -1 \Rightarrow f(-1) = (-1)^{-1} = -1$

If $x = \frac{1}{2} \Rightarrow f\left(\frac{1}{2}\right) = \left(\frac{1}{2}\right)^{\frac{1}{2}} = \frac{1}{\sqrt{2}} \approx = \frac{1}{1.4} = 0.71$

Thus, $f(x) = x^x$ is valid for all real values of x.

(B) We have $g(x) = x \times x \times x \times x \times \cdots \times x$ (x times)

If x is multiplied for x times, the result obtained is x^x.

Thus, apparently, $f(x)$ and $g(x)$ appear to be identical.

However, we are multiplying x for x times, which makes sense only when x is a positive integer. (Multiplying $\frac{1}{2}$ for $\frac{1}{2}$ times or multiplying -1 for -1 times makes no sense)

For example

$g(3) = 3 \times 3 \times 3$ (3 is multiplied for 3 times)

$= 27$

Thus, $g(x)$ is valid only for positive integer values of x.

Thus, $f(x)$ is not the same as $g(x)$.

(4) If $(2, 1)$ are the coordinates of a point on the graph of $f(x)$, what would be the coordi nates of that point for the function $-f(x) + 1$?

Explanation:

To modify $f(x)$ to $-f(x) + 1$, we follow the following steps:

- $f(x) \to -f(x)$: The graph is reflected about the X-axis. Thus, the Y-coordinate of the point would be negated.
  Thus, the coordinates of $-f(x) = (2, \ -1)$

- $-f(x) \to -f(x) + 1$: The graph is shifted 'up' by 1 unit. Thus, the Y-coordinate of the point would increase by '1'.
  Thus, the final coordinates of $-f(x) + 1 = (2, (-1 + 1)) = (2, 0)$

- If (2, 1) are the coordinates of a point on the graph of $f(x)$, what would be the coordinates of that point for the function $f(x - 1)$?

  Explanation:

  We need to modify $f(x) \to f(x - 1)$, i.e. the graph should shift one unit left, hence, value of x should decrease by 1. Thus, (2, 1) gets modified to (1, 1).

  Alternate approach

  If (2, 1) are the coordinates of a point on the graph of $f(x)$, what would be the coordinates of that point for the function $2f(-3x - 1) + 1$? The new x value is obtained by solving $-3x - 1 = 2 \Rightarrow x = -1$. The new y value is obtained by substituting the original y value in the new function: $2(1) + 1 = 3$. Thus, the coordinates are $(-1, 3)$.

Chapter 8

Practice Questions: Algebra

1. If 3 apples, 3 guavas and 4 bananas, together cost \$10 and 3 apples, 2 guavas and 4 bananas together cost \$9, what is the cost of 9 apples, 8 guavas and 12 bananas?

   (A) 26
   (B) 29
   (C) 30
   (D) 32
   (E) 37

2. A person has a few dollars and a few cents such that the total amount with him is a dollars and b cents, where $b < 100$. After spending \$3.50, he was left with $2b$ dollars and 64 cents. What is the value of $(a + b)$?

   (A) 14
   (B) 28
   (C) 32
   (D) 46
   (E) 64

3. If $7x - 2y = 12$, $4x + y = 9$ and $2x + 5y = K$, what is the value of K?

   (A) 9
   (B) 10
   (C) 11
   (D) 12
   (E) 13

4. What is the value of k for the following set of equations?

$$3x + 7y - 11z = 0$$
$$6x - y - 7z = 0$$
$$3x + y - kz = 0$$

   (A) 1
   (B) $\frac{4}{3}$
   (C) 2
   (D) $\frac{7}{3}$
   (E) 5

5. A ball is dropped from the top of a building. The distance, in feet, covered by the ball in t seconds after it was dropped is given by $15t^2$. If the distance covered by the ball in the tth second after it was dropped was 225 feet, what is the value of t?

(A) 3

(B) 4

(C) 7

(D) 8

(E) 10

6. A ball is thrown up from a height of 3 feet above the ground. The distance, in feet, of the ball from the ground is given by $h = 3 + 24t - 4t^2$, where t = time in seconds. What is the maximum height above the ground reached by the ball?

(A) 32 feet

(B) 33 feet

(C) 35 feet

(D) 36 feet

(E) 39 feet

7. What is the minimum positive integer value of p so that $x^2 - px + 8p = 0$ has real and unequal roots?

(A) 12

(B) 24

(C) 27

(D) 32

(E) 33

8. If one of the solutions of $x^2 - px + 12 = 0$ is $\frac{3}{2}$, what is the value of p?

(A) $\frac{3}{2}$

(B) 8

(C) $\frac{19}{2}$

(D) $\frac{57}{4}$

(E) 15

9. What is the minimum integer value of p so that $x^2 - px + 8p = 0$ has real and equal roots?

(A) 0

(B) 1

(C) 2

(D) 32

(E) 33

10. A roller-coaster track is designed in the form of a parabolic arch, whose height, in feet, above the ground is given as $h = -kd(d - 20)$, where k is a positive number and d represents the distance along the length of the track measured from the left-most point where the arch starts. If the arch reaches to a maximum height of 30 feet, what is the value of k?

(A) $\frac{3}{10}$

(B) $\frac{2}{5}$

(C) $\frac{3}{4}$

(D) $\frac{4}{5}$

(E) $\frac{3}{2}$

11. For $f(x) = x^2 + bx + c$, $f(1) = 0$. If $f(3) = 2f(5)$, what is the value of k such that $f(k) = 0, k \neq 1$?

(A) $\frac{10}{3}$

(B) $\frac{37}{7}$

(C) $\frac{17}{3}$

(D) 7

(E) $\frac{19}{2}$

12. If the roots of a fourth degree equation $x^4 - 2x^3 + x^2 + x + 3 = 0$ are p, q, r and s, what is the value of $(2 + p)(2 + q)(2 + r)(2 + s)$?

(A) -37

(B) -9

(C) 9

(D) 37

(E) 40

13. Which of the following is the solution of the inequality $x + \frac{2}{x} \leq 3$?

I. $x > 2$

II. $1 \leq x \leq 2$

III. $x < 0$

(A) Only I

(B) Only II

(C) Only III
(D) Both I and III
(E) Both II and III

14. Which of the following is the correct solution of the inequality $-1 < \frac{x+3}{x+7} < 1$?

I. $x < 7$
II. $x < -5$
III. $x > -5$

(A) Only I
(B) Only II
(C) Only III
(D) Both I and II
(E) Both I and III

15. What is the smallest integer value of x that satisfies the inequality $\frac{x-3}{x^2-9x+18} > \frac{1}{2}$?

(A) 5
(B) 6
(C) 7
(D) 8
(E) 9

16. Which of the following is the correct solution of the inequality $x^3 - 9x^5 > 0$?

I. $0 < x < \frac{1}{3}$
II. $-\frac{1}{3} < x < 0$
III. $x < -\frac{1}{3}$

(A) Only I
(B) Only II
(C) Only III
(D) Both I and II
(E) Both I and III

17. If $-2 \le x \le 3$, $-6 \le y \le 4$, $-1 \le a \le 2$ and $-3 \le b \le 1$, which of the following is the correct range of values of $(a+b)(x+y)$?

(A) $-32 \le (a+b)(x+y) \le 21$

(B) $-28 \leq (a+b)(x+y) \leq 21$
(C) $-24 \leq (a+b)(x+y) \leq 32$
(D) $-24 \leq (a+b)(x+y) \leq 21$
(E) $-28 \leq (a+b)(x+y) \leq 32$

18. If x and y are integers, such that $2 \leq x \leq 10$ and $-12 \leq y \leq -5$, what is the product of the maximum possible value of $\frac{x}{y}$ and the minimum possible value of $\frac{y}{x}$?

(A) -1
(B) $-\frac{2}{3}$
(C) $\frac{2}{3}$
(D) 1
(E) 6

19. A dealer in electronic goods spends \$15,000 on a certain model of TV sets and a certain model of DVD players. The price of each TV set is \$360 and the price of each DVD player is \$240. If he does not wish to purchase greater than 42 items, what is the minimum number of total items the dealer can purchase?

(A) 23
(B) 36
(C) 37
(D) 41
(E) 42

20. A burger store sells chicken burgers, priced at \$8 per burger and vegetable burgers, priced at \$5 per burger. If potato fries are also ordered along with a burger, there is a discount of \$1 on the burger irrespective of the type of burger. The total revenue from selling burgers on a particular day was not greater than \$1,110, and potato fries were ordered along with a burger 50 times. If the total number of burgers purchased is at least 160, what is the maximum number of chicken burgers that were sold?

(A) 41
(B) 58
(C) 120
(D) 159
(E) 169

21. The people working in Company X are planning a trip, for which they want to use sedans and minivans. Each sedan can accommodate 4 people, while each minivan can accommodate 8 people. The rent applicable for each sedan and each minivan is \$60 and \$80, respectively. The total amount available for transportation is \$600. If the sedans and minivans are always filled to capacity, and the number of people interested for the trip is at least 48, what is the minimum number of minivans necessary for the trip?

(A) 2

(B) 3

(C) 5

(D) 6

(E) 8

22. Gift boxes, to be distributed among school children, contain pencils and erasers. Each gift box can hold a maximum of 12 items. If each box must have at least 3 more pencils than the number of erasers, what is the maximum number of erasers in a gift box?

(A) 2

(B) 3

(C) 4

(D) 5

(E) 6

23. The graph of a quadratic function $f(x)$ shown below is symmetric about the line $x = 4$. If $f(1) = 9$, what is the value of $f(7)$?

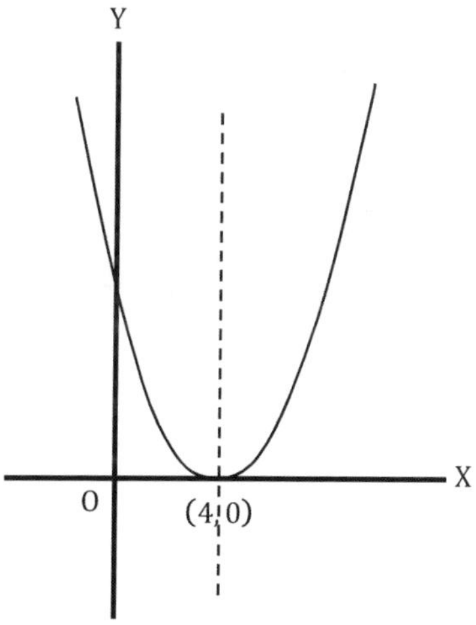

(A) 3

(B) 5

(C) 6

(D) 9

(E) 12

24. In the graph shown below, the functions $f(x) = x^2 - 6x + 8$, which intersects the X-axis at $x = 2$ and $x = 4$ and $g(x) = x - 2$ are shown. For how many integer values of x is $g(x) \geq f(x)$?

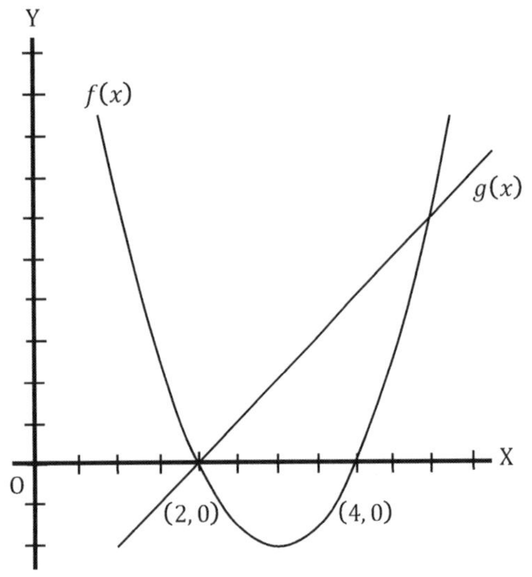

(A) One
(B) Two
(C) Three
(D) Four
(E) Five

25. The graphs of $f(x)$ and $g(x)$ are shown below. Which of the options is correct?

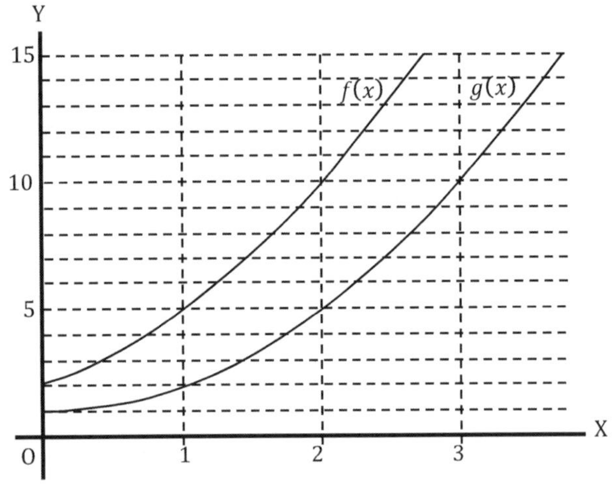

(A) $f(x) = 5x$
(B) $f(x) = x^2 + 4$
(C) $g(x) = (x + 1)^2 - 4$
(D) $g(x) = (x - 1)^2 + 4$
(E) $f(x) - g(x) = 2x + 1$

26. What is the area of the triangle bounded by the lines $f(x) = x - 2$, $g(x) = 2x - 3$ and the Y-axis, as shaded in the graph below?

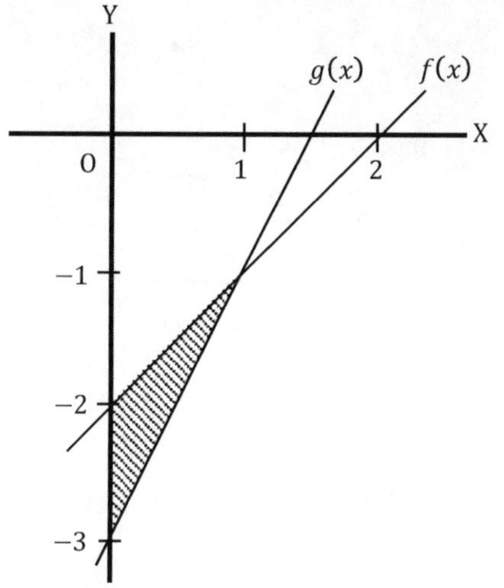

(A) $\frac{1}{4}$

(B) $\frac{1}{2}$

(C) 1

(D) 2

(E) $\frac{3}{2}$

27. The graph of a quadratic function $f(x) = ax^2 + bx + c$ shown below intersects the X-axis at $(2, 0)$ and $(4, 0)$ and intersects the Y-axis at $(0, 4)$. What is the value of $(a + b + c)$?

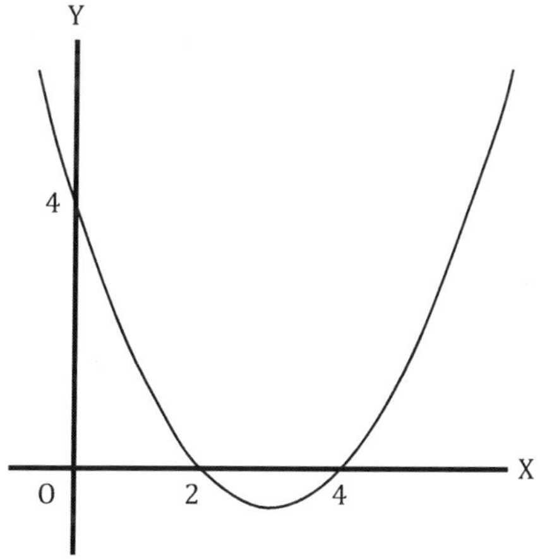

(A) $\frac{1}{2}$

(B) $\frac{3}{2}$

(C) 3

(D) 4

(E) 6

28. For what value of n, the equation below has no possible solution?

$4(2x-1)+3(x-2)=n(x+2)-3(x+1)$

(A) -4

(B) $-\frac{7}{2}$

(C) $\frac{7}{2}$

(D) 7

(E) 14

29. If the equation $x^2-px+12=0$ has exactly one root common with the equation $x^2-6x+9=0$, what is the value of p?

(A) -10

(B) -7

(C) 4

(D) 7

(E) 10

30. If $x+y=2$ and $z=x^2+y^2$, what is the minimum value of z?

(A) $\frac{1}{2}$

(B) 1

(C) 2

(D) 4

(E) 6

31. If q and $2q$ are the roots of the equation $k(x^2-x)+x+1=0$, where $q \neq 0$, which of the following is a correct relation between k and q?

(A) $k=\frac{q}{2}$

(B) $k=\frac{1}{1+3q}$

(C) $k=2q-1$

(D) $k=\frac{1}{2q^2}$

(E) $k=q^2+1$

32. If $f(x)$ is a quadratic polynomial such that $f(1)=1$ and $f(2)=2$. If $f(3)=5$, what is the value of $f(0)$?

(A) -6

(B) -2

(C) 0

(D) 1

(E) 2

33. For all positive integers $x > 2$, function $f(x) = (-1)^x \times f(x-1) \times f(x-2)$. If $f(1) = -f(2) = 1$, what is the value of $f(23)$?

(A) -1

(B) 0

(C) 1

(D) 3

(E) 7

34. For all values of x, let $f(x) = \frac{2x+1}{x-2}$. If $f(p) = q$, what is the value of $f(q)$?

(A) $-\frac{3}{2}$

(B) $\frac{1}{3}$

(C) 1

(D) p

(E) pq

35. What is the value of the area bounded by $f(x) = |x-1| - x$, the X axis and Y axis?

(A) $\frac{1}{4}$

(B) $\frac{1}{2}$

(C) 1

(D) 2

(E) $\frac{3}{2}$

36. If $f(x) = ax^2 + bx$ such that $f(1) = f(-1) + 2$, what is the value of $f(3) - f(-3)$?

(A) 1

(B) 2

(C) 4

(D) 6

(E) 8

37. If $x = 1 + \sqrt{3}$, what is the value of $(x^2 - 2x - 2)$?

(A) -2

(B) $-\sqrt{3}$

(C) 0

(D) 1

(E) $\sqrt{3}$

38. If x is a positive integer such that $24^{\sqrt{x}} + 10^{\sqrt{x}} = 26^{\sqrt{x}}$, what is the value of x?

(A) 0

(B) 1

(C) 2

(D) 4

(E) 6

39. The function $f(x) = 2x^2 - 7x + 6$ is positive for all values of x except when x lies in a particular range. How many integer values of x lie within this particular range?

(A) None

(B) One

(C) Two

(D) Three

(E) Four

40. Which of the following is the correct range of values of x such that $\left|\frac{12 - x}{3}\right| \leq 2$?

(A) $2 \leq x \leq 6$

(B) $4 \leq x \leq 10$

(C) $8 \leq x \leq 16$

(D) $6 \leq x \leq 18$

(E) $4 \leq x \leq 24$

41. How many integer values of x satisfy $x^2 - 4|x| + 3 \leq 0$?

(A) Two

(B) Three

(C) Four

(D) Six

(E) Eight

42. What is the maximum value of $\left\{\frac{12}{x^2 - 12x + 40}\right\}$?

(A) 1

(B) 2

(C) 3

(D) 4

(E) 6

43. If a and b are distinct roots of $x^2 - px - q = 0$, which of the following denotes the value of $(a + b)$?

(A) p

(B) q

(C) $p + q$

(D) $p - q$

(E) pq

44. If $[x]$ denote the greatest integer value less than or equal to x for all positive integers x, such that $1 \leq x \leq 4$, which of the following expressions denotes the value of $\left[\frac{x+1}{2}\right] + \left[\frac{x+2}{4}\right] + \left[\frac{x+4}{8}\right]$?

(A) $2x - 2$

(B) $2x - 1$

(C) x

(D) $x + 1$

(E) $x + 2$

45. If $f(x) = x^3 - kx^2 + 2x$, $x \neq 0$ and $f(-x) = -f(x)$, what is the value of $f(1 - k)$?

(A) -3

(B) 0

(C) 1

(D) 3

(E) 4

46. A function is defined as $f(x) = ax^2 + bx + c$. If $f(2) = f(3) = 0$, and $f(4) = 2$, what is the value of $(a + b + c)$?

(A) 1

(B) 2

(C) 8

(D) 11

(E) 12

47. How many solutions of x are possible, if $x^{-3} + x^{-\frac{3}{2}} = 2$?

(A) One

(B) Two

(C) Four

(D) Five

(E) Six

48. If a function is defined by $f(x) = \frac{x}{x-2}$, what is $f(2x)$ in terms of $f(x)$?

(A) $\frac{f(x)}{3f(x)-1}$

(B) $\frac{2f(x)}{f(x)+1}$

(C) $\frac{f(x)}{f(x)+1}$

(D) $\frac{3f(x)}{2f(x)+1}$

(E) $\frac{2f(x)}{f(x)-1}$

49. If $g(x) = x + 4$ for all x and $f(x) = \frac{2x^2-5}{x}$ for all $x < 0$, for what value of x is $f(x) = g(x)$?

(A) -5

(B) -3

(C) -2

(D) -1

(E) $-\frac{1}{2}$

50. A function is defined such that $f(n) = f(n-1) \times f(n-2)$, for all $n > 2$. If $f(1) = 1$ and $f(2) = -1$, what is the value of $f(33) + f(34)$?

(A) -2

(B) -1

(C) 0

(D) 1

(E) 2

Chapter 9

Answer Key: Algebra

(1) B

(2) D

(3) A

(4) E

(5) D

(6) E

(7) E

(8) C

(9) A

(10) A

(11) C

(12) D

(13) E

(14) C

(15) C

(16) E

(17) E

(18) D

(19) E

(20) C

(21) B

(22) C

(23) D

(24) D

(25) E

(26) B

(27) B

(28) E

(29) D

(30) C

(31) D

(32) E

(33) C

(34) D

(35) A

(36) D

(37) C

(38) D

(39) B

(40) D

(41) D

(42) C

(43) A

(44) C

(45) D

(46) B

(47) A

(48) B

(49) D

(50) C

Chapter 10

Solution: Algebra Practice Questions

1. Let the cost of one apple, one guava and one banana be $\$x$, $\$y$ and $\$z$, respectively.

Thus, we have

$3x + 3y + 4z = 10 \ldots$ (i)

$3x + 2y + 4z = 9 \ldots$ (ii)

We need to determine the value of $(9x + 8y + 12z)$.

Adding (i) and (ii):

$6x + 5y + 8z = 19 \ldots$ (iii)

Adding (i) and (iii):

$9x + 8y + 12z = 29$

Alternate approach

We can see that if (i) is multiplied by 3:

$9x + 9y + 12z = 30$

$=> 9x + 8y + 12z = 30 - y$, i.e. the required value is less than 30.

Similarly, if (ii) is multiplied by 3:

$9x + 6y + 12z = 27$

$=> 9x + 8y + 12z = 27 + 2y$, i.e. the required value is greater than 27.

Thus, the required answer lies between 27 and 30, and hence, must be 29 (the only possible option).

The correct answer is option B.

2. Initial amount = a dollars, b cents.

Amount spent = \$3.50 = 3 dollars, 50 cents.

Final amount = $2b$ dollars, 64 cents.

There are two possibilities:

- Dollars are subtracted from dollars and cents are subtracted from cents, i.e. b cents does not borrow from a.
  Thus, $b - 50 = 64 => b = 114$
  However, we know that $b < 100$
  Hence, this scenario is not possible.
- In order to properly subtract 50 cents from b to get 64 cents as result, we need to borrow 100 cents from a dollars, so the initial sum of money becomes $(a - 1)$ dollars and $(100 + b)$ cents.
  Thus, we have $(100 + b) - 50 = 64 => b = 14$

Considering the subtraction of the dollars:

$(a - 1) - 3 = 2b$

$=> a - 4 = 2 \times 14 = 28$

$=> a = 32$

Thus, we have

Initial amount = \$32.14, amount spent = \$3.50, final amount = \$28.64

Thus, we have

$a + b = 32 + 14 = 46$.

The correct answer is option D.

3. We solve the first two equations and substitute the values of x and y in the third equation.

Thus, we have

$7x - 2y = 12 \ldots \text{(i)}$

$4x + y = 9 \ldots \text{(ii)}$

$2x + 5y = K \ldots \text{(iii)}$

Multiplying (ii) by 2 and adding with (i):

$7x + 8x = 12 + 18 => x = 2$

Substituting $x = 2$ in (ii):

$8 + y = 9 => y = 1$

Substituting $x = 2$ and $y = 1$ in (iii):

$K = 2 \times 2 + 5 \times 1 = 9$.

The correct answer is option A.

4. We have

$3x + 7y = 11z \ldots \text{(i)}$

$6x - y = 7z \ldots \text{(ii)}$

$3x + y = kz \ldots \text{(iii)}$

Multiplying (i) with 2 and subtracting (ii) from it:

$6x - 6x + 14y - (-y) = 22z - 7z$

$=> y = z$

Substituting $y = z$ in (ii):

$6x = 7z + y = 7z + z$

$$=> x = \frac{4}{3}z$$

Substituting $x = \frac{4}{3}z$ and $y = z$ in (iii):

$$3\left(\frac{4}{3}z\right) + z = kz$$

$=> 5z = kz$

$=> k = 5$

The correct answer is option E.

5. The distance covered by the ball in the tth second

= (Distance covered in t seconds) - (Distance covered in $(t-1)$ seconds) $= 15t^2 - 15(t-1)^2$

$= 15\left\{t^2 - (t-1)^2\right\}$

$= 15\,(t - (t-1))\,(t + (t-1))$

$= 15\,(2t - 1)$

Thus, we have

$15\,(2t - 1) = 225$

$=> 2t - 1 = 15$

$=> t = 8$

The correct answer is option D.

6. Height of the ball above the ground

$= 3 + 24t - 4t^2$

$= -4\,(t^2 - 6t) + 3$

$= -4\,(t^2 - 2 \times t \times 3) + 3$

$= -4\,(t^2 - 2 \times t \times 3 + 3^2 - 3^2) + 3$

$= -4\left\{(t-3)^2 - 9\right\} + 3$

$= -4(t-3)^2 + 39$

Since $(t-3)^2$ is a perfect square term, it is always non-negative. Thus, when it is multiplied with (-4), it becomes non-positive.

Thus, the height of the ball from the ground is 39, reduced by the amount $\left\{-4(t-3)^2\right\}$.

Thus, the height of the ball above the ground will be the maximum when the negative term, i.e. $\left\{-4(t-3)^2\right\}$ becomes zero, which happens when $-4(t-3)^2 = 0 => t = 3$

The corresponding value of the height (maximum value) is thus, 39 feet.

The correct answer is option E.

7. In a quadratic $ax^2 + bx + c = 0$, the roots are real and unequal, if the discriminant, i.e. $b^2 - 4ac > 0$.

Thus, for the quadratic $x^2 - px + 8p = 0$, we have

$(-p)^2 - 4\,(1)\,(8p) > 0$

$=> p^2 - 32p > 0$

$=> p^2 - 2 \times p \times 16 + 16^2 - 16^2 > 0$

$=> (p - 16)^2 > 16^2$

Thus, we have

$p - 16 > 16 => p > 32$

OR

$p - 16 < -16 \Rightarrow p < 0$

Thus, the minimum positive integer value of $p = 33$.

The correct answer is option E.

8. We know that $x = \frac{3}{2}$ satisfies the equation $x^2 - px + 12 = 0$

Substituting $x = \frac{3}{2}$ in the equation, we have

$$\left(\frac{3}{2}\right)^2 - p\left(\frac{3}{2}\right) + 12 = 0$$

$$\Rightarrow \frac{3}{2}p = \frac{9}{4} + 12 = \frac{57}{4}$$

$$\Rightarrow p = \frac{2}{3} \times \frac{57}{4} = \frac{19}{2}$$

Alternate approach

Let the other solution be $x = k$.

Thus, we have

$$x^2 - px + 12 \equiv \left(x - \frac{3}{2}\right)(x - k)$$

$$\Rightarrow x^2 - px + 12 \equiv x^2 - x\left(k + \frac{3}{2}\right) + \frac{3}{2}k$$

Comparing coefficients:

$$12 = \frac{3}{2}k \Rightarrow k = 12 \times \frac{2}{3} = 8$$

$$p = k + \frac{3}{2} = 8 + \frac{3}{2} = \frac{19}{2}$$

The correct answer is option C.

9. In a quadratic $ax^2 + bx + c = 0$, the roots are real and equal, if the discriminant, i.e. $b^2 - 4ac = 0$.

Thus, for the quadratic $x^2 - px + 8p = 0$, we have

$$(-p)^2 - 4(1)(8p) = 0$$

$$\Rightarrow p^2 - 32p = 0$$

$$\Rightarrow p(p - 32) = 0$$

$$\Rightarrow p = 0 \text{ or } p = 32$$

Thus, the minimum integer value of $p = 0$.

The correct answer is option A.

10. We know that the height of the arch above the ground is zero at the two ends of the arch, as shown in the diagram below

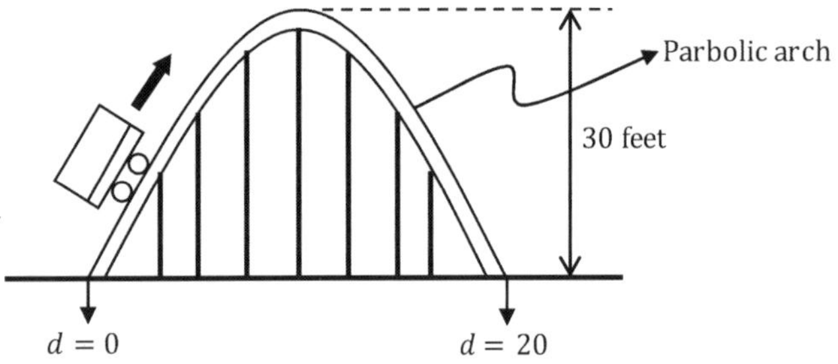

Thus, we have

$h = 0 => -kd(d - 20) = 0$

$=> d = 0$ or 20

The arch reaches its maximum height of 30 feet at the middle, i.e. for $d = \frac{0 + 20}{2} = 10$

Thus, we have

$d = 10,\ h = 30$

$=> h = -kd(d - 20)$

$=> 30 = -k \times 10(10 - 20)$

$=> k = \frac{3}{10}$

Alternate approach

$h = -kd(d - 20)$

$= -k(d^2 - 20d)$

$= -k\left\{(d - 10)^2 - 100\right\}$

$= -k(d - 10)^2 + 100k$

In order to maximize the height, the negative term, i.e. $\left\{-k(d - 10)^2\right\}$ should be '0', i.e. $d = 10$; thus, the corresponding maximum value is $100k$.

Thus, we have

$100k = 30$

$=> k = \frac{3}{10}$

The correct answer is option A.

11. We know that $f(1) = 0$, i.e. $x = 1$ is a solution of $x^2 + bx + c = 0$.

We need to determine k, where $f(k) = 0,\ k \neq 1$, i.e. we need to determine the other solution of $x^2 + bx + c = 0$.

Since 1 and k are the solutions of $x^2 + bx + c = 0$, we have

$x^2 + bx + c \equiv (x - 1)(x - k)$

Since $f(3) = 2f(5)$, we have

$(3-1)(3-k) = 2(5-1)(5-k)$

$=> 2(3-k) = 8(5-k)$

$=> 3-k = 20-4k$

$=> k = \frac{17}{3}$

The correct answer is option C.

12. Since p, q, r and s are the roots of $x^4 - 2x^3 + x^2 + x + 3 = 0$, we have

$x^4 - 2x^3 + x^2 + x + 3 \equiv (x-p)(x-q)(x-r)(x-s)$

Substituting $x = -2$ above, we have

$(-2)^4 - 2(-2)^3 + (-2)^2 + (-2) + 3 = (-2-p)(-2-q)(-2-r)(-2-s)$

$=> 16 + 16 + 4 + 1 = (2+p)(2+q)(2+r)(2+s)$

$=> (2+p)(2+q)(2+r)(2+s) = 37$

The correct answer is option D.

13. We have

$x + \frac{2}{x} \le 3$

$=> x + \frac{2}{x} - 3 \le 0$

$=> \frac{x^2 + 2 - 3x}{x} \le 0$

$=> \frac{(x-2)(x-1)}{x} \le 0$

Thus, the possible scenarios are

- $(x-2)(x-1) \ge 0$ AND $x < 0$
  - $x - 2 \ge 0$ AND $x - 1 \ge 0$ AND $x < 0$
    $=> x \ge 2$ AND $x \ge 1$ AND $x < 0$ - Not possible
  - $x - 2 \le 0$ AND $x - 1 \le 0$ AND $x < 0$

  $=> x \le 2$ AND $x \le 1$ AND $x < 0$

  $=> x < 0 \ldots$(i)
- $(x-2)(x-1) \le 0$ AND $x > 0$
  - $x - 2 \ge 0$ AND $x - 1 \le 0$ AND $x > 0$
    $=> x \ge 2$ AND $x \le 1$ AND $x > 0$ - Not possible
  - $x - 2 \le 0$ AND $x - 1 \ge 0$ AND $x > 0$

  $=> x \le 2$ AND $x \ge 1$ AND $x > 0$

  $=> 1 \le x \le 2 \ldots$(ii)

  Thus, the solution, from (i) and (ii), is:

  $x < 0$

  OR

  $1 \le x \le 2$

The correct answer is option E.

Alternate approach

Using the options: Plug in values of x in I, II and III:

I. Let $x = 3$,
$3 + \frac{2}{3} \nleq 3$
Thus, I does not satisfy.

II. Let $x = 1.5$
$1.5 + \frac{2}{1.5} \leq 3$
Thus, II satisfies.

III. Let $x = -1$
Obviously, sum of two negatives is less than 3.
Thus, III satisfies.

14. We have
$-1 < \frac{x+3}{x+7} < 1$
Let us break this in two parts:

- $-1 < \frac{x+3}{x+7}$
  $=> \frac{x+3}{x+7} + 1 > 0$
  $=> \frac{x+3+x+7}{x+7} > 0$
  $=> \frac{2(x+5)}{x+7} > 0$
  $=> \frac{x+5}{x+7} > 0$
  Thus, we have the following scenarios:
  - $x + 5 > 0$ AND $x + 7 > 0$
    $=> x > -5$ AND $x > -7$
    $=> x > -5$
  - $x + 5 < 0$ AND $x + 7 < 0$
    $=> x < -5$ AND $x < -7$
    $=> x < -7$

  Thus, we have $x > -5$ OR $x < -7 \ldots$(i)
- $\frac{x+3}{x+7} < 1$
  $=> \frac{x+3}{x+7} - 1 < 0$
  $=> \frac{x+3-x-7}{x+7} < 0$
  $=> -4x + 7 < 0$
  $=> \frac{1}{x+7} > 0$
  $=> x + 7 > 0$
  $=> x > -7 \ldots$(ii)

Thus, from (i) and (ii), we finally have

$x > -5$ OR $x < -7$: shown by the black arrows

AND

$x > -7$: shown by the gray arrow

Placing them on a number line, we have

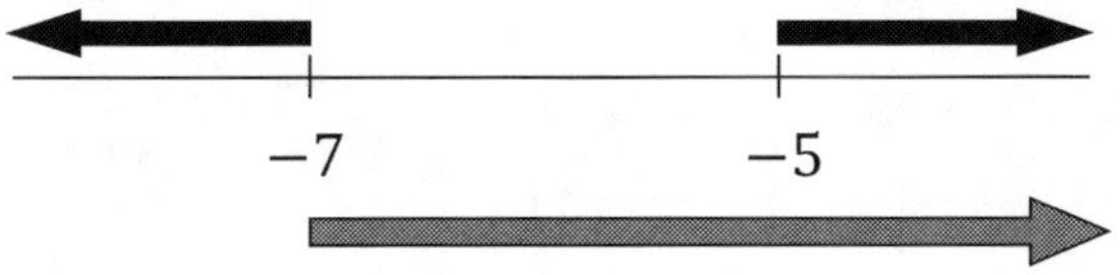

Thus, the overlapping region is

$x > -5$

The correct answer is option C.

Alternate approach

Note that at $x = -7$, the denominator $(x + 7)$ would become 0, and the fraction $\frac{x+3}{x+7}$ will be undefined. Thus, the correct answer must avoid this value.

For Statement I: $x < 7$, we see that $x = -7$ lies in its range; thus, Statement I is not the correct.

For Statement II: $x < -5$, we see that $x = -7$ lies in its range; thus, Statement II is not the correct.

For Statement III: $x > -5$, we see that $x = -7$ does not lie in its range. Since there is no option as "None of the above," Statement III is correct.

15. We have

$$\frac{x-3}{x^2-9x+18} > \frac{1}{2}$$

$$=> \frac{x-3}{(x-3)(x-6)} - \frac{1}{2} > 0$$

Cancelling $(x - 3)$ from numerator and denominator, we have

$$\frac{1}{x-6} - \frac{1}{2} > 0$$

$$=> \frac{2-(x-6)}{2(x-6)} > 0$$

$$=> \frac{-x+8}{x-6} > 0$$

$$=> \frac{x-8}{x-6} < 0$$

Thus, we have the following scenarios:

- $x - 8 < 0$ AND $x - 6 > 0$

  $=> x < 8$ AND $x > 6$

  Thus, the only integer value of $x = 7$

- $x - 8 > 0$ AND $x - 6 < 0$
  $=> x > 8$ AND $x < 6$ - Not possible

Alternate approach

$\frac{x - 3}{(x - 3)(x - 6)} > \frac{1}{2}$

$=> \frac{1}{(x - 6)} > \frac{1}{2}$

Plugging in the options

- Option A: $x = 5$: $\frac{1}{5 - 6} \ngtr \frac{1}{2}$ - Does not satisfy
- Option B: $x = 6$: Division by '0' is not possible - Does not satisfy
- Option C: $x = 7$: $\frac{1}{7 - 6} > \frac{1}{2}$ - Satisfies
- Option D: $x = 8$: $\frac{1}{8 - 6} \ngtr \frac{1}{2}$ - Does not satisfy
- Option E: $x = 9$: $\frac{1}{9 - 6} \ngtr \frac{1}{2}$ - Does not satisfy

The correct answer is option C.

16. We have

$x^3 - 9x^5 > 0$

$=> x^3 (1 - 9x^2) > 0$

$=> x^2 \{x (1 - 9x^2)\} > 0$

Cancelling x^2, since it is always positive:

$x \{(1 - 3x)(1 + 3x)\} > 0$

Thus, we have the following scenarios

- $x > 0$ AND $(1 - 3x)(1 + 3x) > 0$
  - $x > 0$ AND $1 - 3x > 0$ AND $1 + 3x > 0$
    $=> x > 0$ AND $x < \frac{1}{3}$ AND $x > -\frac{1}{3}$
    $=> 0 < x < \frac{1}{3} \ldots$(i)
  - $x > 0$ AND $1 - 3x < 0$ AND $1 + 3x < 0$
    $=> x > 0$ AND $x > \frac{1}{3}$ AND $x < -\frac{1}{3}$ - Not possible
- $x < 0$ AND $(1 - 3x)(1 + 3x) < 0$
  - $x < 0$ AND $1 - 3x > 0$ AND $1 + 3x < 0$
    $=> x < 0$ AND $x < \frac{1}{3}$ AND $x < -\frac{1}{3}$
    $=> x < -\frac{1}{3} \ldots$(ii)
  - $x < 0$ AND $1 - 3x < 0$ AND $1 + 3x > 0$
    $=> x < 0$ AND $x > \frac{1}{3}$ AND $x > -\frac{1}{3}$ - Not possible

Thus, from (i) and (ii), we have

$0 < x < \frac{1}{3}$ OR $x < -\frac{1}{3}$

The correct answer is option E.

17. We have $-2 \leq x \leq 3 \ldots$ (i)

$-6 \leq y \leq 4 \ldots$ (ii)

Thus Adding (i) and (ii):

$\mathbf{-8 \leq x + y \leq 7 \ldots}$ **(iii)**

We also have: $-1 \leq a \leq 2 \ldots$ (iv)

$-3 \leq b \leq 1 \ldots$ (v)

Thus Adding (iv) and (v):

$\mathbf{-4 \leq a + b \leq 3 \ldots}$ **(vi)**

Using (vi) and (iii), for $(a + b)(x + y)$, we have

Maximum possible value occurs by multiplying -4 and -8, i.e. 32

Minimum possible value occurs by multiplying -4 and 7, i.e. -28

Thus, we have

$-28 \leq (a + b)(x + y) \leq 32$

The correct answer is option E.

18. We observe that the values of x are positive while the values of y are negative.

Thus, to maximize the value of $\frac{x}{y}$, we should try to get the least negative value of $\frac{x}{y}$.

Thus, we take x to be the least positive number, i.e. 2 and y to be the most negative number, i.e. -12.

Thus, the maximum possible value of $\frac{x}{y} = -\frac{2}{12} = -\frac{1}{6} \ldots$ (i)

Similarly, to minimize the value of $\frac{y}{x}$, we should try to get the most negative value of $\frac{y}{x}$.

Thus, we take y to be the most negative number, i.e. -12 and x to be the least positive number, i.e. 2.

Thus, the minimum possible value of $\frac{y}{x} = -\frac{12}{2} = -6 \ldots$ (ii)

Thus, from (i) and (ii):

The required product $= \left(-\frac{1}{6}\right) \times (-6) = 1$.

Alternate approach

We can simply use logic to answer this problem.

It is common knowledge that the reciprocal of the maximum value of any term, k, results in the minimum value of the term $\frac{1}{k}$. Thus, the product of the values definitely has to be 1.

The correct answer is option D.

19. Let the number of TV sets and the number of DVD players be x and y, respectively.

Thus, we have

$360x + 240y = 15,000$

$=> 3x + 2y = 125 \ldots$(i)

The least value of $(x + y)$ occurs when x is the largest, i.e. 41 (since 125 when divided by 3 gives 41 as quotient and 2 as remainder). Thus, $2y = 2$, i.e. the value of y is 1.

Alternatively, since the coefficient of x is '3', which is greater than the coefficient of y which is '2', least value of $(x + y)$ would occur if we maximize the value of x or minimize the value of y.

$=> x + y = 41 + 1 = 42$ (least)

The largest value of $(x + y)$ occurs when y is the largest, i.e. 62 (since 125 when divided by 2 gives 62 as quotient and 1 as remainder). However, in that case, x becomes a fraction. However, if we take $y = 61$, then we have $3x = 3$, i.e. the value of x is 1.

$=> x + y = 61 + 1 = 62$ (largest)

Since the dealer does not wish to purchase greater than 42 items, we have $x + y \le 42$

Thus, the only possible solution is

$x + y = 42$.

Thus, the minimum number of total items that the dealer can purchase is 42.

The correct answer is option E.

20. Let the number of chicken burgers and the number of vegetable burgers sold be x and y, respectively.

Thus, total price of the burgers = \$ $(8x + 5y)$.

Since potato fires were ordered with burgers 50 times, each time the discount on a burger being \$1, total discount on the burgers = \$50.

Thus, total revenue generated = \$ $(8x + 5y - 50)$.

Thus, we have

$8x + 5y - 50 \le 1,110$

$=> 8x + 5y \le 1,160 \ldots$(i)

Since the total number of burgers sold was at least 160, we have $x + y \ge 160$

$=> 5x + 5y \ge 800$

$=> -(5x + 5y) \le -800 \ldots$(ii)

Thus, adding (i) and (ii):

$8x + 5y - (5x + 5y) \le 1,160 - 800$

$=> 3x \le 360$

$=> x \le 120$

Thus, the maximum number of chicken burgers sold was 120.

The correct answer is option C.

21. Let the number of sedans and the number of minivans used be s and m, respectively.

Since the number of people is at least 48, we have

$4s + 8m \geq 48$

$=> s + 2m \geq 12 \ldots \text{(i)}$

Since the total amount available is \$600, we have

$60s + 80m \leq 600$

$=> 3s + 4m \leq 30$

$=> -3s - 4m \geq -30 \ldots \text{(ii)}$

From (i) $\times$ 3 + (ii):

$2m \geq 6$

$=> m \geq 3$

Thus, the minimum number of minivans required is 3.

The correct answer is option B.

22. Let the number of pencils and the number of erasers in a gift box be x and y, respectively.

Thus, we have

$x + y \leq 12 \ldots \text{(i)}$

$x \geq y + 3 \ldots \text{(ii)}$

From (ii): Assuming $x = y + 3$ and substituting in (i), we have $(y + 3) + y \leq 12$

$=> 2y \leq 9$

$=> y \leq 4.5$

If x is greater than $(y + 3)$, the value of y will reduce even further to accommodate the higher value of x.

Thus, the maximum value of y is 4 (Since y is an integer).

The correct answer is option C.

23.

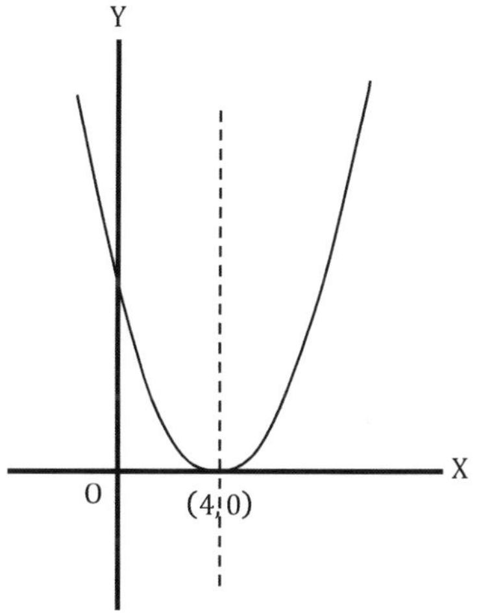

$f(1)$ refers to the value of y at $x = 1$, which is '3' units to the left of $x = 4$.

Since the graph is symmetric about the line $x = 4$, the value of y for a value of x that is '3' units to the left of $x = 4$ is the same as the value of y for a value of x that is '3' units to the right of $x = 4$, i.e. $x = 4 + 3 = 7$

Thus, we have

$f(7) = f(1) = 9$

The correct answer is option D.

24. From the graph, it is clear that the shaded region depicts $g(x) \geq f(x)$, since the line for $g(x)$ is above the line for $f(x)$.

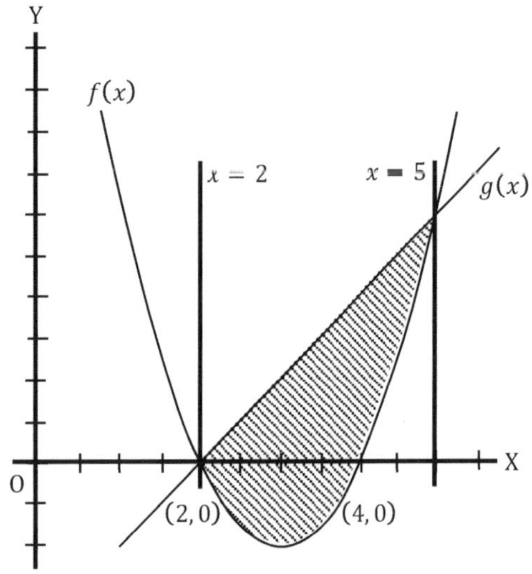

The other intersecting point would be given by equating $g(x)$ with $f(x)$.

$=> x^2 - 6x + 8 = x - 2$

$=> x^2 - 7x - 10 = 0$

$=> x^2 - 5x - 2x - 10 = 0$

$=> x = 2$ or 5

The shaded region is bounded by the lines $x = 2$ and $x = 5$. Note that in the diagram, two dashes depict 1 unit or one dash depicts 0.5 unit.

Since the origin (0, 0) does not fall within the shaded region, the shaded region depicts 'greater than' values for the function $g(x)$.

Thus, the integer values of x are 2, 3, 4 and 5.

Thus, there are 'four' possible values of x.

Alternate approach

We have

$f(x) \leq g(x)$

$=> x^2 - 6x + 8 \leq x - 2$

$=> x^2 - 7x + 10 \leq 0$

=> $(x-2)(x-5) \leq 0$

=> $2 \leq x \leq 5$

Integer values for x are, 2, 3, 4 and 5.

The correct answer is option D.

25.

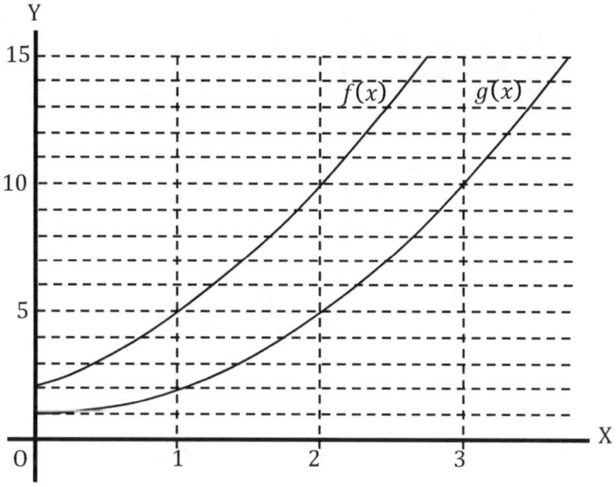

From the graphs, we observe that:

- $f(x)$ satisfies the coordinates: $(0,\ 2)$, $(1,\ 5)$, $(2,\ 10)$
- $g(x)$ satisfies the coordinates: $(0,\ 1)$, $(1,\ 2)$, $(2,\ 5)$, $(3,\ 10)$
- $\{f(x) - g(x)\}$ satisfies the coordinates: $(0,\ 1)$, $(1,\ 3)$, $(2,\ 5)$

Working with the options:

- Option A: Does not satisfy the coordinate $(0,\ 2)$
- Option B: Does not satisfy the coordinates $(0,\ 2)$ or $(2,\ 10)$
- Option C: Does not satisfy the coordinates $(0,\ 1)$ or $(1,\ 2)$ or $(3,\ 10)$
- Option D: Does not satisfy the coordinates $(0,\ 1)$ or $(1,\ 2)$ or $(3,\ 10)$
- Option E: Satisfies the coordinates $(0,\ 1)$, $(1,\ 3)$ and $(2,\ 5)$

The correct answer is option E.

26. From the graph, we have

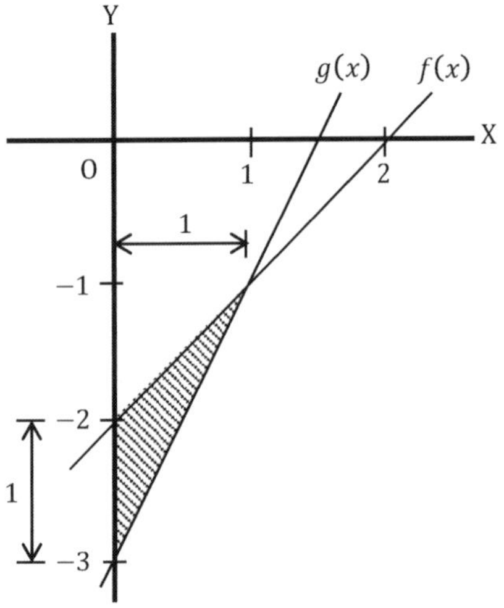

Area of triangle ABC

$= \frac{1}{2} \times \text{Base} \times \text{Height}$

$= \frac{1}{2} \times 1 \times 1 = \frac{1}{2}$

Alternate approach

$g\,(0) = 2 \times 0 - 3 = -3 =>$ Coordinates of C are $(0,\ -3)$

$f\,(0) = 0 - 2 = -2 =>$ Coordinates of B are $(0,\ -2)$

Thus BC $= -2 - (-3) = 1$

Equating $g\,(x)$ and $f\,(x)$:

$2x - 3 = x - 2 => x = 1$

Thus height of the triangle ABC = 1

Thus, area of triangle ABC $= \frac{1}{2} \times 1 \times 1 = \frac{1}{2}$

The correct answer is option B.

27.

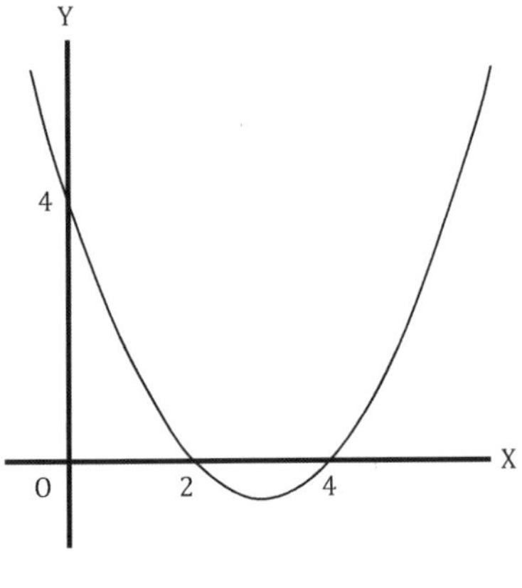

Since the graph intersects at 2 and 4, the quadratic equation $f(x) = ax^2 + bx + c$ has roots 2 and 4.

Thus, we have

$f(x) = ax^2 + bx + c = a(x - 2)(x - 4)$

Since the graph intersects the Y-axis at $(0,\ 4)$, we have

$f(0) = 4$

$=> a(0 - 2)(0 - 4) = 4$

$=> 8a = 4 => a = \frac{1}{2}$

$=> f(x) = \frac{1}{2}(x - 2)(x - 4)$

$=> f(x) = \frac{1}{2}(x^2 - 6x + 8) = \frac{1}{2}x^2 - 3x + 4$

Thus, we have

$ax^2 + bx + c = \frac{1}{2}x^2 - 3x + 4$

$=> a = \frac{1}{2},\ b = -3,\ c = 4$

$=> a + b + c = \frac{3}{2}$

The correct answer is option B.

Alternate approach

We know that c is the y-intercept.

As the graph intersects the y axis at $(0, 4)$, we have $c = 4$

The graph intersects X-axis at (2, 0) and (4, 0) so we have

$0 = a \times 2^2 + b \times 2 + 4$

$=> 4a + 2b = -4$

$=> 2a + b = -2 \ldots$ (i)

Also, we have

$0 = a \times 4^2 + b \times 4 + 4$

$=> 16a + 4b = -4$

$=> 4a + b = -1\ldots$ (ii)

Subtracting (i) from (ii) we get,

$2a = 1 => a = \frac{1}{2}$

Plugging this value in (i) we get,

$b = -3$

Thus, $a + b + c = \frac{1}{2} + (-3) + 4 = \frac{3}{2}$

28. The above is a linear equation. On simplifying the above, we would finally get a form:

$kx + l = nx + m$, where k, l, n and m are constants

In the above, if $k = n$, the terms with x will cancel out.

Thus, if $l = m$, i.e. the constant terms are equal, it implies that the above equation is true for any value of x.

However, if $l \neq m$, the above equation would not be true for any value of x, i.e. there will be no solution.

We have

$4(2x - 1) + 3(x - 2) = n(x + 2) - 3(x + 1)$

$=> n(x + 2) = 4(2x - 1) + 3(x - 2) + 3(x + 1)$

$=> nx + 2n = 14x - 7$

Thus, we should have:

$nx = 14x$ AND $2n \neq -7$

$=> n = 14$ (which satisfies $2n \neq -7$)

The correct answer is option E.

29. We have

$x^2 - 6x + 9 = 0$

$=> (x - 3)^2 = 0$

$=> x = 3$

Thus, the above equation has only one root.

Thus, $x^2 - px + 12 = 0$ must have one root as $x = 3$

Thus, substituting $x = 3$ in $x^2 - px + 12$ would result in '0.'

$=> 3^2 - 3p + 12 = 0$

$=> 3p = 21$

$=> p = 7$

The correct answer is option D.

30. We have

$x + y = 2$

$=> y = 2 - x$

Thus, we have

$z = x^2 + y^2$

$= x^2 + (2 - x)^2$

$= 2x^2 - 4x + 4$

$= 2(x^2 - 2x + 2)$

$= 2(x^2 - 2x + 1) + 2$

$= 2(x-1)^2 + 2$

The minimum value of z occurs when the square term, i.e. $(x-1)^2 = 0$ (which occurs at $x = 1$).

Thus, the minimum value of $z = 0 + 2 = 2$.

Alternate approach

If sum of two variables is a constant, the sum of their squares would be the minimum if they are equal.

Let us see why:

We have $x + y = 2 \Rightarrow x = 2 - y$

Thus, we have $x^2 + y^2$

$= (2-y)^2 + y^2$

$= 4 - 4y + 2y^2$

$= 2(y^2 - 2y) + 4$

$= 2\left\{(y-1)^2 - 1\right\} + 4$

$= 2(y-1)^2 + 2$

The above would have the minimum value if the square term, i.e. $(y-1)^2 = 0$ (which occurs if $y = 1 \Rightarrow x = 2 - y = 1$. The corresponding minimum value is $0 + 2 = 2$).

Thus, we have $x = y \Rightarrow x = 1 \Rightarrow z = 1^2 + 1^2 = 2$

The correct answer is option C.

31. Since q and $2q$ are the roots, the quadratic equation can be represented as: $(x-q)(x-2q) = 0$

$\Rightarrow x^2 - 3qx + 2q^2 = 0 \ldots \text{(i)}$

The given quadratic equation is:

$k(x^2 - x) + x + 1 = 0$

$\Rightarrow kx^2 - x(k-1) + 1 = 0$

$\Rightarrow x^2 - \left(\frac{k-1}{k}\right)x + \frac{1}{k} = 0 \ldots \text{(ii)}$

Comparing (i) and (ii):

$\frac{k-1}{k} = 3q$

$\Rightarrow 1 - \frac{1}{k} = 3q$

$\Rightarrow \frac{1}{k} = 1 - 3q$

$\Rightarrow k = \frac{1}{1-3q} \ldots \text{(iii)}$

$\frac{1}{k} = 2q^2$

$=> k = \frac{1}{2q^2} \ldots$ (iv)

The correct answer is option D.

32. Let $f(x) = ax^2 + bx + c$ - a standard form of a quadratic equation

Thus, we have

$f(1) = 1 => a + b + c = 1 \ldots$ (i)

$f(2) = 2 => 4a + 2b + c = 2 \ldots$ (ii)

$f(3) = 5 => 9a + 3b + c = 5 \ldots$ (iii)

From (ii) - (i): $3a + b = 1 \ldots$ (iv)

From (iii) - (ii): $5a + b = 3 \ldots$ (v)

From (v) - (iv): $2a = 2 => a = 1$

From (iv): $b = 1 - 3 = -2$

From (i): $c - 1 - 1 - (-2) - 2$

Thus, we have

$f(x) = x^2 - 2x + 2$

$=> f(0) = 2$

Alternate approach

We have

$f(1) = 1 => f(1) - 1 = 0$

$f(2) = 2 => f(2) - 2 = 0$

Let the function $g(x) = f(x) - x$

Thus, we see that: $g(1) = f(1) - 1 = 0$ AND $g(2) = f(2) - 2 = 0$

Thus, 1 and 2 are the roots of $g(x)$

$=> g(x) = k(x-1)(x-2)$

$=> f(x) - x = k(x-1)(x-2)$

$=> f(x) = k(x-1)(x-2) + x$

Since $f(3) = 5$, we have

$k(3-1)(3-2) + 3 = 5$

$=> 2k = 2 => k = 1$

$=> f(x) = (x-1)(x-2) + x$

$=> f(0) = (0-1)(0-2) + 0 = 2$

The correct answer is option E.

33. We have $f(1) = -f(2) = 1$

$=> f(1) = 1$ and $f(2) = -1 \ldots$ (i)

Also, we have $f(x) = (-1)^x \times f(x-1) \times f(x-2) \ldots$ (ii)

Substituting $x = 3, 4 \ldots$ etc. in (ii), we have

- $x = 3$: $f(3) = (-1)^3 \times f(2) \times f(1) = 1$
- $x = 4$: $f(4) = (-1)^4 \times f(3) \times f(2) = -1$
- $x = 5$: $f(5) = (-1)^5 \times f(4) \times f(3) = 1$

Thus, we observe that the values of $f(x)$ alternate between 1 and -1, with the even values of x resulting in -1 while the odd values of x resulting in 1.

Thus, we have $f(23) = 1$ (since 23 is odd)

The correct answer is option C.

34. We have

$$f(p) = \frac{2p+1}{p-2} = q$$

$$=> f(q) = \frac{2q+1}{q-2}$$

$$= \frac{2\left(\frac{2p+1}{p-2}\right)+1}{\frac{2p+1}{p-2}-2}$$

$$= \frac{4p+2+p-2}{2p+1-2p+4}$$

$$= \frac{5p}{5} = p$$

The correct answer is option D.

35. The intercept on the X-axis is obtained from $f(x) = 0$

$=> |x-1| - x = 0 => |x-1| = x$

$=> x - 1 = x$ or $-x$

Since $x - 1 \neq x$, we have

$x - 1 = -x => x = \frac{1}{2}$

Thus, $f(x)$ intersects the X-axis at $\left(\frac{1}{2}, 0\right)$.

The intercept on the Y-axis is obtained by substituting $x = 0$ in $f(x)$

$=> f(0) = |0-1| - 0 = 1$

Thus, $f(x)$ intersects the Y-axis at $(0, 1)$.

Thus, the required area is the area of the triangle formed by the points $\left(\frac{1}{2}, 0\right)$, $(0, 1)$ and $(0, 0)$, as shown in the diagram below:

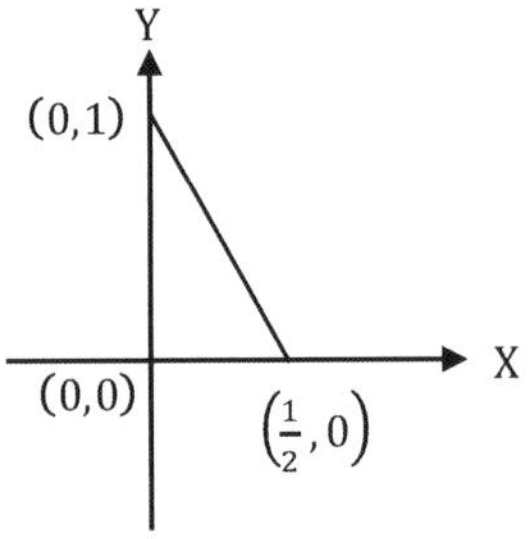

Thus, required area

$= \frac{1}{2} \times 1 \times \frac{1}{2} = \frac{1}{4}$

The correct answer is option A.

36. We have

$f(x) = ax^2 + bx$

Since $f(1) = f(-1) + 2$, we have

$a(1^2) + b(1) = a(-1)^2 + b(-1) + 2$

$=> a + b = a - b + 2$

$=> b = 1$

$=> f(x) = ax^2 + x$

$=> f(3) - f(-3) = \{a(3^2) + 3\} - \{a(-3)^2 + (-3)\}$

$= (9a + 3) - (9a - 3) = 6$

The correct answer is option D.

37. We have

$x = 1 + \sqrt{3}$

$=> x - 1 = \sqrt{3}$

Squaring both sides:

$(x - 1)^2 = (\sqrt{3})^2$

$=> x^2 - 2x + 1 = 3$

$=> x^2 - 2x - 2 = 0$

Alternate approach

Substituting $x = 1 + \sqrt{3}$ in the expression, we have

$x^2 - 2x - 2$

$= (1 + \sqrt{3})^2 - 2(1 + \sqrt{3}) - 2$

$= 1 + 2\sqrt{3} + 3 - 2 - 2\sqrt{3} - 2$

$= 0$

The correct answer is option C.

38. Since the numbers have $\sqrt{x}$ as their exponent, $\sqrt{x}$ must be an integer, i.e. x must be a perfect square (if x is not a perfect square, then each term will be irrational and the equation would never hold).

The only possible options are A, B and D.

Working with the options:

- Option A: $x = 0 => \sqrt{x} = 0$
  Thus, each term becomes 1 and we get: $1 + 1 = 1$, which is not true. - Does not satisfy
- Option B: $x = 1 => \sqrt{x} = 1$
  Thus, the equation becomes: $24 + 10 = 26$, which is not true. - Does not satisfy
- Option D: $x = 4 => \sqrt{x} = 2$
  Thus, the equation becomes: $24^2 + 10^2 = 26^2$, i.e. $576 + 100 = 676$, which is true. - Satisfies.

The correct answer is option D.

Alternate approach

We have

$24^{\sqrt{x}} + 10^{\sqrt{x}} = 26^{\sqrt{x}}$

The numbers 24, 10 and 26 form a Pythagorean triplet, since the numbers are in the ratio 12, 5 and 13, which form a Pythagorean triplet, i.e. $12^2 + 5^2 = 13^2$

Thus, we have

$24^2 + 10^2 = 26^2$

$=> \sqrt{x} = 2$

$=> x = 4$

39. We need to determine the range of values of x for which $f(x) \leq 0$

$=> 2x^2 - 7x + 6 \leq 0$

$=> 2x^2 - 3x - 4x + 6 \leq 0$

$=> x(2x - 3) - 2(2x - 3) \leq 0$

$=> (x - 2)(2x - 3) \leq 0$

$=> 2(x - 2)\left(x - \frac{3}{2}\right) \leq 0$

$=> (x - 2)\left(x - \frac{3}{2}\right) \leq 0$

$=> \frac{3}{2} \leq x \leq 2$

Thus, the only one integer value of $x = 2$

The correct answer is option B.

40. We have

$\left|\frac{12 - x}{3}\right| \leq 2$

$=> \frac{|12-x|}{|3|} \leq 2$

$=> \frac{|12-x|}{3} \leq 2$

$=> |12-x| \leq 6$

$=> |-(x-12)| \leq 6$

$=> |x-12| \leq 6$

$=> -6 \leq x-12 \leq 6$

Adding 12 throughout:

$=> -6+12 \leq x-12+12 \leq 6+12$

$=> 6 \leq x \leq 18$

Alternately, from $|x-12| \leq 6$, we can say that the distance of x from 12 is 6 units, resulting in the extremities 12 + 6 = 18 and 12 - 6 = 6.

The correct answer is option D.

41. We know that

- $|x| = x$ if $x \geq 0$
- $|x| = -x$ if $x < 0$

We have $x^2 - 4|x| + 3 \leq 0$

- Let $x < 0$, i.e. $|x| = -x$:
  $x^2 - 4|x| + 3 \leq 0$
  $=> x^2 - 4(-x) + 3 \leq 0$
  $=> x^2 + 4x + 3 \leq 0$
  $=> (x+1)(x+3) \leq 0$
  $=> -3 \leq x \leq -1$
  Thus, the integer values of x are $-3, -2, -1$, i.e. there are 'three' values ...(i)
- Let $x \geq 0$, i.e. $|x| = x$:
  $x^2 - 4|x| + 3 \leq 0$
  $=> x^2 - 4x + 3 \leq 0$
  $=> (x-1)(x-3) \leq 0$
  $=> 1 \leq x \leq 3$
  Thus, the integer values of x are $1, 2, 3$, i.e. there are 'three' values ...(ii)

Thus, there are 'six' possible values of x.

The correct answer is option D.

42. We have to maximize the value of $\left\{\frac{12}{x^2 - 12x + 40}\right\}$.

The value will be maximized if the value of the denominator is minimized.

Thus, we have

$x^2 - 12x + 40$

$= x^2 - 2 \times x \times 6 + 6^2 + 4$

$= (x - 6)^2 + 4$

The above value is minimized if the square term, i.e. $(x - 6)^2 = 0$ (which occurs at $x = 6$).

Thus, the minimum value of the denominator is '4'.

Thus, the maximum value of the expression

$= \frac{12}{4} = 3$

The correct answer is option C.

43. We have

Since a is a root of $x^2 - px - q = 0$, we have

$a^2 - ap - q = 0 \ldots \text{(i)}$

Since b is a root of $x^2 - px - q = 0$, we have

$b^2 - bp - q = 0 \ldots \text{(ii)}$

Subtracting (i) from (ii):

$b^2 - a^2 - bp + ap = 0$

$=> (b - a)(b + a) - p(b - a) = 0$

$=> (b - a)(b + a - p) = 0$

$=> b - a = 0$ OR $b + a - p = 0$

$=> b = a$ OR $b + a = p$

Since a and b are distinct roots, $a \neq b$

$=> a + b = p$

The correct answer is option A.

44. We know that $[x]$ denotes the greatest integer less than or equal to x.

Thus, for example, $[2.3]$ equals the greatest integer less than or equal to 2.3:

The integers less than 2.3 are: 2, 1, 0, −1, etc.

Among these, the greatest is 2.

Thus, $[2.3] = 2$.

Since x is a positive integer and $1 \leq x \leq 4$, we have

- $x = 1$: $\left[\frac{x+1}{2}\right] + \left[\frac{x+2}{4}\right] + \left[\frac{x+4}{8}\right] = \left[\frac{1+1}{2}\right] + \left[\frac{1+2}{4}\right] + \left[\frac{1+4}{8}\right] = [1] + \left[\frac{3}{4}\right] + \left[\frac{5}{8}\right] = 1 + 0 + 0 = 1$
- $x = 2$: $\left[\frac{x+1}{2}\right] + \left[\frac{x+2}{4}\right] + \left[\frac{x+4}{8}\right] = \left[\frac{2+1}{2}\right] + \left[\frac{2+2}{4}\right] + \left[\frac{2+4}{8}\right] = \left[\frac{3}{2}\right] + [1] + \left[\frac{6}{8}\right] = 1 + 1 + 0 = 2$

- $x = 3$: $\left[\frac{x+1}{2}\right] + \left[\frac{x+2}{4}\right] + \left[\frac{x+4}{8}\right] = \left[\frac{3+1}{2}\right] + \left[\frac{3+2}{4}\right] + \left[\frac{3+4}{8}\right] = [2] + \left[\frac{5}{4}\right] + \left[\frac{7}{8}\right] = 2 + 1 + 0 = 3$
- $x = 4$: $\left[\frac{x+1}{2}\right] + \left[\frac{x+2}{4}\right] + \left[\frac{x+4}{8}\right] = \left[\frac{4+1}{2}\right] + \left[\frac{4+2}{4}\right] + \left[\frac{4+4}{8}\right] = \left[\frac{5}{2}\right] + \left[\frac{6}{4}\right] + \left[\frac{8}{8}\right] = 2 + 1 + 1 = 4$

Thus, we see that the value of the expression is the same as the value of x assumed.

Thus, the value of $\left[\frac{x+1}{2}\right] + \left[\frac{x+2}{4}\right] + \left[\frac{x+4}{8}\right] = x$

The correct answer is option C.

Alternate approach

Check for $x = 1$: $\left[\frac{x+1}{2}\right] + \left[\frac{x+2}{4}\right] + \left[\frac{x+4}{8}\right] = 1$

Plug in $x = 1$ in all answer options and we get the answer as 1 in only B and C, rest all options gets eliminated.

Check for $x = 2$: $\left[\frac{x+1}{2}\right] + \left[\frac{x+2}{4}\right] + \left[\frac{x+4}{8}\right] = 2$

Plug in $x = 2$ in B and C and we get the answer as 2 in only C.

45. $f(x) = x^3 - kx^2 + 2x$

$=> f(-x) = (-x)^3 - k(-x)^2 + 2(-x)$

$= -x^3 - kx^2 - 2x$

Since $f(-x) = -f(x)$, we have

$-x^3 - kx^2 - 2x = -(x^3 - kx^2 + 2x)$

$=> 2kx^2 = 0$

$=> k = 0$

Thus, we have

$f(x) = x^3 + 2x$

$=> f(1-k) = f(1) = 1 + 2 = 3$

The correct answer is option D.

46. Since $f(2) = f(3) = 0$, the roots of $f(x) = ax^2 + bx + c$ must be $x = 2$ or 3.

Thus, we have

$f(x) = ax^2 + bx + c$

$= a(x-2)(x-3)$

Since $f(4) = 2$, we have

$f(4) = a(4-2)(4-3) = 2a$

$=> 2a = 2$

$=> a = 1$

Thus, we have

$f(x) = 1 \times (x-2)(x-3) = x^2 - 5x + 6$

Comparing coefficients:

$ax^2 + bx + c \equiv x^2 - 5x + 6$

$=> a = 1,\ b = -5,\ \&\ c = 6$

$=> a + b + c = 2$

The correct answer is option B.

47. We have

$x^{-3} + x^{-\frac{3}{2}} = 2$

Let $x^{-\frac{3}{2}} = a$

$=> x^{-3} = \left(x^{-\frac{3}{2}}\right)^2 = a^2$

$=> a^2 + a = 2$

$=> a^2 + a - 2 = 0$

$=> (a+2)(a-1) = 0$

$=> a = -2$ OR 1

$=> x^{-\frac{3}{2}} = -2$ OR 1

$=> \frac{1}{x^{\frac{3}{2}}} = -2$ OR 1

Taking reciprocal on both sides:

$=> x^{\frac{3}{2}} = -\frac{1}{2}$ OR 1

$=> (x^3)^{\frac{1}{2}} = -\frac{1}{2}$ OR 1

Since $(x^3)^{\frac{1}{2}}$ represents the square root of a number (x^3), it must be positive (since the square root, by definition only takes the positive root)

Thus, we have

$(x^3)^{\frac{1}{2}} = 1$

Squaring both sides:

$=> x^3 = 1$

Taking cube-root on both sides:

$=> x = 1$

Thus, there is only one solution for x.

The correct answer is option A.

48. $f(x) = \frac{x}{x-2} \ldots$ (i)

Let $f(x) = y$

$=> y = \frac{x}{x-2}$

$=> xy - 2y = x$

$=> xy - x = 2y$

$=> x(y - 1) = 2y$

$=> x = \frac{2y}{y-1} \ldots \text{(ii)}$

From (i), we have

$f(2x) = \frac{2x}{2x-2}$

$= \frac{x}{x-1}$

Substituting the value of x from (ii), we have

$f(2x) = \frac{\left(\frac{2y}{y-1}\right)}{\left(\frac{2y}{y-1} - 1\right)}$

$= \frac{\left(\frac{2y}{y-1}\right)}{\left(\frac{2y-y+1}{y-1}\right)}$

$= \frac{2y}{y+1}$

Substituting $y = f(x)$, we have

$f(2x) = \frac{2f(x)}{f(x)+1}$

The correct answer is option B.

49. $f(x) = \frac{2x^2 - 5}{x}$

$g(x) = x + 4$

Since $f(x) = g(x)$, we have

$\frac{2x^2 - 5}{x} = x + 4$

$=> 2x^2 - 5 = x^2 + 4x$

$=> x^2 - 4x - 5 = 0$

$=> (x + 1)(x - 5) = 0$

$=> x = -1$ OR 5

However, we know that $f(x)$ is only defined for $x < 0$

Thus, the only possible value of $x = -1$

The correct answer is option D.

50. We have

$f(n) = f(n - 1) \times f(n - 2)$, for all $n > 2$

We also have:

- $f(1) = 1$
- $f(2) = -1$

Thus, using the above, we have

- For $n = 3:\ f(3) = f(2) \times f(1) = (-1) \times 1 = -1$
- For $n = 4:\ f(4) = f(3) \times f(2) = (-1) \times (-1) = 1$
- For $n = 5:\ f(5) = f(4) \times f(3) = 1 \times (-1) = -1$
- For $n = 6:\ f(6) = f(5) \times f(4) = (-1) \times 1 = -1$

Thus, we observe the different values of $f(n)$ follow the sequence: $(1, -1, -1)$

Thus, when k is an integer, we have

- If $n = 3k + 1 => f(n) = 1$
- If $n = 3k + 2 => f(n) = -1$
- If $n = 3k => f(n) = -1$

$=> f(33) + f(34)$

$\equiv f(3k) + f(3k + 1)$

$= -1 + 1$

$= 0$

The correct answer is option C.

Chapter 11

Word Problems

11.1 Percents

Percent means 'out of hundred'. A fraction having 100 in the denominator is called a percent.

To convert a fraction into a percent, we need to multiply it by 100.

For example, the fraction $\frac{1}{4} = \frac{1\times 25}{4\times 25} = \frac{25}{100} = 25\%$, which is equivalent to doing $\frac{1}{4} \times 100 = 25\%$.

Similarly, if we were to compare the scores of a candidate in two tests, in one of which he obtained 60 out of 80 and in the other 75 out of 90, we can use percent values:

- In the first test: $\frac{60}{80} \times 100 = 75\%$
- In the second test: $\frac{75}{90} \times 100 = 83.33\%$

Thus, the candidate fared better in the second test.

There are two aspects of percent calculation:

- **Calculating a given percent of some number:**

  If we were to calculate: What is 20% of 120?

  This implies that we need to determine a number, which as a percent of 120 represents 20%.

  Following unitary method:

  If the value is 100, then the required number is 20.

  => If the value is 120, the required number is $\frac{20}{100} \times 120 = 24$

  Thus, we have

  $$\boldsymbol{x\% \text{ of } y = \frac{x}{100} \times y}$$

- **Calculating the percent value of a number with respect to another:**

  If we were to calculate: 30 represents what percent of 150?

  This implies that if we have 30 out of 150, then what is the number that we have out of 100?

Following unitary method:

If the value is 150, then the required number is 30.

=> If the value is 100, the required number is $\frac{30}{150} \times 100 = 20$

Thus, we have

$$\boldsymbol{x} \textbf{ is what \% of } \boldsymbol{y} = \frac{\boldsymbol{x}}{\boldsymbol{y}} \times \mathbf{100}$$

Some fractions and their percent equivalents are shown below:

Fraction	Percent	Fraction	Percent	Fraction	Percent	Fraction	Percent
$\frac{1}{1}$	100%	$\frac{1}{6}$	16.66%	$\frac{1}{11}$	9.09%	$\frac{1}{16}$	6.25%
$\frac{1}{2}$	50%	$\frac{1}{7}$	14.28%	$\frac{1}{12}$	8.33%	$\frac{1}{17}$	5.88%
$\frac{1}{3}$	33.33%	$\frac{1}{8}$	12.50%	$\frac{1}{13}$	7.69%	$\frac{1}{18}$	5.55%
$\frac{1}{4}$	25%	$\frac{1}{9}$	11.11%	$\frac{1}{14}$	7.14%	$\frac{1}{19}$	5.26%
$\frac{1}{5}$	20%	$\frac{1}{10}$	10%	$\frac{1}{15}$	6.66%	$\frac{1}{20}$	5.00%

Percent change: Percent change is calculated as the change between the initial and final values calculated as a percent of the initial value.

Thus, we have

$$\textbf{Percentage change} = \frac{\textbf{Final value (F)} - \textbf{Initial value (I)}}{\textbf{Initial value (I)}} \times \mathbf{100}$$

Given that the percent change be $n\%$, we have $n\% = \left(\frac{F-I}{I}\right) \times 100$.

Rearranging the terms, we get:

$$\frac{F-I}{I} = \frac{n}{100} => \frac{F}{I} - 1 = \frac{n}{100}$$

$$=> \frac{F}{I} = 1 + \frac{n}{100}$$

$$=> \boldsymbol{F} = \boldsymbol{I}\left(\boldsymbol{1} + \frac{\boldsymbol{n}}{\boldsymbol{100}}\right)$$

Thus, in order to get the final value, the initial value is to be multiplied with $\left(1 + \frac{n}{100}\right)$.

Some examples are shown below:

Percent change	Fraction multiplied	Percent change	Fraction multiplied
100% increase	$1 + \frac{100}{100} = 2$	100% decrease	$1 - \frac{100}{100} = 0$
50% increase	$1 + \frac{50}{100} = \frac{3}{2}$	50% decrease	$1 - \frac{50}{100} = \frac{1}{2}$
33.33% increase	$1 + \frac{33.33}{100} = \frac{4}{3}$	33.33% decrease	$1 - \frac{33.33}{100} = \frac{2}{3}$
25% increase	$1 + \frac{25}{100} = \frac{5}{4}$	25% decrease	$1 - \frac{25}{100} = \frac{3}{4}$
20% increase	$1 + \frac{20}{100} = \frac{6}{5}$	20% decrease	$1 - \frac{20}{100} = \frac{4}{5}$

Let us take a few examples:

(1) If the price of an article becomes \$150 after increasing it by 20%, what was the initial price of the article?

Explanation:

Since the price is increased by 20%, we have

$$F = I\left(1 + \frac{20}{100}\right)$$

$$=> I = \frac{F}{1 + \frac{20}{100}} = \frac{150}{1 + \frac{1}{5}} = \frac{150}{\frac{6}{5}} = 150 \times \frac{5}{6} = \$125$$

Alternately, we can say that the final price of the article is $(100 + 20)\,\% = 120\%$ of the initial price.

Thus, we have

$150 = 120\%$ of I

$$=> I = 150 \times \frac{100}{120} = \$125$$

(2) After two successive discounts of 20% each, the price of a television becomes \$12,800. What was the original price?

Explanation:

Let original price = $\$x$.

Thus, we have

$$x\left(1-\frac{20}{100}\right)\left(1-\frac{20}{100}\right)=12,800$$

$$\Rightarrow x = 12,800 \times \frac{5}{4} \times \frac{5}{4} = \$20,000$$

Successive percent change: When back-to-back percent changes happen, then the overall percent change can be calculated using the following relation:

Let there be successive percent changes of $x\%$ and $y\%$. If the overall percent change be $z\%$, then:

$$\boldsymbol{z} = \left(\boldsymbol{x} + \boldsymbol{y} + \frac{\boldsymbol{xy}}{\mathbf{100}}\right)\%$$

Here x and y have to be taken with their sign, positive for increase and negative for decrease.

Let us take an example:

If there is a 20% decrease followed by a 30% increase in the price of a commodity, what is the overall percent change in price of the commodity?

Explanation:

Let the price of the commodity be \$100.

Thus, after 20% decrease in price, the price = $(100 - 20)\%$ or 80% of \$100 = \$80.

Thus, after a 30% increase in price, the final price

$$= (100 + 30)\% \text{ or } 130\% \text{ of } \$80 = \$\left(\frac{130}{100} \times 80\right) = \$104$$

Thus, overall percent change (from \$100 to \$104) = $\frac{104-100}{100} \times 100 = 4\%$

Alternate approach:

We can use the successive percent change formulae shown above above, as following:

We have: $x = -20,\ y = 30$

Thus, overall percent change

$$= -20 + 30 - \frac{20 \times 30}{100} = -20 + 30 - 6 = 4\%.$$

Percent change in product of two terms: The net percent change in the product of two terms, when one changes by x% and the other changes by y%, is also given by the same formula:

$$\left(\boldsymbol{x} + \boldsymbol{y} + \frac{\boldsymbol{xy}}{\mathbf{100}}\right)\%$$

Here x and y have to be taken with their sign, positive for increase and negative for decrease.

Let us take a few examples:

(1) If the length of a rectangle increases by 20% and the breadth decreases by 30%, what is the percent change in area of the rectangle?

Explanation:

Let the length and breadth of the rectangle be $100x$ and $100y$, respectively.

Thus, the area = Length × Breadth = $10,000xy$

New length = $(100 + 20)\%$ of $100x = 120x$

New breadth = $(100 - 30)\%$ of $100y = 70y$

New area = $120x \times 70y = 8,400xy$

Thus, percent change in area = $\frac{10,000xy - 8,400xy}{10,000xy} \times 100 = 16\%$ (decrease)

Alternate approach:

We can use the successive percent change rule shown above as follows:

We have: $x = 20,\ y = -30$

Thus, overall percent change

$$= 20 - 30 - \frac{20 \times 30}{100} = -10 - 6 = -16\%.$$

(2) The price of an item is increased by 20%. As a result, the number of items sold decreases by 10%. What is the percent change in the revenue generated as a result?

Explanation:

Revenue = Price × Number of items sold

Thus, percent change in revenue (a product of two commodities) $= 20 - 10 - \frac{20 \times 10}{100} = 8\%$

Percent change in the sum of two commodities: Let there be two commodities whose values are A and B, which have percent changes of $x\%$ and $y\%$, the net percent change in the sum of the two is given by:

$$\left(\frac{Ax + By}{A + B}\right) \times 100\%$$

Here x and y have to be taken with their sign, positive for increase and negative for decrease.

Let us take an example:

The monthly salaries of a couple are \$1,000 for the husband and \$1,500 for the wife. If the husband's salary increases by 10% and the wife's salary increases by 20%, what is the overall percent increase in their combined salary?

Explanation:

Increased salary of the husband = (100 + 10) % of \$1,000 = \$1,100

Increased salary of the wife = (100 + 20) % of \$1,500 = \$1,800

Thus, new combined salary = \$(1,100 + 1,800) = \$2,900

Initial combined salary = \$(1,000 + 1,500) = \$2,500

Thus, percent increase $= \frac{2,900 - 2,500}{2,500} \times 100\% = 16\%$

Alternate approach:

We can use the percent change formulae shown above as follows:

We have: $A = 1,000,\ B = 1,500,\ x = 10,\ y = 20$

Thus, net percent change

$$= \left(\frac{Ax + By}{A + B}\right) \times 100\%$$
$$= \frac{1,000 \times 10\% + 1,500 \times 20\%}{1,000 + 1,500} \times 100\% = \frac{100 + 300}{2,500} \times 100\%$$
$$= \frac{400}{25}\% = 16\%$$

Change of base: If the percent difference between two values, expressed in terms of one is to be calculated in terms of the other, we use the change of base concept.

- **A is x% more than B => B is $\left(\frac{100x}{100 + x}\right)$% less than A**
- **A is x% less than B => B is $\left(\frac{100x}{100 - x}\right)$% more than A**

Let us take an example:

If A is 20% more than B, by what percent is B less than A?

Explanation:

We can say that: If B is 100, A = (100 + 20)% of 100 = 120.

Thus, B is $\left(\frac{120 - 100}{120}\right) \times 100 = 16.66\%$ less than A.

Alternate approach:

We can use the change of base rule shown above as follows:

Since A is 20% more than B, we have $x = 20$

Thus, the required percent value $= \left(\frac{100 \times 20}{100 + 20}\right)\% = \frac{2,000}{120}\% = 16.66\%$

11.2 Profit, Loss, and Discount

Some important points regarding profit, loss and discount are:

- The price at which an article is purchased or manufactured is the Cost Price
- The price at which the article is sold is the Selling Price
- Profit or Loss is the difference between the Cost Price and Selling Price:
    - Profit = Selling Price − Cost Price
    - Loss = Cost Price − Selling Price
- Percent Profit or Percent Loss are calculated on the Cost Price
- Marked price or List price is the price on which discounts are applicable resulting in the final Selling Price
- Percent Discount is calculated on the Marked (List) Price

Some important relations:

- If an article having cost C is sold at $P\%$ profit, then the selling price S is

  $S = (100 + P)\%$ of C

  $=> S = \frac{100 + P}{100} \times C$

  $=> \boldsymbol{S = C \times \left(1 + \frac{P}{100}\right)}$

- If an article having cost C is sold at $L\%$ loss, then the selling price S is

  $S = (100 - L)\%$ of C

  $=> S = \frac{100 - L}{100} \times C$

  $=> \boldsymbol{S = C \times \left(1 - \frac{L}{100}\right)}$

- If an article having marked (list) price M is sold at $D\%$ discount, then the selling price S is

  $S = (100 - D)\%$ of M

  $=> S = \frac{100 - D}{100} \times M$

  $=> \boldsymbol{S = M \times \left(1 - \frac{D}{100}\right)}$

- If an article having cost C is sold at a price S, the percent profit or loss is $\left(\frac{S-C}{C}\right) \times 100\%$
  If the above value comes positive, it is a profit, else it is a loss

- If an article having marked price M is sold at a price S $(S < M)$, the percent discount is $\left(\frac{M-S}{M}\right) \times 100\%$

Let us take a few examples:

(1) A shopkeeper has a stock of pens, each having the same cost price. If he sells 4 pens at the price of the cost of 5 pens and he sells all the pens he has with him, what is his percent of profit?

Explanation:

We know that

Cost price of 5 pens = Selling price of 4 pens = \$20 (assumed)

(Note: We assumed the price as \$20 since 20 is divisible by both 4 and 5)

Thus, the cost of 1 pen = $\$\frac{20}{5} = \$4$

Selling price of 1 pen = $\$\frac{20}{4} = \$5$

Thus, percent profit = $\frac{5-4}{4} \times 100\% = 25\%$

Alternately, we can say that since the shopkeeper sells 4 pens for the cost of 5, he makes a profit of $(5 - 4) = 1$ pen for every 4 pens.

Thus, his percent profit = $\frac{1}{4} \times 100\% = 25\%$

Point to be noted:

If Selling Price of n articles = Cost price of m articles

Percent profit or loss

$$= \left(\frac{m-n}{n}\right) \times 100\%$$

If the above is positive, then it is a profit, else it is a loss

(2) A shopkeeper has a stock of pens, each having the same cost price. If for every 6 pens that he sells at cost price, he gives away 1 pen for free, and he sells all the pens he has with him, what is his percent loss?

Explanation:

Since the shopkeeper gives away 1 pen for free for every 6 pens sold at cost price, we have

Cost price of 6 pens = Selling price of 7 pens = \$42 (assumed)

(Note: We assumed the price as \$42 since 42 is divisible by both 6 and 7)

Thus, the cost of 1 pen = $\$\frac{42}{6} = \$7$

Selling price of 1 pen = $\$\frac{42}{7} = \$6$

Thus, percent loss = $\frac{6-7}{7} \times 100\% = 14.28\%$

Point to be noted:

If y articles are given for free for every x articles sold at cost price, we can say that $(x+y)$ articles are sold at the cost price of x articles.

Thus, percent loss

$$= \left(\frac{y}{x+y}\right) \times 100\%$$

(3) A shopkeeper has a stock of pens, each having the same cost price. If for every 10 pens that he sells at a price 20% higher than the cost, he gives away 1 pen for free. If he sells all the pens he has with him, what is his percent profit?

Explanation:

Let the cost price of a pen be \$10.

Thus, selling price of a pen = $(100+20)\%$ of \$10 = \$12.

The man gives 1 pen for free for every 10 pens sold at a higher price.

Thus, selling price of $(10+1) = 11$ pens = $\$(10 \times 12) = \$120$

Cost price of 11 pens = $\$\,(\mathbf{11 \times 10}) = \mathbf{\$110}$

Thus, percent profit = $\dfrac{\mathbf{120-110}}{\mathbf{110}} \times \mathbf{100\%} = \mathbf{9.09\%}$

Alternate approach:

Apparently, the shopkeeper sells each pen at a price 20% above the cost price.

Thus, we have

The selling price of 1 pen = $(\mathbf{100} + \mathbf{20})\,\%$ of the cost price = $\dfrac{\mathbf{6}}{\mathbf{5}}$ times the cost price

=> The selling price of 5 pens = Cost price of 6 pens ... (i)

The man gives 1 pen for free for every 10 pens sold at a higher price.

Thus, from (i):

The selling price of 10 pens = Cost price of 12 pens

Thus, the man sells 10 pens for the cost of 12 pens and gives 1 pen for free.

Thus, the selling price of $(\mathbf{10} + \mathbf{1}) = \mathbf{11}$ pens is the same as the cost of 12 pens = \$132 (assumed).

Thus, the selling price of each pen = $\$\dfrac{\mathbf{132}}{\mathbf{11}} = \mathbf{\$12}$

The cost price of each pen = $\$\dfrac{\mathbf{132}}{\mathbf{12}} = \mathbf{\$11}$

Thus, percent profit = $\dfrac{\mathbf{12-11}}{\mathbf{11}} \times \mathbf{100\%} = \mathbf{9.09\%}$

(4) An article is sold at 20% profit. Had it been sold at \$60 less than the selling price, there would have been a loss of 5%. What is the cost price of the article?

Explanation:

Difference in selling prices = \$60

Difference in % profit = $\mathbf{20} - (\mathbf{-5}) = \mathbf{25\%}$ of the Cost price

Thus, we have

25% of Cost price = \$60

$$\Rightarrow \text{ Cost price} = \$\left(60\times\frac{100}{25}\right) = \$240$$

(5) A man sold two tables for \$96 each, gaining 20% on one and losing 20% on the other. What is his overall profit or loss percent?

Explanation:

For the first table:

Selling price = \$96 = (100 + 20) % of Cost price

$$\Rightarrow \text{ Cost price} = \$\left(96\times\frac{100}{120}\right) = \$80$$

For the second table:

Selling price = \$96 = (100 − 20) % of Cost price

$$\Rightarrow \text{ Cost price} = \$\left(96\times\frac{100}{80}\right) = \$120$$

Thus, total cost price = \$(80 + 120) = \$200

Total selling price = \$(96 × 2) = \$192

$$\text{Thus, overall percent loss} = \frac{200-192}{200}\times 100 = 4\%$$

Point to be noted:

If two items are sold, **each at the same price**, one at a **profit of *k*%** and the other at a **loss of *k*%**, there is an overall **loss** given by $\left(\frac{k^2}{100}\right)\%$.

(6) In which of the following cases is the discounted price the greatest?

(A) Successive discounts of 10% and 20%

(B) Successive discounts of 20% and 10%

(C) Successive discounts of 25% and 5%

(D) Single discount of 30%

Explanation:

We know that the overall percent change for successive percent changes of x% and y% is given by $\left(x + y + \frac{xy}{100}\right)\%$

Thus, going by options:

(A) Net percent change = $-10 - 20 + \frac{(-10) \times (-20)}{100\%} = -28\%$

(B) Net percent change = $-20 - 10 + \frac{(-20) \times (-10)}{100\%} = -28\%$

(C) Net percent change = $-25 - 5 + \frac{(-25) \times (-5)}{100\%} = -28.75\%$

Since the minimum discount is 28% for both options A and B, the greatest price would also be for options A and B.

11.3 Ratio, Proportion and Variation

Ratio: A ratio is defined as the multiple or part a quantity of one kind bears with another quantity of the same kind.

For example:

Let us have two quantities $\boldsymbol{a}$ and $\boldsymbol{b}$, such that $\boldsymbol{a} = \mathbf{30}$ and $\boldsymbol{b} = \mathbf{45}$

Since $\mathbf{30}\times\frac{3}{2} = \mathbf{45}$, we have

$\boldsymbol{a}\times\frac{3}{2} = \boldsymbol{b} \Rightarrow \frac{\boldsymbol{a}}{\boldsymbol{b}} = \frac{2}{3}$

$\Rightarrow \boldsymbol{a} : \boldsymbol{b} = 2 : 3$

Thus, the ratio of a and b is 3 : 2

Some important relations

- If $A : B = p : q$, we have
  - $A = \left(\frac{p}{p+q}\right) \times (A+B)$
  - $B = \left(\frac{q}{p+q}\right) \times (A+B)$

- Ratio remains same when the two quantities of a ratio are multiplied or divided by the same number. Thus, we have

  $\frac{\boldsymbol{a}}{\boldsymbol{b}} = \frac{\boldsymbol{a}\times\boldsymbol{k}}{\boldsymbol{b}\times\boldsymbol{k}}$
  $\Rightarrow \boldsymbol{a} : \boldsymbol{b} = (\boldsymbol{a} \times \boldsymbol{k}) : (\boldsymbol{b} \times \boldsymbol{k})$

- If the ratio of two quantities is less than '1', adding the same positive number to both quantities of the ratio increases its value. Thus, we have

  - If $\frac{a}{b} < 1$, and k is a positive number:
    $\frac{\boldsymbol{a}+\boldsymbol{k}}{\boldsymbol{b}+\boldsymbol{k}} > \frac{\boldsymbol{a}}{\boldsymbol{b}}$
    For example: $\frac{a}{b} = \frac{2}{3} \Rightarrow \frac{a+1}{b+1} = \frac{3}{4} > \frac{2}{3}$
  - Similarly, if $\frac{a}{b} < 1$, and k is a positive number:

    $\frac{\boldsymbol{a}-\boldsymbol{k}}{\boldsymbol{b}-\boldsymbol{k}} < \frac{\boldsymbol{a}}{\boldsymbol{b}}$

For example: $\frac{a}{b} = \frac{2}{3} => \frac{a-1}{b-1} = \frac{1}{2} < \frac{2}{3}$

- If the ratio of two quantities is greater than '1', adding the same positive number to both quantities of the ratio decreases the value. Thus, we have

  - If $\frac{a}{b} > 1$, and k is a positive number:
    $\boldsymbol{\frac{a+k}{b+k} < \frac{a}{b}}$
    For example: $\frac{a}{b} = \frac{3}{2} => \frac{a+1}{b+1} = \frac{4}{3} < \frac{3}{2}$
  - Similarly, if $\frac{a}{b} > 1$, and k is a positive number:

    $\boldsymbol{\frac{a-k}{b-k} > \frac{a}{b}}$
    For example: $\frac{a}{b} = \frac{3}{2} => \frac{a-1}{b-1} = \frac{2}{1} > \frac{3}{2}$

- If $\frac{a}{b} = \frac{c}{d}$, we have
  Let $\frac{a}{b} = \frac{c}{d} = k$
  $=> a = b \times k$ and $c = d \times k$
  $=> a + c = k\,(b + d)$
  $=> \frac{a+c}{b+d} = k$
  $=> \boldsymbol{\frac{a}{b} = \frac{c}{d} = \frac{a+c}{b+d}}$

- If $\frac{a}{b} = \frac{c}{d}$, we have

  Let $\frac{a}{b} = \frac{c}{d} = k$
  $=> a = b \times k$ and $c = d \times k$
  Thus, we have
  - $a + b = bk + b = b\,(k+1)$, $a - b = bk - b = b\,(k-1)$
  - $c + d = dk + d = d\,(k+1)$, $c - d = dk - d = d\,(k-1)$

  $=> \boldsymbol{\frac{a+b}{a-b} = \frac{c+d}{c-d}}$

- If we have $\boldsymbol{k \times a = l \times b = m \times c}$, we have

  $\boldsymbol{a : b : c = \frac{1}{k} : \frac{1}{l} : \frac{1}{m}}$

Proportion: If the ratio of two quantities is equal to the ratio of other two quantities then the four quantities are said to be in proportion.

Thus, if $a : b = c : d$, then a, b, c, d are in proportion.

Variation:

- If two quantities a & b are such that any increase in b produces a proportionate increase in a, and vice-versa, then a is 'directly proportional' to b.

  Thus, we have

  $\boldsymbol{a \propto b}$

  => $a = kb$, where k is a constant

  => $\boldsymbol{\frac{a}{b}}$ = **constant**

- If two quantities a and b are such that any increase in b will bring about a proportionate decrease in a and vice-versa, then a is inversely proportional to b.

  Thus, we have

  $\boldsymbol{a \propto \frac{1}{b}}$

  => $a = \frac{k}{b}$, where k is a constant

  => $a \times b$ = constant

- If A, B and C are three quantities, such that $A \propto B$ and $B \propto C$

  => $\boldsymbol{A \propto C}$

- Joint variation: If A, B and C are three quantities, such that $A \propto B$ if C is constant and $A \propto C$ if B is constant, we have

  $\boldsymbol{A \propto (B \times C)}$, where both B and C vary

Let us take a few examples:

(1) If $\frac{p}{3} = \frac{q}{8} = \frac{7p - 2q}{k}$, what is the value of k?

Explanation:

We know that

$$\frac{p}{3} = \frac{q}{8} = \frac{7p - 2q}{k}$$

Let $\frac{p}{3} = \frac{q}{8} = \frac{7p - 2q}{k} = r$

$=> p = 3r$ and $q = 8r$

Thus, we have

$7p = 21r$ and $2q = 16r => 7p - 2q = 5r$

Thus, we have

$$\frac{7p - 2q}{k} = r$$

$$=> \frac{5r}{k} = r => k = 5$$

(2) An alloy has iron and copper mixed in the ratio 2 : 3. If 21 pounds of iron is mixed with 15 pounds of the above alloy, what would be the percent of copper in the final alloy?

Explanation:

Amount of copper present in 15 pounds of the original alloy $= \frac{3}{2+3} \times 15 = 9$ pounds

Since the quantity of copper remains unchanged, while the total alloy quantity becomes $(15 + 21) = 36$ pounds, we have

Percent of copper $= \frac{9}{36} \times 100 = 25\%$

(3) The cost of manufacturing an article is partly fixed and partly varies with the number of articles to be manufactured. If the cost of manufacturing 10 articles is \$100 and the cost of manufacturing 20 articles is \$180, what is the fixed cost involved?

Explanation:

Let the cost of manufacturing an article be given by C.

Let the fixed cost be f and the variable cost be v.

Thus, we have

$v \propto n$, where n is the number of articles

$=> v = kn$, where k is a constant of proportionality

Thus, we have

$C = f + v => C = f + kn$

For $n = 10 : C = 100 = f + 10k \ldots$ (i)

For $n = 20 : C = 180 = f + 20k \ldots$ (ii)

From (ii) − (i):

$80 = 10k => k = 8$

Thus, from (i):

$100 = f + 80 => f = 20$

Thus, the fixed cost is \$20.

Alternate approach:

The cost of manufacturing 10 articles is \$100 and that for 20 articles is \$180

Thus, the incremental cost for $(20 - 10) = 10$ articles = \$ $(180 - 100) =$ \$80

Since the cost of manufacturing 10 articles is \$100, the balance amount, i.e. \$ $(100 - 80)$ = \$20 is the fixed cost.

(4) A telecom operator charges a fixed amount $\$f$ for the first $\frac{1}{4}$ minutes of a call. Thereafter, for every $\frac{1}{2}$ minutes, it charges $\$v$. If the total charge for a x minute call is $\$c$, what is the expression for x in terms of f, v and c?

Explanation:

Let the duration of the call be x minutes.

Charges for the first $\frac{1}{4}$ minutes = $\$f$

Duration of call remaining = $\left(x - \frac{1}{4}\right)$ minutes.

Call charge every $\frac{1}{2}$ minute = $\$v$

Thus, call charge for $\left(x - \frac{1}{4}\right)$ minutes =

$$\$\left\{\left(\frac{v}{\frac{1}{2}}\right) \times \left(x - \tfrac{1}{4}\right)\right\} = \$\left\{2v\left(x - \tfrac{1}{4}\right)\right\}$$

Thus, total call charge = $\$\left\{f + 2v\left(x - \tfrac{1}{4}\right)\right\}$

Thus, we have

$$f + 2v\left(x - \tfrac{1}{4}\right) = c$$

$$=> 2v\left(x - \tfrac{1}{4}\right) = c - f$$

$$=> x - \frac{1}{4} = \frac{c - f}{2v}$$

$$=> x = \frac{c - f}{2v} + \frac{1}{4}$$

11.4 Mixtures

Mixture problems involve forming a mixture from two or more things, and determining some quantity (percentage, price, etc.) of the final mixture.

Mixture problems can be seen in any of (or similar to) the following scenarios:

- Combining multiple varieties of commodities having different prices and determining the average price of the mixture
- Combining multiple alloys with different percent purity and determining the percent purity of the mixture
- Combining multiple solutions with different components and determining the ratio of the components in the mixture

To find the mean value of a mixture:

- When X and x quantities of A and B, costing C and cost c, respectively, are mixed, the average cost, M, of mixture is given by:

$$M = \frac{CX + cx}{X + x}$$

- When two solutions A and B, having volume V and v, containing $R\%$ concentration of ingredient X and $r\%$ of ingredient X, respectively, are mixed, the concentration, $M\%$, of ingredient X in the mixture is given by:

$$M = \frac{V\left(\frac{R}{100}\right) + v\left(\frac{r}{100}\right)}{V + v} \times 100\% = \left[\frac{VR + vr}{V + v}\right]\%$$

Let us take a few examples:

(1) Three varieties of tea priced at \$10 per pound, \$15 per pound and \$24 per pound are mixed together in the ratio 2 : 1 : 5, respectively. What is the average price of the resulting mixture?

Explanation:

We know that the three varieties are mixed in the ratio 2 : 1 : 5

Let the quantities of the three varieties be 2 pounds, 1 pound and 5 pounds, respectively.

Thus, total prices of the three varieties of tea are:

$\$(2 \times 10) = \$20$, $\$(1 \times 15) = \$15$ and $\$(5 \times 24) = \$120$

Thus, combined price of $(2 + 1 + 5) = 8$ pounds of tea $= \$(20 + 15 + 120) = \$155$

Thus, average price of the mixture $= \$\left(\frac{155}{8}\right) = \$19\frac{3}{8}$

(2) Three varieties of tea priced at \$20 per pound, \$15 per pound and \$24 per pound are mixed. If the average price of the resulting mixture is \$20 per pound, which of the following is a possible ratio in which the three varieties of tea were mixed?

(A) 1 : 4 : 5
(B) 2 : 3 : 8
(C) 3 : 8 : 12
(D) 5 : 8 : 10

Explanation:

Let the quantities of the three varieties of tea used are x pounds, y pounds and z pounds.

Thus, total prices of the three varieties of tea are:

$\$\,(x \times 20) = \$20x$, $\$\,(y \times 15) = \$15y$ and $\$\,(z \times 24) = \$24z$

Thus, combined price of $(x + y + z)$ pounds of tea $= \$\,(20x + 15y + 24z)$

Thus, average price of the mixture $= \$\left(\frac{20x + 15y + 24z}{x + y + z}\right)$

Thus, we have

$$\frac{20x + 15y + 24z}{x + y + z} = 20$$

$$=> 20x + 15y + 24z = 20x + 20y + 20z$$

$$=> 5y = 4z => \frac{y}{z} = \frac{4}{5}$$

Since x is not there in the above equation, any value of x can be used. Observe that the average price of the mixture is the same as the price per pound of the variety having quantity x.

Thus, the presence or the absence of that particular variety will not affect the average price of the mixture.

Thus, we need to choose those options where $y : z = 4 : 5$

The correct answers are options A and D.

(3) Iron ore mined from Site A has 72% iron, while that from Site B has 56% iron. A certain quantity of both ores are taken and melted to form a substance having 65% iron. If the quantity of iron ore taken from Site A is 5 pounds more than that from Site B, what is the weight of the ore used from Site B?

(A) 12
(B) $16\frac{1}{2}$
(C) $17\frac{1}{2}$
(D) 20

Explanation:

Let the quantity of ore taken from Site B be x pounds and that taken from Site A be $(x + 5)$ pounds.

Thus, quantity of iron (in pounds) present in the mixture

$= 56\%$ of $x + 72\%$ of $(x + 5)$

$$= \frac{56x}{100} + \frac{72(x+5)}{100} = \frac{128x + 360}{100}$$

Total quantity of the mixture $= x + (x + 5) = (2x + 5)$ pounds

Thus, the percent of iron in the mixture

$$= \frac{128x + 360}{100(2x + 5)} \times 100\%$$

Thus, we have

$$\frac{128x + 360}{100(2x + 5)} \times 100\% = 65\%$$

$=> 128x + 360 = 130x + 325$

$=> 2x = 35 => x = 17\frac{1}{2}$

Note: Percent of iron present in ores from Site A and Site B = 72% and 56%

If the two ores were mixed in equal ratio, percent of iron in the mixture $= \frac{72 + 56}{2} = 64\%$

Since the actual percent of iron in the mixture is 65%, i.e. slightly greater than the above value, there should be slightly higher quantity of ore from Site A.

Alternate approach:

Let the quantities of ores from Site A and Site B be a and b pounds, respectively.

Thus, we have

$$\frac{72a + 56b}{a + b} = 65$$

$=> 72a + 56b = 65a + 65b => 7a = 9b$

$=> a = \frac{9}{7}b$

But, we also know that $a = b + 5$

$=> b + 5 = \frac{9}{7}b => \frac{2}{7}b = 5$

$=> b = 17\frac{1}{2}$

Method of alligation:

Alligation is the method by which we calculate the ratio in which two different ingredients of known values are mixed to produce a mixture of a given value. The value of the ingredient may refer to the price, composition, etc.

In the alligation method, the amount of working required to get to the answer is considerably reduced.

Let us look into the method for different scenarios:

- When P quantity of dearer ingredient A of cost C and p quantity of cheaper ingredient B of cost c (where $C > c$) are mixed to get an average cost of mixture M:

$$\frac{\textbf{Weight of cheaper ingredient}}{\textbf{Weight of dearer ingredient}} = \frac{\textbf{Cost of dearer} - \textbf{Cost of mixture}}{\textbf{Cost of mixture} - \textbf{Cost of cheaper}}$$

The rule can also be represented in the following way:

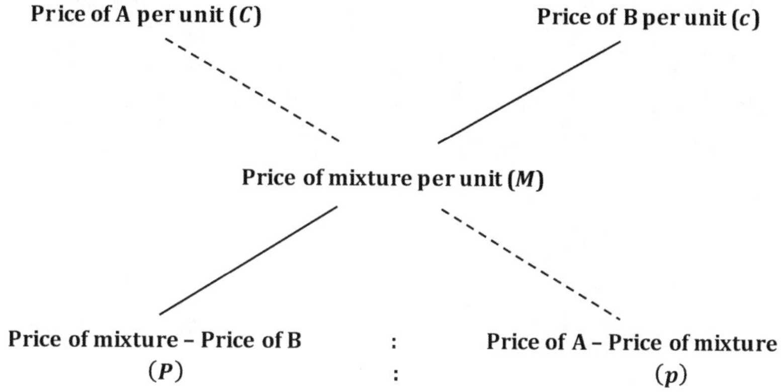

- When P quantity of a solution A having $R\%$ of constituent X and p quantity of a solution B having $r\%$ of constituent X (where $R > r$) are mixed to get an average concentration of constituent X as $M\%$:

$$\frac{\textbf{Weight of higher conc. solution}}{\textbf{Weight of lower conc. solution}} = \frac{\textbf{Conc. in mixture} - \textbf{Conc. in lower}}{\textbf{Conc. in higher} - \textbf{Conc. in mixture}}$$

The rule can also be represented in the following way:

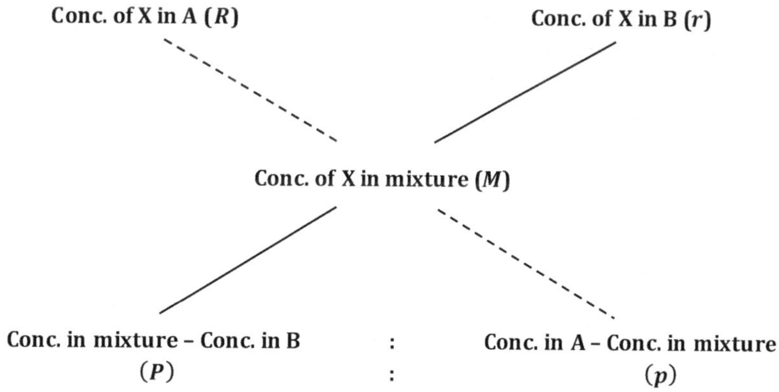

Let us take a few examples:

(1) A merchant mixes x kg of tea costing \$12 per kg and y kg of another variety costing \$20 per kg. If the price of the mixture is \$15 per kg, what is the ratio of x and y?

Explanation:

We have: x kg of tea priced at \$12 per kg and y kg of tea priced at \$20 per kg.

Thus, the average price of the mixture

$$= \$\left(\frac{12x + 20y}{x + y}\right)$$

Thus, we have

$$\frac{12x + 20y}{x + y} = 15$$

$$=> 12x + 20y = 15x + 15y$$

$$=> 3x = 5y => \frac{x}{y} = \frac{5}{3}$$

Alternate approach:

Let us solve using the method of alligation:

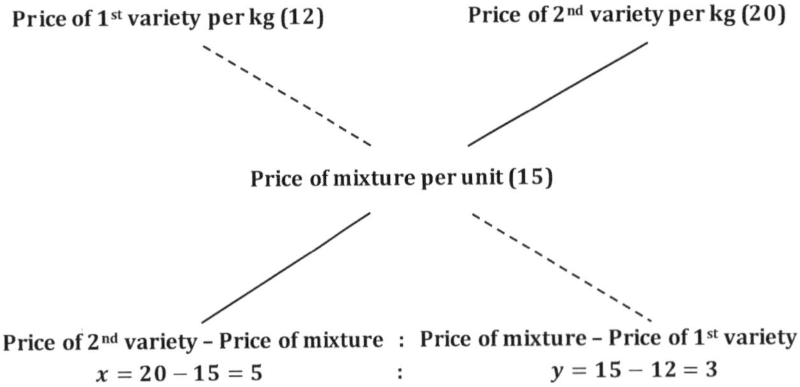

Thus, the required ratio of $x : y = 5 : 3$

(2) Solution A contains alcohol and spirit in the ratio 4 : 1 and Solution B contains alcohol and spirit in the ratio 2 : 3. In what proportion should the solutions to be taken from A and B to form a mixture having alcohol and spirit in the ratio of 1 : 1?

Explanation:

Let us consider the concentration of alcohol in the solutions.

Concentration of alcohol in solution A = $\left(\frac{4}{4+1}\right) \times 100 = 80\%$

Concentration of alcohol in solution B = $\left(\frac{2}{2+3}\right) \times 100 = 40\%$

Let the quantity taken from solution A and solution B be x and y, respectively.

Thus, average concentration of alcohol in the mixture

$$= \frac{80x + 40y}{x + y}$$

Concentration of alcohol in the final mixture = $\left(\frac{1}{1+1}\right) \times 100 = 50\%$

Thus, we have

$$\frac{80x + 40y}{x + y} = 50$$

$$=> 80x + 40y = 50x + 50y$$

$$=> 30x = 10y => \frac{x}{y} = \frac{1}{3}$$

Alternate approach:

Let us solve using the method of alligation:

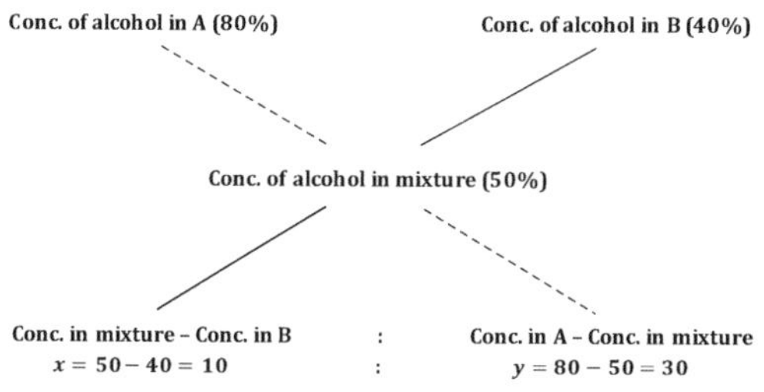

Thus, the required ratio of $x : y = 1 : 3$

Removal and Replacement

Say, a vessel contains a solution of different ingredients in a ratio. A part of the solution is removed and some quantity of one of the existing ingredients is added to the solution.

Let us take an example:

A container has 100 liters of spirit. 10 liters of the contents are taken out and replaced with petrol. This process is repeated one more time. What is ratio of spirit and petrol finally in the container?

Explanation:

The idea behind the problem is as follows:

If a vessel having capacity V has two constituents A and B in the ratio $a : b$, and a part of the contents of the vessel, say $p\%$ are removed, the remaining vessel would still have A and B in the same ratio $a : b$. Also, the contents that are removed, would also have A and B in the ratio $a : b$.

Since $p\%$ of the contents of the vessel are removed, the part remaining is $(100 - p)\%$ of the vessel. Thus, the remaining parts of the constituents are also $(100 - p)\%$ of their initial values.

Applying the above concept in the problem, we have

Initial contents of the vessel = 100 liters spirit.

Since 10 liters are removed and replaced with petrol, we have 90 liters spirit and 10 liters petrol.

After this, again 10 liters of the above solution of 100 liters is removed.

Thus, $\frac{10}{100} \times 100\% = 10\%$ of the solution is removed.

Thus, spirit remaining = $(100 - 10) = 90\%$ of $90 = \frac{90}{100} \times 90 = 81$ liters.

Similarly, petrol remaining = $(100 - 10) = 90\%$ of $10 = \frac{90}{100} \times 10 = 9$ liters.

Since 10 liters petrol is added again, the final quantity of petrol = $10 + 9 = 19$ liters.

Thus, ratio of spirit and petrol finally = 81 : 19

11.5 Computation

Computational problems are typically those which cannot be categorized into any fixed category.

Such problems are usually solved using basic arithmetic skills applied along with logical reasoning skills.

There can be various types of computational problems. A few of the typical ones are mentioned below:

Alphanumeric problems: An alphanumeric problem is an arithmetic problem involving addition, subtraction, multiplication, and in some rare cases, division, where the digits of the numbers are replaced with letters.

Careful observation of the patterns is necessary to determine the numeric value of each letter.

Let us take an example:

In the multiplication shown below, $1ABCDE$ is a six-digit number. What is the value of $(A + B + C + D + E)$?

$$\begin{array}{r} 1\ A\ B\ C\ D\ E \\ \text{x} \qquad\qquad 3 \\ \hline A\ B\ C\ D\ E\ 1 \\ \hline \end{array}$$

Explanation:

In the above multiplication, we observe:

- $E \times 3$ ends in 1 => E must be 7 => There is a carry of 2
- $D \times 3 + 2$ ends in E, i.e. 7 => D must be 5 => There is a carry of 1
- $C \times 3 + 1$ ends in D, i.e. 5 => C must be 8 => There is a carry of 2
- $B \times 3 + 2$ ends in C, i.e. 8 => B must be 2 => There is no carry
- $A \times 3$ ends in B, i.e. 2 => A must be 4 => There is a carry of 1
- $1 \times 3 + 1 = 4$ => $A = 4$, as obtained above

Thus, we have

$A + B + C + D + E = 4 + 2 + 8 + 5 + 7 = 26$

Symbol based problems: In symbol based problems, a symbol is defined, based on which, the calculations are to be done.

Let us take an example:

If $r \# s = r^s + s^r$ and $p @ 0 = p$, where @ implies one of $+$, $-$, $\times$, $\div$, what is the value of $(2 @ 3) \# 5$?

Explanation:

We have $p @ 0 = p$, where @ implies one of $+$, $-$, $\times$, $\div$

Thus, we may have

- @ implies '+': $p @ 0 = p + 0 = p$
- @ implies '−': $p @ 0 = p - 0 = p$

Working with the above scenarios:

- $(2 @ 3) \# 5 = (2 + 3) \# 5 = 5 \# 5 = 5^5 + 5^5 = 2 \times (5^5)$
- $(2 @ 3) \# 5 = (2 - 3) \# 5 = (-1) \# 5 = (-1)^5 + 5^{(-1)} = -1 + \frac{1}{5} = -\frac{4}{5}$

Thus, the possible solutions are: $2 \times (5^5)$ OR $-\frac{4}{5}$

Calculation intensive: Certain problems may be simply calculation based, where the calculation may be simplified using some basic identities or rules.

Let us take an example:

What is the approximate value of $\left\{(0.11 + 0.07)^2 - (0.11 - 0.07)^2\right\}$ % of $\left\{\frac{100}{(0.11)^2 - (0.07)^2}\right\}$?

(A) 2.4

(B) 3.6

(C) 4.3

(D) 8.1

Explanation:

We have

$\left\{(0.11 + 0.07)^2 - (0.11 - 0.07)^2\right\}$

$= \left(0.11^2 + 2 \times 0.11 \times 0.07 + 0.07^2\right) - \left(0.11^2 - 2 \times 0.11 \times 0.07 + 0.07^2\right)$

$= 4 \times 0.11 \times 0.07$

Again, we have

$$\frac{100}{(0.11)^2 - (0.07)^2}$$
$$= \frac{100}{(0.11 + 0.07)\,(0.11 - 0.07)}$$
$$= \frac{100}{0.18 \times 0.04}$$

Thus, the required value

$$= (4 \times 0.11 \times 0.07)\% \text{ of } \frac{100}{0.18 \times 0.04}$$
$$= \frac{4 \times 0.11 \times 0.07}{100} \times \frac{100}{0.18 \times 0.04}$$
$$= 4 \times \frac{0.11 \times 0.07}{0.18 \times 0.04}$$
$$= 4 \times \frac{11 \times 7}{18 \times 4}$$
$$= 4 \times \frac{77}{72}$$

Since $2 > \frac{77}{72} > 1$, the value should be slightly greater than 4 (but definitely less than 8).

The correct answer is Option C.

11.6 Averages

The average of a set of numbers is a number expressing the central or typical value of the set. It is calculated as the ratio of the sum of the set of numbers and the number of terms in the set.

Thus, for a set of n numbers $x_1, x_2, x_3, \ldots x_n$, the average (arithmetic mean) is calculated as

$$\overline{x} = \frac{\text{Sum of the } n \text{ numbers}}{n} = \frac{x_1 + x_2 + x_3 + \cdots + x_n}{n}$$

Thus, we have

Sum of the n numbers = $\overline{x} \times n$

Some important points:

- If a constant number k is added to or subtracted from each term of a set of numbers, the average of the set would also increase by k or decrease by k, respectively.

  Thus, we have

  If the average of $x_1, x_2, x_3, \ldots x_n$ is $\overline{x}$, then we have

  - Average of $x_1 + k,\ x_2 + k,\ x_3 + k,\ \ldots\ x_n + k$ is $\overline{x} + k$
  - Average of $x_1 - k,\ x_2 - k,\ x_3 - k,\ \ldots\ x_n - k$ is $\overline{x} - k$

- If each term of a set of numbers is multiplied by or divided by a constant number k, the average of the set would also be multiplied by k or divided by k, respectively.

  Thus, we have

  If the average of $x_1, x_2, x_3, \ldots x_n$ is $\overline{x}$, then we have

  - Average of $x_1 \times k,\ x_2 \times k,\ x_3 \times k,\ \ldots\ x_n \times k$ is $\overline{x} \times k$
  - Average of $\frac{x_1}{k}, \frac{x_2}{k}, \frac{x_3}{k}, \ldots \frac{x_n}{k}$ is $\frac{\overline{x}}{k}$

Weighted average:

If the average value of a set of n observations is x_n, the average value of a set of m observations is x_m, and so on, the combined average of all $(n + m + \ldots)$ observations is given as

$$\overline{x} = \frac{x_n \times n + x_m \times m + \ldots}{n + m + \ldots}$$

Let us take an example:

In a class of 5 students, in a test, 3 students scored 20 and 2 students scored 15. What is the average score of the class?

Explanation:

Simply speaking, the sum of the scores is 20 + 20 + 20 + 15 + 15 = 90

Thus, the average score = $\frac{90}{5} = 18$

Using the concept of weighted average, we have

Average score = $\frac{20 \times 3 + 15 \times 2}{3 + 2} = \frac{90}{5} = 18$

Additional rules:

- If n numbers having an average x_n are added to an existing set of m numbers having an average x_m, we have

  Final average = Initial average + $\frac{n(x_n - x_m)}{m + n}$

  Let us take an example:

  The average of four numbers is 30. What is the new average if the numbers 66 and 64 are included in the set?

  Explanation:

  Initial total = $4 \times 30 = 120$

  Final total = $120 + 66 + 64 = 250$

  Thus, final average = $\frac{250}{6} = 41\frac{2}{3}$

  Alternate approach

  The average of the two new numbers = $\frac{66 + 64}{2} = 65$

  Using the relation above, we have

  Final average = $30 + \frac{2(65 - 30)}{4 + 2} = 30 + \frac{2 \times 35}{6} = 30 + 11\frac{2}{3} = 41\frac{2}{3}$

Note: This approach offers a quick calculation for the difference between the initial and final average values.

- If n numbers having an average x_n, are removed from an existing set of m numbers having an average x_m, we have

  Final average = Initial average $- \dfrac{n\,(x_n - x_m)}{m - n}$

  Let us take an example:

  Average of six numbers is 30. What is the new average if the numbers 24 and 26 are removed from the set?

  Explanation:

  Initial total = $6 \times 30 = 180$

  Final total = $180 - 24 - 26 = 130$

  Thus, final average = $\dfrac{130}{6-2} = \dfrac{130}{4} = 32\dfrac{1}{2}$

  Alternate approach

  The average of the two numbers which are removed = $\dfrac{24 + 26}{2} = 25$

  Using the relation above, we have

  Final average = $30 - \dfrac{2\,(25 - 30)}{6 - 2} = 30 - \left(-\dfrac{10}{4}\right) = 32\dfrac{1}{2}$

- When a few incorrect numbers are taken while calculating the average, we have

  Actual average = Incorrect average + $\dfrac{\text{Sum of correct terms} - \text{Sum of incorrect terms}}{\text{Total number of terms}}$

Let us take an example:

The average of 10 numbers is 48. It was later found out that two numbers had been taken wrongly: 36 was taken as 54 and 46 was taken as 32. What is the actual average?

Explanation:

$$\text{Actual average} = 48 + \frac{(36 + 46) - (54 + 32)}{10} = 48 + \left(-\frac{4}{10}\right) = 47\frac{3}{5}$$

- When the average of a group changes as a result of one term being replaced by another, we have

New term = Replaced term + (Increase in average × Number of terms)

Let us take an example:

The average of a group of six increases by 3 when 72 is replaced by another number. What is the value of the new number included in place of 72?

Explanation:

Let the new number be x and the initial average be y.

Thus, initial total = $6y$ => Final total = $6y - 72 + x$

Thus, final average = $\frac{6y - 72 + x}{6}$

Thus, we have $\frac{6y - 72 + x}{6} = 3 + y \Rightarrow 6y - 72 + x = 18 + 6y$

$\Rightarrow x = 72 + 18 = 90$

Alternate approach

Using the above relation:

Value of the new number = $72 + 3 \times 6 = 90$

11.7 Speed/Time/Distance

Speed refers to the rate of travel, or, in other words, the distance covered in unit time.

Thus, we have the basic rule:

$$\text{Speed } (S) = \frac{\text{Distance traveled } (D)}{\text{Time taken } (T)} \Rightarrow D = S \times T$$

A few important relations:

- If the ratio of speeds of two bodies A and B is $a : b$, the ratio of the times taken by A and B to cover the same distance d is $\frac{d}{a} : \frac{d}{b} = \frac{1}{a} : \frac{1}{b} = b : a$.
- If the ratio of speeds of A and B is $a : b$, the ratio of the distances traveled by them in the same time t is $t \times a : t \times b = a : b$.
- If the ratio of time of travel of A and B is $a : b$, the ratio of the distances traveled by them with the same speed s is $s \times a : s \times b = a : b$.

Let us take an example:

Walking at $\left(\frac{3}{4}\right)^{\text{th}}$ of his usual speed, a person reaches his destination 10 minutes late. What is the usual time it would take him to travel at his usual speed to cover the same distance?

Explanation:

Let the usual time be x minutes.

In both situations, the distance covered is the same.

Thus, the ratio of time taken is the reciprocal of the ratio of distance.

Thus, assuming his usual speed to be s, we have

$$\frac{\frac{3}{4}s}{s} = \frac{x}{x + 10}$$

$$\Rightarrow \frac{3}{4} = \frac{x}{x + 10}$$

$$\Rightarrow 3x + 30 = 4x \Rightarrow x = 30$$

Average Speed: If a man covers different distances at different speeds, the average speed is given by

$$\text{Average Speed} = \frac{\text{Total Distance traveled}}{\text{Total Time taken}}$$

We have the possible scenarios:

- If $d_1,\ d_2,\ \ldots\ d_n$ are distances traveled, taking time $t_1,\ t_2,\ \ldots\ t_n$, respectively, we have

  $$\text{Average speed} = \frac{d_1 + d_2 + \cdots + d_n}{t_1 + t_2 + \cdots + t_n}$$

- If a body travels at speeds $s_1,\ s_2,\ \ldots\ s_n$, for time $t_1,\ t_2,\ \ldots\ t_n$, respectively, we have

  $$\text{Average speed} = \frac{s_1t_1 + s_2t_2 + \cdots + s_nt_n}{t_1 + t_2 + \cdots + t_n}$$

  - If the time of travel is t in each case, we have

  $$\text{Average speed} = \frac{s_1t + s_2t + \cdots + s_nt}{t + t + \cdots + t} = \frac{s_1 + s_2 + \cdots + s_n}{n}$$

  If $n = 2$, we have

  $$\text{Average speed} = \frac{s_1 + s_2}{2}$$

- If $d_1,\ d_2,\ \ldots\ d_n$ are distances traveled at speeds $s_1,\ s_2,\ \ldots\ s_n$ respectively, we have

  $$\text{Average speed} = \frac{d_1 + d_2 + \cdots + d_n}{\frac{d_1}{s_1} + \frac{d_2}{s_2} + \cdots + \frac{d_n}{s_n}}$$

  - If the same distance d is traveled at the above speeds, we have

  $$\text{Average speed} = \frac{d + d + \cdots + d}{\frac{d}{s_1} + \frac{d}{s_2} + \cdots + \frac{d}{s_n}} = \frac{n}{\frac{1}{s_1} + \frac{1}{s_2} + \cdots + \frac{1}{s_n}}$$

  If $n = 2$, we have

  $$\text{Average speed} = \frac{2}{\frac{1}{s_1} + \frac{1}{s_2}} = \frac{2s_1s_2}{s_1 + s_2}$$

Let us take an example:

A man travels from City A to City B. If he covers the first half of the distance at 60 miles per hour and the remaining at 40 miles per hour, what is the average speed of the man?

How does the answer change if the man had covered the first half of the total time taken at 60 miles per hour and the remaining at 40 miles per hour?

Explanation:

Let the total distance be $2d$ miles.

Thus, he traveled d miles at 60 miles per hour and the remaining d miles at 40 miles per hour.

Thus, total time he took for the journey $= \frac{d}{60} + \frac{d}{40} = \frac{5d}{120} = \frac{d}{24}$ hours.

Thus, average speed $= \frac{2d}{\frac{d}{24}} = 48$ miles per hour.

Alternate approach 1

We can assume any suitable value of the distance such that it is divisible by 60 and 40.

Let the distance for each case be 120 miles.

Thus, the man traveled 120 miles at 60 miles per hour and the remaining 120 miles at 40 miles per hour.

Thus, total time he took for the journey $= \frac{120}{60} + \frac{120}{40} = 5$ hours

Total distance = 120 + 120 = 240 miles.

Thus, average speed $= \frac{240}{5} = 48$ miles per hour.

Alternate approach 2

Using the relation given above, we have

Average speed $= \frac{2s_1s_2}{s_1 + s_2} = \frac{2 \times 60 \times 40}{60 + 40} = \frac{4,800}{100} = 48$ miles per hour.

Looking at the second part of the problem:

Let the time he traveled at 60 miles per hour and 40 miles per hour be t hours each.

Thus, total distance $= 60t + 40t = 100t$ miles.

Total time $= t + t = 2t$ hours.

Thus, average speed $= \frac{100t}{2t} = 50$ miles per hour.

Alternate approach

Using the relation given above, we have

Average speed $= \frac{s_1 + s_2}{2} = \frac{60 + 40}{2} = 50$ miles per hour.

Stoppage Time: Let the distance covered by a body be d, the average speed of the body excluding stoppage time be s_e, and the average speed including stoppage time be s_i. We need to determine the average stoppage time per hour.

Time taken to cover the distance d had there been no stoppages $= \frac{d}{s_e} = t_e$

Actual time taken to cover the distance d since there were stoppages $= \frac{d}{s_i} = t_i$

Thus, total stoppage time during the journey $= t_i - t_e$

$$= \frac{d}{s_i} - \frac{d}{s_e} = l_s$$

Thus, average stoppage time per hour

$$= \frac{t_s}{t_i} \times 60 = \left(\frac{\frac{d}{s_i} - \frac{d}{s_e}}{\frac{d}{s_i}} \times 60 \right) \text{ minutes}$$

$$= \left(\frac{s_e - s_i}{s_e} \times 60 \right) \text{ minutes}$$

Let us take an example:

A train maintains an average speed of 60 miles per hour while traveling between two cities, A and B. However, since the train stops at multiple stations between the two cities, its average speed comes down to 40 miles per hour. What is the average stoppage time per hour?

Explanation:

We can assume a suitable distance such that it is divisible by 60 and 40.

Let the distance be 120 miles.

Time taken, assuming the train does not stop between the cities $= \frac{120}{60} = 2$ hours.

Actual time taken by the train (including stoppages) $= \frac{120}{40} = 3$ hours.

Thus, total stoppage time $= 3 - 2 = 1$ hour.

Thus, the train had a stoppage time of 1 hour in a total of 3 hours.

Thus, stoppage time per hour = $\frac{1}{3}$ hours = $\frac{1}{3} \times 60 = 20$ minutes

Relative Speed: Relative speed refers to the speed of a moving body with respect to another moving body.

- If two bodies A and B move in the same direction with speeds S_A and S_B $(S_A > S_B)$, their relative speed is given by $(S_A - S_B)$

  If B is d distance ahead of A, the time taken by A to catch up with B = $\frac{d}{S_A - S_B}$

- If two bodies A and B are moving in opposite directions with speeds S_A and S_B, their relative speed is given by $(S_A + S_B)$.

  If A and B are separated by distance d, the time taken by them to meet each other $= \frac{d}{S_A + S_B}$

A few important points:

- Time taken by a body of length x to pass a stationary object of negligible length, such as a signal post or a pole or a stationary man equals the time taken by the body to cover a distance equaling its own length of x.
- Time taken by a body moving at speed S of length x to pass a moving object of negligible length, such as the object is running at speed s equals $\left(\frac{x}{S - s}\right)$.
- Time taken by a body of length x to pass a stationary object of length y equals the time taken by the body to cover a distance of $(x + y)$.

Effective or Resultant Speed: Resultant speed refers to the speed at which a body moves in a moving medium.

Let the speed of a body without the effect of any medium be S, and the speed of the medium be M.

Thus, we have

- If the body travels against the flow of the medium (also known as 'Upstream motion' in the context of a man swimming/rowing against a flowing river):

  Resultant speed (Upstream) = $S - M = U$

- If the body travels with the flow of the medium (also known as 'Downstream motion' in the context of a man swimming/rowing along a flowing river):

Resultant speed (Downstream) = $S + M = D$

- Speed of the body without the effect of medium (also known as speed in 'Still water' in the context of a man swimming/rowing along a flowing river):

$$S = \frac{U + D}{2}$$

- Speed of the medium (or 'River' in the context of a man swimming/rowing along a flowing river):

$$M = \frac{D - U}{2}$$

The diagram depicting the situation of a man rowing in a river is shown below:

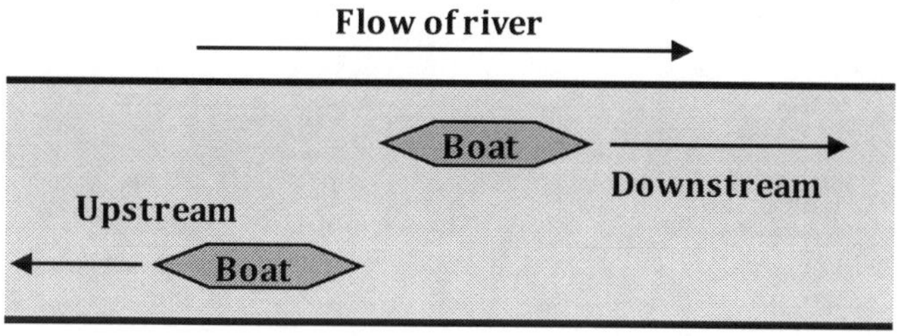

Circular track movement: If two objects P and Q move around a circular track of length L with speeds S_P and S_Q, we have the following relations:

- Time taken to meet anywhere on the track for the first time $(t) = \dfrac{L}{\text{Relative speed of P and Q}}$
- Time taken to meet at the starting point for the first time $(T) = \text{LCM}\left(\dfrac{L}{S_P}, \dfrac{L}{S_Q}\right)$

Let us take an example:

If P and Q move along a circular track of length 1,200 meters with speeds of 4 meters per second and 6 meters per second, determine

- Time taken by them to meet on the track for the first time
- Time taken by them to meet at the starting point for the first time

Explanation:

To determine the time taken for P and Q to meet on the track, we need to determine their relative speed, for which we need to know their directions of motion.

Since their directions are not known, let us assume:

- **Same direction**: Relative speed = 6 – 4 = 2 meters per second.

  For them to meet again, Q has to cover one lap, i.e. 1,200 meters more than what P covers in the same time.

  Thus, distance = 1,200 meters

  => Time taken to meet = $\frac{\text{Distance}}{\text{Relative speed}} = \frac{1,200}{2}$ = 600 seconds.

- **Opposite direction**: Relative speed = 6 + 4 = 10 meters per second

  For them to meet again, P and Q together have to cover one lap, i.e. 1,200 meters.

  Thus, distance = 1,200 meters => Time taken to meet = $\frac{1,200}{10}$ = 120 seconds.

To determine the time taken by P and Q to meet at the starting point for the first time, the directions of motion are not required.

Time taken by P to reach the starting point = $\frac{1,200}{4}$ = 300 seconds

Time taken by Q to reach the starting point = $\frac{1,200}{6}$ = 200 seconds

Thus, time taken by both to reach the starting point simultaneously

= LCM of 300 and 200 = 600 seconds.

Let us take a few additional examples:

(1) A man travels for 6 hours. He covers the first half of the distance at 20 miles per hour and the rest at 30 miles per hour. What is the total distance?

Explanation:

Let the distance be $2d$ miles.

Thus, total time taken $= \frac{d}{20} + \frac{d}{30} = \frac{d}{12}$ hours.

Thus, we have

$\frac{d}{12} = 6 \Rightarrow 2d = 144$

Alternate approach

Since the man covers equal distances at speeds of 20 miles per hour and 30 miles per hour, we have

Average speed $= \frac{2 \times 20 \times 30}{20 + 30} = 24$ miles per hour.

Total time of travel = 6 hours.

Thus, total distance covered = Average speed $\times$ Total distance = $24 \times 6 = 144$ miles.

(2) A person goes around the three sides of an equilateral triangle shaped field at speeds of 10, 12, and 20 miles per hour and reaches back to his starting point. What is his average speed during the journey?

Explanation:

Let us assume a suitable length for each side of the triangle so that it is divisible by 10, 12, and 20.

Let the distance be 60 miles.

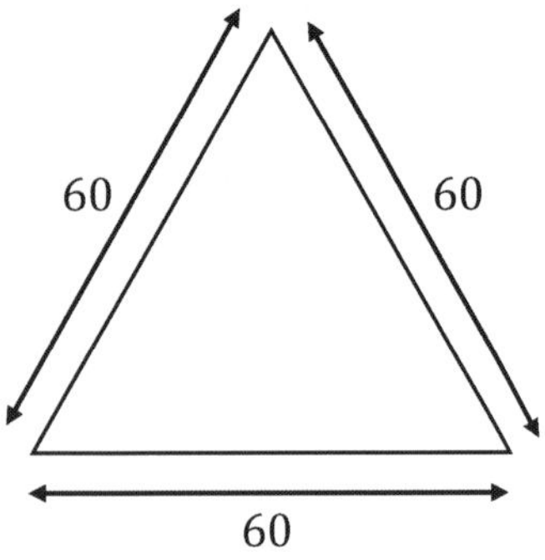

Thus, time taken to cover the perimeter of the triangle

$$= \frac{60}{10} + \frac{60}{12} + \frac{60}{20} = 14 \text{ hours.}$$

Total distance = $3 \times 60 = 180$ miles.

Thus, average speed = $\frac{180}{14} = 12\frac{6}{7}$ miles per hour.

(3) A is twice as fast as B, and B is thrice as fast as C. If C covers a distance in 54 minutes, how long will A take to cover the same distance?

Explanation:

We know that B is thrice as fast as C.

Since C takes 54 minutes to cover a particular distance, time taken by B to cover the same distance

$$= \frac{54}{3} = 18 \text{ minutes}$$

We also know that A is twice as fast as B.

Since B takes 18 minutes to cover a particular distance, time taken by A to cover the same distance

$$= \frac{18}{2} = 9 \text{ minutes}$$

(4) A man covers a certain distance at his normal speed. Had he moved 3 miles per hour faster, it would have taken him 40 minutes less to reach the destination. However, had he moved 2 miles per hour slower, it would have taken him 40 more minutes to reach the destination. What is the normal speed of a man?

Explanation:

Let the distance be x miles.

Time taken to cover the distance at 30 miles per hour = $\frac{x}{30}$ hours.

Time taken to cover the distance at 40 miles per hour = $\frac{x}{40}$ hours.

Thus, time difference = $\frac{x}{30} - \frac{x}{40} = \frac{x}{120}$ hours.

Since the man was 10 minutes late in the first case and was 5 minutes early in the second, the time difference = $5 - (-10) = 15$ minutes = $\frac{1}{4}$ hour.

Thus, we have

$$\frac{x}{120} = \frac{1}{4} => x = 30$$

For the above type of problems, we may remember a simple rule:

$$\text{Distance} = \frac{\text{Product of the speeds}}{\text{Difference of the speeds}} \times \text{Difference in time}$$

Thus, applying the above rule, we have

$$\text{Distance} = \frac{30 \times 40}{40 - 30} \times \left(\frac{5}{60} - \left(\frac{10}{60}\right)\right) = \frac{1,200}{10} \times \frac{15}{60} = 30$$

(5) A man covers a certain distance at his normal speed. Had he moved 3 miles per hour faster, he would have taken 40 minutes less to reach the destination. However, had he moved 2 miles per hour slower, he would have taken 40 more minutes to reach the destination. What is the normal speed of the man?

Explanation:

$$\text{Distance} = \frac{\text{Product of the speeds}}{\text{Difference of the speeds}} \times \text{Difference in time}$$

Let the normal speed be S.

Case 1: New speed = $(S + 3)$

$$\text{Distance} = \frac{S(S+3)}{(S+3) - S} \times \frac{40}{60} = \frac{2S\,(S+3)}{9} \ldots \text{(i)}$$

Case 2: New speed = $(S - 2)$

Distance = $\frac{S(S-2)}{S-(S-2)} \times \frac{40}{60} = \frac{S(S-2)}{3} \ldots \text{(ii)}$

Thus, from (i) and (ii), we have

$\frac{2S(S+3)}{9} = \frac{S(S-2)}{3} \Rightarrow 2S + 6 = 3S - 6$

$\Rightarrow S = 12$

Alternate approach

Let the distance be d miles and the normal speed be s miles per hour.

Thus, we have

Case 1:

$\frac{d}{s} - \frac{d}{s+3} = \frac{40}{60} \Rightarrow \frac{3d}{s(s+3)} = \frac{2}{3} \ldots \text{(i)}$

Case 2:

$\frac{d}{s-2} - \frac{d}{s} = \frac{40}{60} \Rightarrow \frac{2d}{s(s-2)} = \frac{2}{3} \ldots \text{(ii)}$

From (i) and (ii), we have

$\frac{3d}{s(s+3)} = \frac{2d}{s(s-2)} \Rightarrow s = 12$

(6) A man takes 8 hours to walk to a certain place and ride back. However, he could have taken 4 hours less if he had covered both ways by riding. How long would it have taken him to walk both ways?

Explanation:

Walking time + Riding time = 8 hours

Riding time (both ways) = 8 − 4 = 4 hours.

$\Rightarrow$ Riding time (one way) = $\frac{4}{2}$ = 2 hours.

=> Walking time (one way) = 8 – 2 = 6 hours

Hence, walking time (both ways) = 6 + 6 = 12 hrs.

Alternate approach

Two-way riding saves 4 hours. It implies that one-way riding takes 4 hours less than one-way walking. Thus, one-way walking takes 4 hours more than one-way riding.

Thus, walking both ways will take 8 + 4 = 12 hours.

11.8 Time and Work

Let us look at the basic rules:

- If M_1 men do a work W_1 in D_1 days and M_2 equally efficient men do a work W_2 in D_2 days, we have

  $$\frac{M_1 D_1}{W_1} = \frac{M_2 D_2}{W_2}$$

  Let us take an example:

  If 20 men can complete a work in 10 days, how long would 40 men take to complete a work twice as large?

  Explanation:

  1 man would be able to complete the work in $20 \times 10 = 200$ days.

  Hence, 40 men would complete the work in $\frac{200}{40} = 5$ days.

  Thus, 40 men would complete double the work in $5 \times 2 = 10$ days.

  Alternately, substituting the values in the above relation:

  $$\frac{20 \times 10}{W} = \frac{40 \times D_2}{2W} \Rightarrow D_2 = 10$$

- If the number of hours (or days) required to complete a work is t, the fraction of the work completed in one hour (or one day) $= \frac{1}{t}$

  For example: If A can do a work in 20 days, part of the work done in 1 day $= \frac{1}{20}$

- If the fraction of work completed in one hour (or one day) is f, the time taken to complete the entire work $= \frac{1}{f}$

  For example: If A can complete $\frac{1}{4}$ of a work in 1 day, time taken by A to complete the entire work $= \frac{1}{\frac{1}{4}} = 4$ days

- If two men take x and y days to do the same work, the ratio of their efficiencies

  $$= \frac{1}{x} : \frac{1}{y} = y : x$$

Let us take a few examples:

(1) A and B can do a piece of work in 15 days and 20 days, respectively. In how many days will they finish the work if both of them work together?

Explanation:

Fraction of work completed by A in 1 day = $\frac{1}{15}$

Fraction of work completed by B in 1 day = $\frac{1}{20}$

Thus, the fraction of work completed by A and B together in 1 day = $\frac{1}{15} + \frac{1}{20} = \frac{7}{60}$

Thus, the time taken by them to complete the entire work = $\frac{1}{\frac{7}{60}} = \frac{60}{7} = 8\frac{4}{7}$ days.

Alternate approach

In such questions, it is best to assume the total work to be the LCM of the number of days each person takes to complete the work (rather than working with fractions).

Let the work = LCM of 15 & 20 = 60 units

Work completed by A in 1 day = $\frac{60}{15}$ = 4 units

Work completed by B in 1 day = $\frac{60}{20}$ = 3 units

Thus, the work completed by A and B together in 1 day = 4 + 3 = 7 units

Hence, the number of days that both of them working together will take = $\frac{60}{7} = 8\frac{4}{7}$ days

(2) A can complete a work in 24 days. B is 60% more efficient than A. What is the number of days it would take A and B to complete the same piece of work working together?

Explanation:

Let B take x days to complete the work.

Ratio of efficiency of A and B = 100 : 160

Thus, ratio of time taken by A and B = $\frac{1}{100} : \frac{1}{160}$ = 160 : 100 = 8 : 5

Thus, we have

$\frac{8}{5} = \frac{24}{x}$ => $x = 15$

Let the total work = LCM of 15 and 24 = 120 units

Thus, the work done by A per day = $\frac{120}{24}$ = 5 units and

work done by B per day = $\frac{120}{15}$ = 8 units

Thus, the total work done by A and B per day = 5 + 8 = 13 units

Thus, the time taken = $\frac{120}{13} = 9\frac{3}{13}$ days

Alternate approach

Let the required time be t days.

Ratio of efficiency of A and B = 100 : 160

=> Ratio of efficiency of A and A-B together = 100 : (100 + 160) = 100 : 260

Thus, ratio of time taken by A and A-B together = $\frac{1}{100} : \frac{1}{260}$ = 260 : 100

=> $\frac{260}{100} = \frac{24}{t}$

=> $t = 24 \times \frac{100}{260} = \frac{120}{13} = 9\frac{3}{13}$

(3) 2 men and 3 boys can do a piece of work in 10 days, while 3 men and 2 boys can do the same work in 8 days. What is the ratio of efficiency between a man and a boy?

Explanation:

We know that:

$\frac{M_1 D_1}{W_1} = \frac{M_2 D_2}{W_2}$

The work in both cases is the same.

Also, we have

M_1 = (2 men + 3 boys), M_2 = (3 men + 2 boys), D_1 = 10 days, D_2 = 8 days

We know that $M_1 \times D_1 = M_2 \times D_2$; note that $W_1 = W_2$

=> (2 men + 3 boys) × 10 = (3 men + 2 boys) × 8

=> 20 men + 30 boys = 24 men + 16 boys

=> 4 men = 14 boys

=> Ratio of efficiencies of a man and a boy = $\frac{14}{4} = \frac{7}{2}$

(4) A and B finish a work in t days. A alone takes $(t + 12)$ days and B alone takes $(t + 3)$ days. What is the value of t?

Explanation:

Part of work done by A in 1 day = $\left(\frac{1}{t + 12}\right)$

Part of work done by B in 1 day = $\left(\frac{1}{t + 3}\right)$

Part of work done by A and B together in 1 day = $\left(\frac{1}{t}\right)$

Thus, we have

$$\frac{1}{t} = \frac{1}{t + 12} + \frac{1}{t + 3}$$

$$=> \frac{1}{t} = \frac{2t + 15}{t^2 + 15t + 36}$$

$$=> t^2 + 15t + 36 = 2t^2 + 15t$$

$$=> t^2 = 36$$

$$=> t = 6$$

Pipes and cisterns: A pipe connected to a tank, a cistern, or a reservoir that fills it is known as an inlet. Similarly, a pipe connected to a tank, a cistern, or a reservoir and emptying it is known as an outlet.

Let us take a few examples:

(1) A pipe can fill a tank in 3 hours. Because of a leak in the tank, it took 4 hours to fill. How long will it take for the leak to drain out the entire filled-up tank?

Explanation:

Let the capacity of the tank = LCM of 3 & 4 = 12 liters

Quantity filled by the filling pipe per hour = $\frac{12}{3}$ = 4 liters

Quantity filled by both in 1 hour = $\frac{12}{4}$ = 3 liters

Hence, the leak empties 4 – 3 = 1 liter per hour.

Thus, the time taken by the leak to empty the tank = $\frac{12}{1}$ = 12 hours.

(2) A leak can empty a completely filled cistern in 4 hours. If a pipe that admits 12 liters of water per hour is opened, the cistern is emptied in 6 hours. What is the capacity of the cistern?

Explanation:

Let the capacity of the cistern be x liters.

Since the leak can empty the cistern in 4 hours, the rate of emptying = $\frac{x}{4}$ liters per hour.

After the filling pipe, which fills at the rate of 12 liters per hour, is opened, the effective rate of emptying = $\left(\frac{x}{4} - 12\right)$ liters per hour.

Thus, time, in hours, taken to empty the cistern

$$= \frac{x}{\frac{x}{4} - 12} = \frac{4x}{x - 48}$$

Thus, we have

$$\frac{4x}{x - 48} = 6$$

$=> 4x = 6x - 288$

$=> x = 144$ liters

(3) Pipes P and Q are inlet pipes that can fill a tank in 3 hours and 5 hours, respectively. Pipe R is an outlet pipe. Starting with an empty cistern, if pipes P and Q are opened for 2 hours and 3 hours, respectively, and pipe R is opened for 4 hours, the entire cistern gets filled. How long does pipe R take to empty a cistern that is completely filled?

Explanation:

Let the capacity of the tank be the LCM of the time taken by pipes P and Q

= LCM of 3 and 5 = 15 liters.

Thus, filling rate of pipe P per hour = $\frac{15}{3}$ = 5 liters per hour

Filling rate of pipe Q per hour = $\frac{15}{5}$ = 3 liters per hour

Let the emptying rate of pipe R = x liters per hour

We know that to fill the cistern, the pipes P, Q and R are opened for 2 hours, 3 hours and 4 hours, respectively.

Thus, we have

(Amount of water filled by P) + (Amount of water filled by Q) – (Amount of water drained by R) = Capacity of the tank

$(5 \times 2) + (3 \times 3) - (x \times 4) = 15$

$=> x = 1$

Thus, the pipe R empties at the rate of 1 liter per hour.

Thus, time taken by pipe R to empty the cistern = $\frac{15}{1}$ = 15 hours.

11.9 Simple & Compound Interest

If a person borrows some amount of money for a certain time, then

- Principal (P) is the amount of money borrowed
- Interest (I) is the additional money to be paid
- Amount (A) is the sum of principal and interest
- Time (T) is period for which the sum is borrowed
- Rate $(R\%)$ is the interest that the borrower has to pay for every \$100 per year

Simple Interest: If the interest on a certain period is calculated only on the principal borrowed, it is simple interest.

Compound Interest: If the interest is calculated on the amount after each year (period) instead of only on the principal, it is compound interest.

The table below gives an example of how Simple Interest and Compound Interest operate. The principal at the beginning of 1st year is \$1,000, and the rate of interest is 10% per annum. The details for 3 years are shown:

	Simple Interest				**Compound Interest**			
Year	Principal	Interest for the year	Cumulative interest	Amount	Principal	Interest for the year	Cumulative interest	Amount
1	1,000	100	100	1,100	1,000	100	100	1,100
2	1,000	100	200	1,200	1,000	110	210	1,210
3	1,000	100	300	1,300	1,000	121	331	1,331
	A few important points							
	The interest for every year (not cumulative interest) is a constant				The interest for every year increases by *R*% of the interest in the previous year			
	The amount every year increases by a constant value				The amount every year increases by *R*% of the interest in the previous year			
	$I = \frac{P \times R \times T}{100}$ $A = P + I = P\left(1 + \frac{R \times T}{100}\right)$				$A = P\left(1 + \frac{R}{100}\right)^T$ $I = A - P = P\left\{\left(1+\frac{R}{100}\right)^T - 1\right\}$			

Concept of compounding: In compound interest, the interest does not always need to be collected annually. Interest may be collected semi-annually, quarterly, etc. In such cases, the formula for the amount under compound interest is modified to:

$$A = P\left(1 + \frac{\frac{R}{n}}{100}\right)^{(T \times n)}$$

Here, n represents the number of compounding, for example: $n = 2$ implies compounding is done semi-annually, while $n = 4$ implies compounding is done quarterly.

Difference between Compound Interest and Simple Interest after 2 years:

Let us consider Principal P and Rate of interest R%. The difference can be easily observed from the table below:

Year	Simple Interest for the year	Compound Interest for the year
1	$R\%$ of $P = x$	$R\%$ of $P = x$
2	$R\%$ of $P = x$	$(R\%$ of $P) + (R\%$ of $x) = x + R\%$ of x
Total	$\mathbf{2x}$	$\mathbf{2x + R\%}$ **of** $\mathbf{x}$

Compound Interest and Simple Interest is the same for a certain sum of money at the same rate for the first year.

Difference after 2 years with interest calculated yearly:

Compound Interest – Simple Interest

$= R\%$ of $x = R\%$ of $(R\%$ of $P)$

$$=P\left(\frac{R}{100}\right)^2$$

Let us take a few examples:

(1) In how many years will a principal of \$400 yield an interest of \$100, given that the rate of interest and the number of years are numerically equal?

Explanation:

Let Rate of interest (%) = Time = k.

Since the principal (P) is \$400, we have

$$\text{Simple Interest} = \frac{P \times R \times T}{100}$$

$$=> 100 = \frac{400 \times k \times k}{100}$$

$=> k = 5$

Thus, the required time is 5 years.

(2) In how many years, at 10% simple interest per year, will the amount become 5 times the principal?

Explanation:

Since the amount becomes 5 times the principal, the interest accumulated must be 4 times the principal.

Let the principal be P and the time taken be T.

Thus, we have

$$\frac{P \times 10 \times T}{100} = 4P$$

$=> T = 40$

(3) A sum of money invested under simple interest triples in 6 years. In how many years would the sum of money become 27 times its value?

What would be the answer if the sum of money was invested at compound interest?

Explanation:

Let the sum of money (Principal) be P.

Thus, the amount after 6 years = $3P$.

Thus, interest accumulated in 6 years = $3P - P = 2P$

Thus, for the interest to be $2P$, time taken is 6 years.

If the sum of money has to be 27 times, i.e. the amount is $27P$, interest accumulated $= 27P - P = 26P$

Thus, for the interest to be $26P$, time taken

$$= \frac{6}{2} \times 26 = 78 \text{ years.}$$

Thus, time taken is 78 years.

Let us now see the scenario under compound interest:

We know that the value of the sum becomes 3 times in 6 years.

Thus, we have

P becomes $3P$ in 6 years

=> $3P$ becomes $3 \times 3P = 9P$ in another 6 years, i.e. $6 + 6 = 12$ years

=> $9P$ becomes $3 \times 9P = 27P$ in another 6 years, i.e. $12 + 6 = 18$ years.

Thus, the time taken is 18 years.

(4) A sum of money invested under simple interest amounts to \$1,500 after 5 years and \$1,600 after 7 years. What is the rate of interest?

Explanation:

Let the sum of money invested (Principal) be \$$P$.

Let the rate of interest be R%.

Thus, amount after 5 years:

$$P + \frac{5PR}{100} = 1,500 \ldots \text{(i)}$$

Amount after 7 years:

$$P + \frac{7PR}{100} = 1,600 \ldots \text{(ii)}$$

From (ii) – (i):

$$\frac{2PR}{100} = 100 \Rightarrow \frac{PR}{100} = 50 \ldots \text{(iii)}$$

Substituting the above in (i):

$$P + 5 \times 50 = 1,500 \Rightarrow P = 1,250$$

Thus, from (iii):

$$\frac{1,250 \times R}{100} = 50 => R = 4\%$$

Alternate approach

The interest accumulated in $7 - 5 = 2$ years = $\$1,600 - \$1,500 = \$100$

Thus, interest accumulated each year = $\$\left(\frac{100}{2}\right) = \$50$

Thus, interest accumulated in 5 years = $\$\,(5 \times 50) = \$250$

Thus, Principal = $\$1,500 - \$250 = \$1,250$

Thus, interest per year on $1,250 is $50.

$$=> \text{Rate of interest} = \frac{50}{1,250} \times 100 = 4\%$$

(5) A sum of money invested under compound interest gives $240 in interest in the second year and $288 in interest in the third year. What is the value of the sum invested?

Explanation:

We know that in compound interest, the interest for every year increases by $R\%$ of the interest in the previous year (where $R\%$ is the rate of interest).

Thus, percent increase in the interest from the second year to the third year

$$= \frac{288 - 240}{240} \times 100 = \frac{48}{240} \times 100 = 20\%$$

Thus, the percent increase in the interest from the first year to the second year is also 20%.

Thus, we have

$(100 + 20)\,\%$ = 120% of the Interest in the first year = Interest in the second year

=> 120% of the Interest in the first year = $240

=> Interest in the first year = $\$\left(\frac{240}{120} \times 100\right) = \$200$

Since the first year interest is simply *R*% of the invested sum, we have

20% of Invested sum = $200

=> Invested sum of money = $\$\left(\frac{200}{20} \times 100\right) = \$1,000$

11.10 Combinatorics

Combinatorics can be called the **science of counting**. This is the branch of mathematics in which we study the selection and/or the arrangement of the elements of finite sets, having certain characteristics.

We seek to find out the answers to the questions like:

"In how many ways can a team of 4 members be selected from 10 people?"

or

"In how many possible ways can different photographs be taken of a group of people standing in a line, considering that a different order results in a new photograph?"

or

"How many ten-digit phone numbers starting with "98" are possible?"

There are two parts to Combinatorics, namely, **Permutation**, and **Combination**. These can also be called **Counting Methods**.

Combinatorics as a topic has attracted a fair amount of attention to the GMAT recently. It is also a vital topic from the perspective of probability, which is frequently tested on the GMAT.

11.11 Permutation

The notion of permutation implies the act of permuting or arranging of a few objects. Permutation of a set of objects is an act of arranging them in a particular order. For example, there are six possible permutations (arrangements) of the set (A, B, C), namely (A, B, C), (A, C, B), (B, A, C), (B, C, A), (C, A, B), and (C, B, A).

11.12 Combination

The notion of combination is a way of selecting one or several objects out of a set, where the order or the arrangement of objects does not matter. In the above example, the selection of all the three objects from the set (A, B, C), namely (A, B, C) is possible only in one way. Arranging the elements as (A, C, B), (B, A, C), (B, C, A), (C, A, B), and (C, B, A) does not make any difference as the same objects are there in each set.

However, if we select only two objects from the set (A, B, C), then there are three possible selections, namely (A, B), (A, C), and (B, C). Sets (B, A), (C, A), and (C, B) do not make a different selection as the order is not important.

Similarly, if we select only one object from the set, then there are three possible selections, namely (A), (B), and (C).

11.13 What's the difference?

We use the word "combination" casually in daily lives, without bothering whether the order of objects is important. Let's see examples.

If I say that the vehicle registration number of my car is formed out of four digits: 0, 2, 5, and 6, can you guess what could be the number? Well, it's not possible to guess correctly as there are many possible numbers out of 0, 2, 5, and 6. If we arrange 0, 2, 5, and 6, there are, for example, 2056, 5602, 6205, and many more numbers. Just for your curiosity, there are as many as 24 possible numbers out of these 4 digits. This calls for **permutation.**

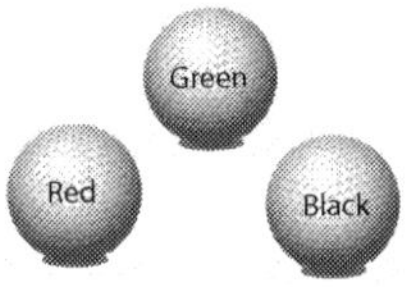

If I say to you that I have three colored balls—red, green, and black. Had it made any difference if I had rather said that I have three colored balls—green, red, and black?—No; as the order in not important. This calls for **combination.**

In other words—***A Permutation is an ordered Combination.***

Combinatorics	**Selection is important**	**Order/Arrangement is important**
• Permutation	Yes	Yes
• Combination	Yes	No

11.14 Permutation

Let us develop a formulaic approach to solve lengthy questions.

Question: In how many ways can Alex, Betty, and Chris be photographed?

The foremost decision in the Permutation & Combination based question is to ask which of the two concepts needs to be applied in the question? The answer is: ask yourself whether the arrangement of the objects will render a unique way or a presentation. If the answer is yes, apply permutation, else combination.

In the above question, the arrangement is important, as the photograph of Betty, Alex, and Chris (standing in that order) will be different from the photograph of Alex, Betty, and Chris (standing in that order), so we will apply permutation.

Develop a formula

There are a total of three places to be occupied by Alex, Betty, and Chris. They can stand in any order to get photographed. The first place can be occupied in three ways: any one of Alex, Betty, and Chris will stand on the first place; similarly, the second place can be occupied in only two ways as one of Alex, Betty, and Chris has already occupied the first place; and the last place can be occupied in only one way as two of Alex, Betty, and Chris have already occupied the first, and the second place.

So, the **total number of ways = 3 × 2 × 1 = 6 ways.**

What if there are n number of friends who want to get photographed?

In a similar way, the first place can be occupied in n number of ways: any one of n friends will stand here; correspondingly, the second place can be occupied in $(n - 1)$ ways as one of n friends has already occupied the first place; comparably, the third place can be occupied in $(n - 2)$ ways as two of n friends have already occupied the first and the second place; and the last place can be occupied in only one way.

So, **the total number of ways** = $\boldsymbol{n \times (n-1) \times (n-2) \times (n-3) \times (n-4) \times \ldots\ldots\ldots\ldots 1 = n!}$.

$\boldsymbol{n!}$ is called factorial n. Its value equals to the product of all the integers from '1' to 'n'.

Selecting few objects from all the objects

Say, if any **two** of Alex, Betty, and Chris are to be photographed, then in how many ways can they be photographed?

Now, there are only two places to be occupied by any two of Alex, Betty, and Chris. The first place can be occupied by 3 ways, and the second place can be occupied by 2 ways.

So, the **total number of ways** = $3 \times 2 = \mathbf{6}$ **ways.**

What if there are r friends out of n friends to get photographed?

On a similar way, the first place can be occupied in n ways; the second place can be occupied in $(n-1)$ ways; the third place can be occupied in $(n-2)$ ways; and similarly, the rth place can be occupied in $(n-r+1)$ or $[n-(r-1)]$ ways.

So, the **total number of ways** = $n \times (n-1) \times (n-2) \times (n-3) \times \ldots\ldots\ldots [n-(r-1)]$

In mathematical notations, it is written as P_r^n or $P(n,r)$.

So, $\boldsymbol{P_r^n}$ or $P(n,r) = n \times (n-1) \times (n-2) \times (n-3) \times \ldots\ldots\ldots [n-(r-1)]$;

If $r = n$, then P_n^n or $P(n,n) = n \times (n-1) \times (n-2) \times (n-3) \times \ldots\ldots\ldots\ldots 1 = n!$.

If we need to find the permutations of r distinct objects out of n distinct objects, we apply:

$$P_r^n = \frac{n!}{(n-r)!}\text{; and } P_n^n = n!$$

Some most common values of factorials:

- **5! = 5 × 4 × 3 × 2 × 1 = 120**
- **4! = 4 × 3 × 2 × 1 = 24**
- **3! = 3 × 2 × 1 = 6**
- **2! = 2 × 1 = 2**
- **1! = 1**
- **0! = 1** (you may by surprised that the value of 0! is '1', but it is so.)

Question: In how many ways can 3 out of 4 friends be photographed?

Solution: The number of ways = $4 \times 3 \times 2 = 24$ ways.

Formula approach: $n = 4$, and $r = 3$; so, $P_3^4 = \frac{4!}{(4-3)!} = \frac{4!}{1!} = 4 \times 3 \times 2 \times 1 = 24$ ways.

Question: In how many ways can all 4 friends be photographed?

Solution: The number of ways = $4! = 4 \times 3 \times 2 \times 1 = 24$ ways.

Formula approach: $n = 4$, and $r = 4$; so, $P_4^4 = \frac{4!}{(4-4)!} = \frac{4!}{0!} = \frac{4 \times 3 \times 2 \times 1}{1} = 24$ ways.

11.14.1 Permutation with repetition of objects

In the examples above, we tried to find out the number of ways 3 friends can be photographed. In the example, you must have noticed that each successive place has one less way to get filled. It is because if, say, Betty occupied the first place, she cannot occupy the second or the third place.

Let us see the following example.

Question: How many 3-digit combination codes with digits 1, 2, and 3 are possible for a lock?

If your answer is 3! = 6 combinations, you assumed that NO digit can be repeated. Say, if the digit '2' occupied the first place, it cannot reappear in the second and the third place.

However, in a real life, there can be a combination code such as '333', which means that the digits can be repeated.

So, if the digits can be repeated, the possible number of combinations $= 3 \times 3 \times 3 = 27$.

If there are n digits, and n to be filled, the **total number of ways = n^n**, if repetition is allowed.

Let us see following example.

Question: How many 3-digit passwords are possible with the digits 0, 1, 2 .... 9? Repetition is allowed.

Since each place can be filled in 10 ways (any digit among 0 to 9 may be used), the total number of possible passwords = $10 \times 10 \times 10 = 10^3 = 1,000$.

Similarly, for a question: how many r-digit passwords are possible with n digits, if repetition of digits is allowed?

The answer would be $n \times n \times n \times ....r$ times = n^r passwords.

Number of arrangements out of selecting r objects from n objects = n^r; Repetition is allowed.

Remember that it is important to know beforehand whether repetition is allowed. If it is NOT allowed, then for the above example, the answer would be $P_r^n = \frac{n!}{(n-r)!}$.

11.14.2 Circular permutation

We have already seen the question: In how many ways can Alex, Betty, and Chris get seated?

The answer to the question is 3! = 6 ways. It is implied that all three will stand in a row. What if the question were:

In how many distinct ways can Alex, Betty, and Chris get seated around a round table?

Well, the answer would be $(3-1)! = 2! =$ only 2 ways! This is because, unlike the arrangement in a line, merely moving from one place to the other does not make a distinct seating arrangement with a round table.

If there are n objects to be arranged in a **circle**, in which all the positions are identical, we need to assign one of the objects to a seat to mark it as the starting point (since in a circle, there is no distinct starting point). Thus, there are $(n-1)$ objects left to be arranged. So, these n objects can be arranged in $(n-1)!$ ways.

Let us see how.

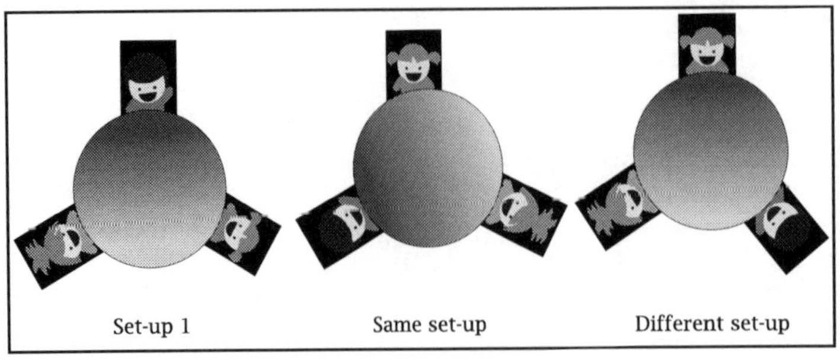

So, the number of distinct ways n objects can be arranged in a straight line = $n!$;

but,

The number of distinct ways n objects can be arranged around a circle = $(n - 1)!$

11.15 Combination

Let us develop a formulaic approach to solve lengthy questions.

Question: In how many ways can two balls out of three balls (a red, a green, and a black) be selected?

As stated earlier, the foremost decision in the Permutation & Combination based questions is to ask: Which of the two concepts needs to be applied? Ask yourself whether the arrangement of two balls selected will render a unique selection. If the answer is yes, apply permutation, otherwise apply combination.

In the above question, the arrangement is not important as, if the two selected balls are, say, red and green; it is does not matter whether they are **red and green** or **green and red**. So, we will apply combination.

Develop the formula

Two balls selected can either be **red and green**, **red and black** or **green and black**, so there are total three ways (RG/RB/GB).

What if the order was given importance, though conceptually wrong in this case, the answer would have been $P_2^3 = 3 \times 2 = 6$ ways (wrong answer).

So, the correct answer can be written as $\frac{P_2^3}{2!} = \frac{6}{2} = 3$ ways.

So, if r different objects are to be selected out of n different objects, then the total number of ways would be $C_r^n = \frac{P_r^n}{r!} = \frac{n!}{r! \times (n-r)!}$

$$C_r^n = \frac{n!}{r! \times (n-r)!}$$

$$C_r^n = \frac{n \times (n-1) \times (n-2) \times (n-3).....[n-(r-1)]}{[r \times (r-1) \times (r-2).....1]}$$

Question: In how many ways can 4 out of 6 balls be drawn from an urn?

Total number of ways = $C_4^6 = \frac{6!}{4! \times (6-4)!} = \frac{6!}{4! \times 2!} = \frac{6 \times 5 \times 4 \times 3 \times 2 \times 1}{(2 \times 1) \times (4 \times 3 \times 2 \times 1)} = 15$ ways.

Alternate approach 1:

The above approach to calculate the value of C_4^6 is lengthy. It could have been calculated using the following approach. See the next question.

Question: In how many ways can 5 out of 7 balls be drawn from an urn?

Total number of ways = $C_5^7 = \frac{7 \times 6 \times \cancel{5} \times \cancel{4} \times \cancel{3}}{\cancel{5} \times \cancel{4} \times \cancel{3} \times 2 \times 1} = 21$ ways.

Alternate approach 2:

The calculation of C_5^7 can further be simplified; you would have observed that 5, 4, and 3 are getting cancelled from the numerator and the denominator.

However, C_5^7 can be rewritten as $C_{(7-2)}^7$ and is equal to $C_2^7 = \frac{7 \times 6}{2 \times 1} = 21$ ways. So,

$$\boxed{C_r^n = C_{(n-r)}^n}$$

But, remember that

$$\boxed{P_r^n \neq P_{(n-r)}^n}$$

Note: Apply Alternate approach 2, if $r > (n - r)$

Question: In how many ways can 2 out of 7 balls be drawn from an urn?

Total number of ways = $C_2^7 = \frac{7 \times 6}{2 \times 1} = 21$ ways; however, in this case, we must not transform C_2^7 to $C_{(7-5)}^7$, as this will rather lengthen the calculation, however, the answer would be the same.

11.16 Concepts Revisited

Question	Permutation or Combination	Should we apply $C_r^n = C_{(n-r)}^n$?	Solution
In how many ways can 4 out of 6 friends be seated in a row?	**Permutation;** Order is important	Not applicable for permutation	# of ways = $6 \times 5 \times 4 \times 3$ = 360 ways
In how many ways can 12 out of 14 players be selected for a team?	**Combination;** Order is not important; only selection the matters.	Yes; as $r > n - r$ $r = 12$ & $n = 14$	# of ways = $C_{12}^{14} = C_2^{14} = \frac{14 \times 13}{1 \times 2}$ = 91 ways
In how many ways can 5 different colored balls be arranged in a row to show distinguishable patterns?	**Permutation;** Order is important. Remember that not every question on balls is a combination question.	Not applicable for permutation	# of ways = 5! = $5 \times 4 \times 3 \times 2 \times 1$ = 120 ways
In how many ways can 5 different colored balls be arranged around a circle to show distinguishable patterns?	**Circular Permutation;** Order is important	Not applicable for permutation	# of ways = $(5-1)! = 4! = 4 \times 3 \times 2 \times 1$ =24 ways
In how many ways can 2 out of 6 players be selected for a team?	**Combination;** Order is not important; only the selection matters.	No; $r < n - r$ $r = 2$ & $n = 6$	# of ways = $C_2^6 = \frac{6 \times 5}{1 \times 2}$ = 15 ways
In how many ways can all players out of 6 players be selected for a team?	**Combination;** Order is not important; only the selection matters.	Though yes, it is not needed.	For a combination question, if $r = n$ or $r = 0$, # of ways is ALWAYS equal to 1.

11.17 Permutation of multiple indistinguishable objects

Question: How many meaningful or meaningless words can be formed out of the word 'book'?

The answer '4! = 24 words' is a wrong answer as two O's were counted as different letters, but they are the same.

If there are multiple indistinguishable objects, we must exclude the number of ways formed because of them as they are basically the same.

Total number of words = $\frac{4!}{2!}$ = 12 words; the 2! in the denominator is for two O's. The 12 words would be **book, boko, bkoo, kboo, kobo, koob, oobk, ookb, okob, obok, obko, & okbo**.

Develop a formula

If there are total n balls, and among these, there are p number of identical green, q number of identical red, and r number of identical black balls, in how many ways can these balls be arranged in a row to have unique patterns?

$$\text{Total number of ways} = \frac{n!}{p! \times q! \times r!}$$

Question: How many words with or without meaning can be formed out of the word 'Mississippi'?

There are total 11 letters, so $n = 11$; there are four Is, so $p = 4$; four Ss, so $q = 4$; two Ps, so $r = 2$.

So, total number of words = $\frac{11!}{4! \times 4! \times 2!} = \frac{11!}{2 \times (4!)^2}$ words

11.18 Restricted Permutation

Question: How many 4-digit even numbers can be formed out of digits 0–9? Repetition is allowed.

> This is a question on restricted permutation. We not only have to find the count of 4-digit numbers, but also assure that the numbers are even. It implies that the unit digits must be either 0, 2, 4, 6, or 8—only 5 ways; the unit place can be filled in five ways.
>
> Also, the digit at the thousandth place cannot be '0', as allowing '0' at the thousandth place will render 3-digit numbers. It implies that the thousandth place digit must be any digit among (1-9): 9 possibilities.
>
> So, total # of 4-digit numbers

= [# of ways to fill thousandth place] × [# of ways to fill hundredth place] × [# of ways to fill tenth place] × [# of ways to fill unit place]

= 9 × 10 × 10 × 5 = 4,500 numbers

11.19 Restricted Combination

Question: In how many ways can 3 out of 11 players of a team be selected if the captain must be included in the selection?

Since the captain must be included, the question becomes: In how many ways can 2 out of 10 players of a team be selected?, as one selection (the captain) is already done.

$$\# \text{ of ways} = C_2^{10} = \frac{10 \times 9}{1 \times 2} = 45 \text{ ways.}$$

If the question is changed to: In how many ways can 3 out of 11 players of a team be selected, provided the captain should NOT be included in the selection?

Since the captain is not to be included, the question becomes: In how many ways can 3 out of 10 players of a team be selected?, as one player (the captain) is out of the selection process.

$$C_3^{10} = \frac{10 \times 9 \times 8}{1 \times 2 \times 3} = 120 \text{ ways.}$$

Let us see another question with two different set of objects.

Question: In how many ways can a man choose 2 trousers and 3 shirts out of 3 trousers and 4 shirts?

Let's choose trousers first. # of ways to choose trousers = $C_2^3 = C_1^3 = 3$ ways; $[C_r^n = C_{(n-r)}^n]$

Let's choose shirts now. # of ways to choose shirts = $C_3^4 = C_1^4 = 4$ ways;

Total number of ways of choosing 2 trousers and 3 shirts = 3 × 4 = 12 ways.

11.20 Rule of Sum

If there are 4 boys and 3 girls, in how many ways can a team of 2 be formed such that there are either both the boys or both the girls in the team?

# of ways 2 boys can be selected = $C_2^4 = \frac{4 \times 3}{1 \times 2} = 6$ ways;

# of ways 2 girls can be selected = $C_2^3 = C_1^3 = 3$ ways;

Since either of the ways can occur, the total number of ways either two boys or two girls can be selected = $6 + 3 = 9$ ways.

The key word is EITHER.

If an event can be done in two or more ways, the numbers of ways are added. This is called **'Rule of Sum'**.

11.21 Rule of Product

If there are 4 boys and 3 girls, in how many ways can a team of 4 be formed such that there must be 2 boys and 2 girls in the team?

# of ways 2 boys can be selected = $C_2^4 = \frac{4 \times 3}{1 \times 2} = 6$ ways;

# of ways 2 girls can be selected = $C_2^3 = C_1^3 = 3$ ways;

Since selecting 4 members is composed of selecting 2 boys and 2 girls, so the total number of ways 4 members would be selected = $6 \times 3 = 18$ ways.

The key word is AND.

If an event is composed of two or more sub-events, total number of ways of doing the event equals to the product of the total number of ways of doing each of the sub-event. This is called **'Rule of Product'**.

Key word	**Total # of ways**
• Either or	Add the number of ways (+)
• And	Multiply the number of ways (×)

11.22 Probability

What are the chances of Messi winning the toss in a FIFA world cup match? Well, ideally it's 50%! He has an equal chance of winning and losing the toss, provided the coin and the trial are fair.

It is very important that the coin tossed is fair, or else the coin will be called a biased coin, the outcomes of 'Head and Tail' would be influenced, and the chances of getting a Head and getting a Tail will not be 50-50.

What do we mean when we say that the chance of getting a Head is 50%?

It means that if we toss a coin many number of times, say 100 times, it is likely that the fair coin will fall Head 50 times and fall Tail 50 times. But can this occur practically? The answer is no. Just to share a fact that Carolina Panthers has lost 13 tosses consecutively; however, it is so rare that it happens in 1 out of 8,192 cases!

So, is there a significance of 50-50? Yes, there is. If, say, the experiment of throwing a fair coin is done over 100,000 times (Read: very large number of trials), it is very likely that you get Head close to 50,000 times.

11.23 Chance vs. Probability

Chance is a term used by common people; it is measured on a scale of '100'. The count of any event to occur out of 100 possibilities is a chance percent.

Probability is a term used by mathematicians; which is measured on a scale of '1' instead of '100'.

11.24 A few definitions

11.24.1 Experiment

Tossing a coin or throwing a die is an experiment. It is an activity that can be repeated an infinite number of times and produces a well-defined set of outcomes. In an experiment, only one of all the possible outcomes is the result. The repetition of an experiment is called a **Trial**.

11.24.2 Event

In an experiment, getting a Head upon tossing a coin or getting a '3' on the upper face upon throwing a die is an event. Events are a set of all possible outcomes of an experiment. We may also talk about a favorable or an unfavorable event. Tossing a coin renders two events: {Heads and Tails}, while throwing a die renders six events: {1, 2, 3, 4, 5, 6}. In an experiment with a die, if you were to have '5' as an outcome, then getting a '5' is a favorable event, and not getting a '5' are unfavorable events.

11.24.3 Equally likely events

If the chances of occurrence of some events are equal, they are called equally likely events. For a fair die, the chance of the occurrence of '3' equals that of '5'. So, these events are equally likely. Similarly, for a coin, the chance of getting a 'Head' equals that of a 'Tail'.

11.24.4 Sample set

A set of all possible outcomes of an experiment is called the sample set. For a fair die, the sample set is {1, 2, 3, 4, 5, 6}; whereas for a coin, the sample set is {Head, Tail}.

$$\text{Probability of an event} = \frac{\text{Number of favorable events}}{\text{Total number of events}}$$

Question: What is the probability of getting an outcome as Head in an experiment of tossing a fair coin?

The sample set is {Head, Tail}, so only two events are possible; out of these two events, only one event—Head is the favorable event.

$$\text{P(Head)} = \frac{\text{Number of favorable events}}{\text{Total number of events}} = \frac{n(\text{favorable events})}{n(\text{total events})} = \frac{1}{2}$$

Question: In an experiment of throwing a die, what is the probability of getting an outcome as '2' or '4' on the upper face of an unbiased die?

The sample set is {1, 2, 3, 4, 5, 6}, so there are a total six events; out of these six events, two events: {2, 4} are the favorable events.

$$\text{Probability of '2' or '4'} = \frac{n(\text{favorable events})}{n(\text{total events})} = \frac{n\{2,4\}}{n\{1,2,3,4,5,6\}} = \frac{2}{6} = \frac{1}{3}$$

Sum of probabilities of all events = '1'. Event with probability = 1 is a **Certain event.**

$$\text{P(H or T)} = \frac{n(\text{favorable events})}{n(\text{total events})} = \frac{2}{2} = 1$$

Probability cannot be less than '0'. Event with probability = 0 is an **Impossible event.**

$$\text{Probability (neither Head nor Tail)} = \frac{n(\text{favorable events})}{n(\text{total events})} = \frac{0}{2} = 0$$

So, we can deduce that:

$\mathbf{0 \le P(A) \le 1}$

Question: There are 2 red, 4 green, and 5 black balls in a bag. If a ball is drawn randomly, what is the probability that it is a green color ball?

$$\text{P(G)} = \frac{n(\text{ favorable events})}{n(\text{total events})} = \frac{4}{(2+4+5)} = \frac{4}{11}$$

11.25 An experiment with two dice

Die is singular and **Dice** is plural. Two dice will have $6 \times 6 = 36$ sample points or possible events. The sample set of SUM would be $\{2, 3, 4, 5, 6, 7, 8, 9, 10, 11, 12\}$; each element in the set is the sum of outcomes on the upper face of each die.

Let us list down the number ways each sum can be formed.

Sum	Ways	Number of ways	Frequency distribution & Probabilities
2	{1,1}	1	$p = \frac{1}{36}$
3	{2,1}, {1,2}	2	$p = \frac{1}{18}$
4	{3,1}, {2,2}, {1,3}	3	$p = \frac{1}{12}$
5	{4,1}, {3,2}, {2,3}, {1,4}	4	$p = \frac{1}{9}$
6	{5,1}, {4,2}, {3,3}, {2,4}, {1,5}	5	$p = \frac{5}{36}$
7	{6,1}, {5,2}, {4,3}, {3,4}, {2,5}, {1,6}	6	$p = \frac{1}{6}$
8	{6,2}, {5,3}, {4,4}, {3,5}, {2,6}	5	$p = \frac{5}{36}$
9	{6,3}, {5,4}, {4,5}, {3,6}	4	$p = \frac{1}{9}$
10	{6,4}, {5,5}, {4,6}	3	$p = \frac{1}{12}$
11	{6,5}, {5,6}	2	$p = \frac{1}{18}$
12	{6,6}	1	$p = \frac{1}{36}$
	Total number of ways	**36**	

Memorize...

1. If Sum $\leq$ 7, then Number of ways = Sum – 1

2. If Sum $\geq$ 8, then Number of ways = 13 – Sum

Question: What is the probability of getting a sum of '5' on the upper faces of two unbiased dice?

The sample set is $\{2, 3, 4, 5, 6, 7, 8, 9, 10, 11, 12\}$; and there are a total 36 possible events.

Probability of getting a sum of '5' = $\frac{n(\text{ favorable events})}{n(\text{total events})} = \frac{4}{36} = \frac{1}{9}$

11.26 A few more definitions

11.26.1 Mutually Exclusive Events

If two events cannot occur together, they are called **Mutually Exclusive** events or **Disjoint** events.

Example: Getting a Head and a Tail on the upper face of a coin; another example, getting a '3' and a '5' in a throw of a die.

With a coin, only one of the two events—Head or Tail will occur at a time, hence they are mutually exclusive events; similarly, if '3' appears on the upper face of the die, it is obvious that '5' will not appear.

Following Venn diagram represents the scenario.

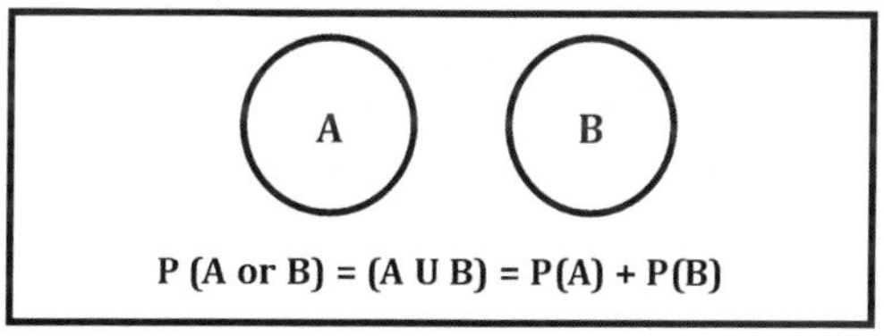

However, an event of getting '3' and 'prime number' in a throw of a die is not mutually exclusive as '3' is an element of the set of prime numbers.

If the events are joint or non-mutually exclusive, then the following formula applies:

P(A or B) = P(A $\cup$ B) = P(A) + P(B) – P(A & B)

Following Venn diagram represents the scenario.

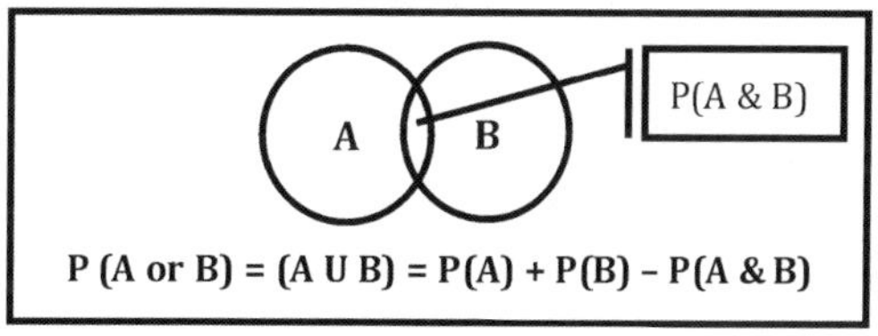

The table below depicts the difference between mutually exclusive and joint events with an application of an example.

Event	Example	Solution
Disjoint (Mutually exclusive)	A die is rolled. What is the probability of getting a '3' or a '5' on the upper face of the die?	P('3') = 1/6; P('5') = 1/6; So, P('3' or '5') = (1/6) + (1/6) = 1/3.
Joint	A die is rolled. What is the probability of getting a '3' or a 'Prime number' on the upper face of the die?	P('3') = 1/6; P('Prime') = 3/6 = 1/2; P('3' & 'P') = 1/6 ('3' is common in Prime numbers: 2, 3, and 5); So, P('3' or 'P') = (1/6) + (1/2) - (1/6) = 1/2.

11.26.2 Independent Events

Two events are called independent if the occurrence of one event does not affect the occurrence of the other.

Example: In a throw of two dice, the occurrence of a '3' on the first die does not influence the occurrence of any number on the second die; hence these events are independent events.

For independent events, **P(A & B) = P(A $\cap$ B) = P(A) $\times$ P(B)**

If the events are not independent, then the concept of 'conditional probability' is involved; however, this concept is not frequently tested on the GMAT.

11.26.3 Exhaustive event

A set of events is called **Exhaustive** if there is no possibility of any other event to occur other than the events in the set; in other words, one of those events must occur. For example: In the toss of a coin, Events of a Head and a Tail are exhaustive events. (assuming, of course, that the coin does not land on its side!)

11.27 AND vs. OR rule

In a nutshell, you must know rules for probability calculations on the GMAT: "AND" means MULTIPLY, and "OR" means ADDITION.

Concept	Application	Example	Solution
AND	×; multiply the probabilities	A coin is tossed twice. What is the probability of getting a Head in first trial, and Tail in the second trial?	P(Head in I trial) = 1/2; P(Tail in II trial) = 1/2; So, P(HT) = (1/2) × (1/2) = 1/4.
OR	+; add the probabilities	A die is rolled. What is the probability of getting a '3' **or** an 'Even number' on the upper face of the die?	P('3') = 1/6; P('E') = 3/6 = 1/2; So, P('3' or 'E') = (1/6) + (1/2) = 2/3.

11.28 Conditional Probability

The concept of conditional probability relates to deducing the probability of an event given that another event has already occurred. This concept is important since the occurrence of one event can affect the probability of a dependent event.

The conditional probability of A given B is denoted by $P(A|B)$, it means probability of A if event B has occurred.

$$\mathbf{P(A|B) = \frac{P(A \cap B)}{P(B)}}$$

Question: What is the probability of getting a '3' on the upper face of a die given that the number appeared is a 'Prime number'?

Say, P('3') = probability of getting a '3';
and P('P') = probability of getting a 'Prime number'

We know that P('P') = 1/2;
and P('3'∩'P') = 1/6

So, $P(3|P) = \frac{P(3 \cap P)}{P(P)} = \frac{1/6}{1/2} = 1/3$

Logical deduction approach:

Since there are only three prime numbers: 2, 3, & 5 and '3' is one of them, hence the chance of getting '3' out of '2', '3' and '5' is 33.33% = 1/3.

11.29 Independent and Dependent Events

- For **Independent Events: P(A|B) = P(A);**

- For **Dependent Events: P(A|B) $\neq$ P(A)**

Question: A bag contains 3 yellow and 7 blue marbles. What is the probability of drawing a yellow marble in the first draw and then drawing a blue marble in the second draw?

Let us understand the concept of dependent event.

The second event–drawing a blue marble–is a dependent event since the draw of a yellow marble in the first draw will increase the probability of drawing a blue marble in the second draw; while the draw of a blue marble in the first draw will decrease the probability of drawing another blue marble in the second draw. So, the second event is a dependent event.

Let us see how.

(1) If a yellow marble is drawn in the first draw, then the bag will contain ~~3~~2 yellow and 7 blue marbles; this makes the probability of drawing the blue marble in the second draw equals to 7/(2 + 7) = 7/9.

(2) If a blue marble is drawn in the first draw, then the bag will contain 3 yellow and ~~7~~6 blue marbles; this makes the probability of drawing the blue marble in the second draw equals to 6/(3 + 6) = 6/9 = 2/3.

Clearly, 7/9 $\neq$ 2/3 implies that the probability of the second event is dependent on that of the first event.

Coming back to the question...

Drawing a yellow marble in the first draw = 3/10.
Drawing a blue marble in the second draw = 7/9.

So, the probability of drawing a yellow marble in the first draw, and blue marble in the second draw = (3/10) × (7/9) = 7/30.

11.30 Complement of an Event

If the occurrence of an event is denoted as A, then its non-occurrence is denoted as A' or A^C.

Probability of occurrence of an event + Probability of non-occurrence of an event = 1.

So, $\boldsymbol{P(A')}$ or $\boldsymbol{P(A^C) = 1 - P(A)}$;

or $\boldsymbol{P(A) + P(A') = 1}$

Question: If a coin is tossed 3 times, what is the probability that it lands up on heads at least once?

Traditional approach:

P (Head ≥ 1) = P (Head = 1; Tail = 2) + P (Head = 2; Tail = 1) + P (Head = 3; Tail = none);

=> (Head = 1; Tail = 2) can occur in 3 ways; HTT, THT, and TTH;

For each way, P (Head = 1; Tail = 2) = (1/2) × (1/2 × 1/2) = 1/8;

So, for all 3 ways, P (Head = 1; Tail = 2) = 3 × (1/8) = 3/8.

=> Similarly, (Head = 2; Tail = 1) can occur in 3 ways; HHT, THH, and HTH;

Again, for each way, P (Head = 2; Tail = 1) = (1/2) × (1/2 × 1/2) = 1/8;

So, for all 3 ways, P (Head = 2; Tail = 1) = 3 × (1/8) = 3/8.

=> At last, (Head = 3; Tail = none) can occur in only one way; HHH;

P (Head = 3; Tail = none) = (1/2 × 1/2 × 1/2) = 1/8.

So, P (Head ≥ 1) = P (Head = 1; Tail = 2) + P (Head = 2; Tail = 1) + P (Head = 3; Tail = none);
= 3/8 + 3/8 + 1/8
= 7/8.

This approach is time consuming. Let us see an alternate approach.

Alternate approach:

P (Head ≥ 1) = 1 – P (Head < 1)' = 1 – P (Head = 0; Tail = 3)
= 1 - (1/2 × 1/2 × 1/2)
= 7/8.

So, the learning is that we can make use of the concept of complement of an event to shorten the calculations.

11.31 Application of combination

What if the question is like the following?

Question: If a coin is tossed 3 times, what is the probability that it lands up on Head at least twice?

In this question, the alternate approach and the traditional approach will take almost the same time.

Alternate approach:

P (Head $\geq$ 2) = 1 – P (Head < 2)′ = 1 – P (Head = 0; Tail = 3) – P (Head = 1; Tail = 2)

You will have to do the calculations twice–once for P (Head = 0; Tail = 3), and then for P (Head = 1; Tail = 2).

Traditional approach:

P (Head $\geq$ 2) = P (Head = 2; Tail = 1) + P (Head = 3; Tail = none)

Again, you will have to do the calculations twice–once for P (Head = 2; Tail = 1), and then for P (Head = 3; Tail = none). Practically there is no saving of time!

However, with the application of the concept of combination, we can save some time.

Let us take a seemingly time-consuming question.

Question: If a coin is tossed 5 times, what is the probability that it lands up on Head at least thrice?

P (Head $\geq$ 3) = P (Head = 3; Tail = 2) + P (Head = 4; Tail = 1) + P (Head = 5; Tail = none)

The meaning of (Head = 3; Tail = 2) is that we want 3 Heads occurring in 3 out of 5 trials, so it can be deduced that we want to choose 3 events out of 5 events, which is equal to C_3^5.

So, P (Head = 3; Tail = 2) = $C_3^5 \times [(1/2 \times 1/2 \times 1/2) \times (1/2 \times 1/2)]$
$= C_2^5 \times (1/2)^5 = (5.4)/(1.2) \times (1/2)^5 = (5.4)/(1.2) \times (1/2)^5 = 5/2^4$.

(Note that $C^5_3 = C^5_2$)

Similarly, P (Head = 4; Tail = 1) = $C^5_4 \times [(1/2 \times 1/2 \times 1/2 \times 1/2) \times (1/2)]$
$= C^5_1 \times (1/2)^5 = 5/2^5$.

And, P (Head = 5; Tail = none) = $C^5_5 \times (1/2 \times 1/2 \times 1/2 \times 1/2 \times 1/2) = 1 \times (1/2)^5$
$= 2^5$.

So, P (Head $\geq$ 3) = $5/2^4 + 5/2^5 + 1/2^5 = 16/2^5 = 1/2$.

P (Head $\geq$ 3) = 1/2.

Let us see one more question.

Question: A bag contains 4 yellow and 5 blue marbles. If three marbles are drawn simultaneously, what is the probability of drawing a yellow and two blue marbles?

P(Y = 1; B = 2)

$$= \frac{(\text{\# of ways one yellow marble can be drawn}) \times (\text{\# of ways two blue marbles can be drawn})}{(\text{\# of ways three marbles can be drawn})}$$

$$= \frac{(C^4_1) \times (C^5_2)}{C^9_3} = 4 \times \left[\frac{\frac{5\times4}{1\times2}}{\frac{9\times8\times7}{1\times2\times3}}\right] = \frac{10}{21}.$$

11.32 Sets

A Set is a collection of well-defined things. The objects of a set are called its elements or members.

Nomenclature:

- **Universal Set:** All sets are assumed to be contained in a set called the universal set.
- **Null Set:** An empty set having no elements in it is called a null set.
- **Disjoint Set**: Two sets are disjoint if they have no elements in common.
  For example: $A = \{1, 2, 3\}$ and $B = \{4, 5, 6\}$ are two disjoint sets.
- **Cardinal Number of a Set:** The Cardinal number of a finite set A is the number of elements of the set, denoted by $n(A)$. For example: For $A = \{1, 2, 3\}$, $n(A) = 3$

Venn Diagrams: A Venn diagram is a pictorial representation of sets represented by enclosed areas in a plane. The universal set is represented by a rectangle, and the other sets are represented by areas lying within it.

Set Operations:

- **Union:** The Union of two sets A and B, i.e. $A \cup B$, is a set that contains all the elements contained in set A or set B.

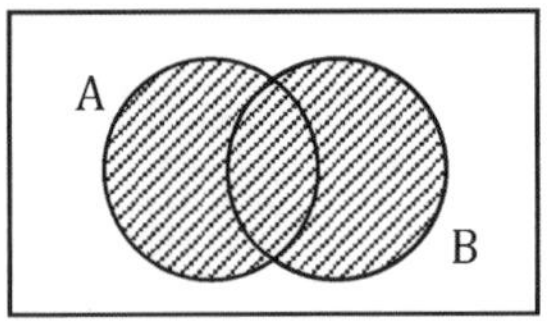

For example: If $A = \{2, 3\}$, $B = \{1, 3, 5\}$

$\Rightarrow A \cup B = \{1, 2, 3, 5\}$

- **Intersection:** Intersection of two sets A and B, i.e. $A \cap B$, that contains the elements common to both sets A and B.

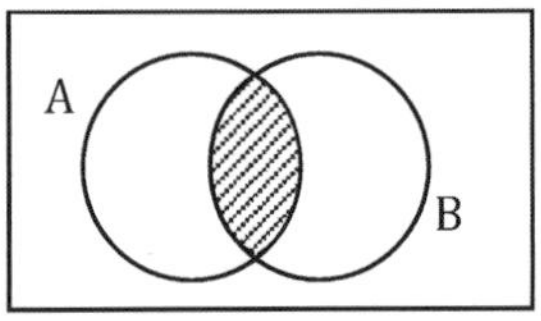

For example: For $A = \{2, 3\}$, $B = \{1, 3, 5\}$, $A \cap B = \{3\}$

- **Difference of Two Sets:** Difference of two sets A and B, i.e. $A - B$ is a set of elements present in set A but not in set B.

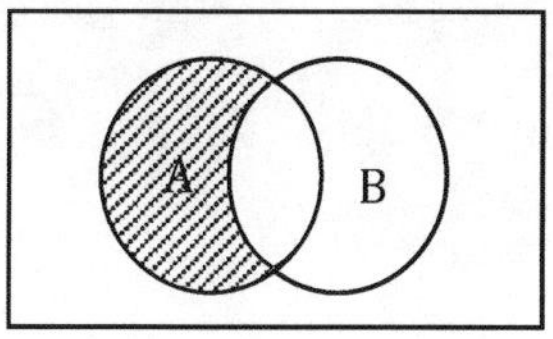

For example: For $A = \{2, 4\}$, $B = \{2, 6, 5\}$, $A - B = \{4\}$, $B - A = \{5, 6\}$

- **Complement:** Complement of set A, i.e. A^{-1} or A' is a set that contains the elements outside set A.

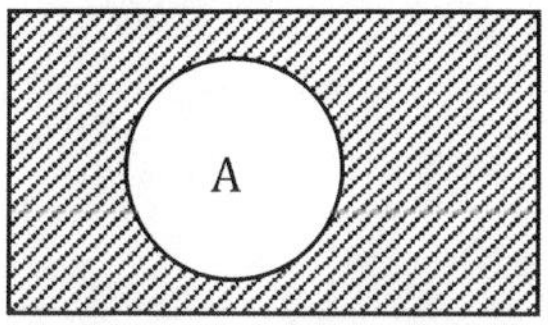

For example: If $A = \{1, 2\}$ and Universal Set U $= \{1, 2, 7, 8\}$, $A' = \{7, 8\}$

A few important rules:

- For two sets A and B:
  - $A \cup B = A + B - A \cap B \ldots$(i)
  - $(A \cup B)' = A' \cap B' \ldots$(ii)
  - $(A \cap B)' = A' \cup B' \ldots$(iii)

  Let us see how:

  We refer to the diagram given below:

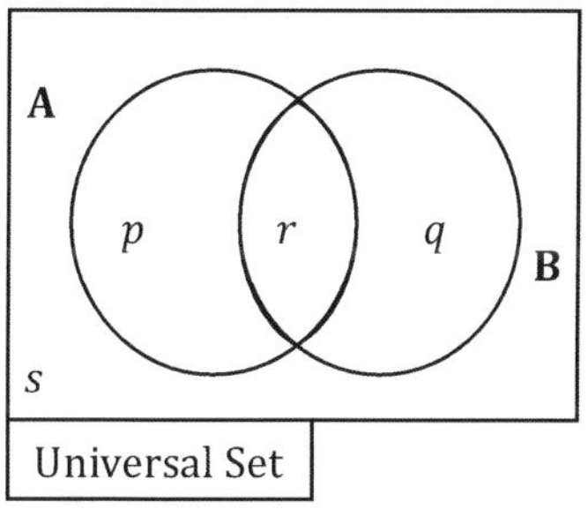

The cardinal numbers of the regions are shown as p, q, r and s.

we have

$A = p + r,\ B = q + r,\ A \cap B = r,\ A \cup B = p + r + q$

Thus, we have

$A \cup B = p + q + r = (p + r) + (q + r) - r$

$= A + B - A \cap B \ldots$(i)

$=> (A \cup B)' = s$

We also have: $A' = q + s,\ B' = p + s => A' \cap B' = s$

$=> (A \cup B)' = A' \cap B' \ldots$(ii)

Similarly

$(A \cap B)' = p + q + s$

Also, $A' \cup B' = A' + B' - A' \cap B' = (q + s) + (p + s) - s = p + q + s$

$=> (A \cap B)' = A' \cup B' \ldots$(iii)

- For three sets $A,\ B$ and C:

  $A \cup B \cup C = A + B + C - (A \cap B + B \cap C + C \cap A) + A \cap B \cap C$

  Let us see how:

  The cardinal numbers of the regions are shown:

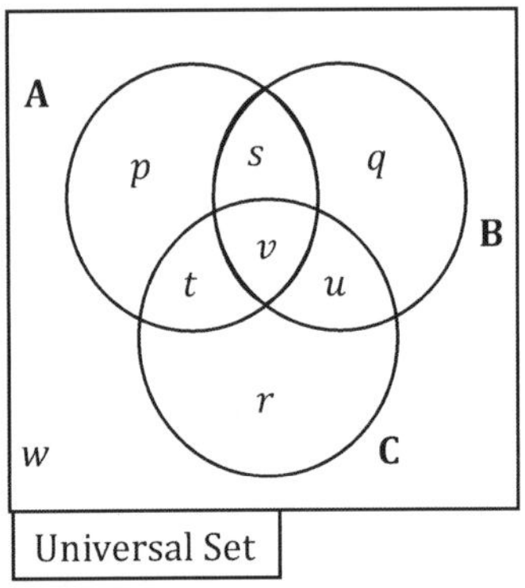

$A = p + s + t + v$

$B = q + s + u + v$

$C = r + t + u + v$

$A \cap B = s + v,\ A \cap C = t + v,\ B \cap C = u + v$

$A \cap B \cap C = v$

$A \cup B \cup C = p + q + r + s + t + u + v$

Also, $A \cap B + B \cap C + C \cap A = (s + v) + (t + v) + (u + v) = s + t + u + 3v$

$=> A + B + C - (A \cap B + B \cap C + C \cap A) + A \cap B \cap C$

$= (p + s + t + v) + (q + s + u + v) + (r + t + u + v) - (s + t + u + 3v) + v$

$= p + q + r + s + t + u + v$

$= A \cup B \cup C$

Let us take a few examples:

(1) In a survey, it was found that 65 people keep dogs as pets, 70 people keep cats, 40 people keep birds, 20 people keep both dogs and cats, 10 people keep both cats and birds, 5 people keep both birds and dogs and 3 people keep all three. How many people participated in the survey if it is known that each person has at least one pet?

Explanation:

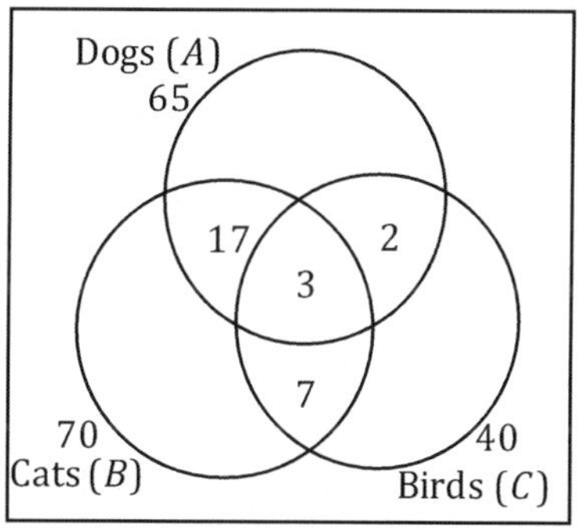

$A = 65,\ B = 40,\ C = 70$

$A \cap B = 20,\ B \cap C = 10,\ C \cap A = 5$

$A \cap B \cap C = 3$

The total number of people surveyed

$= A \cup B \cup C$

$= A + B + C - (A \cap B + B \cap C + C \cap A) + A \cap B \cap C$

$= 65 + 40 + 70 - (20 + 10 + 5) + 3 = 143$

(2) In a class of 50 boys, some play cricket, some football, and some hockey. The number of boys who play cricket is higher than those who play football, which is higher than those who play hockey, which is higher than those who play only two games, which is higher than those who play all three games. If each boy plays at least one game and there is at least one boy who plays all three games, what is the minimum number of boys who play football?

Explanation:

We know that the boys play cricket (C), football (F) or hockey (H).

As the information,

50 = C + F + H − Only two games + All three games

In order to minimize F, we need to minimize the number of students in the regions a, b, c, and d, as shown in the diagram below.

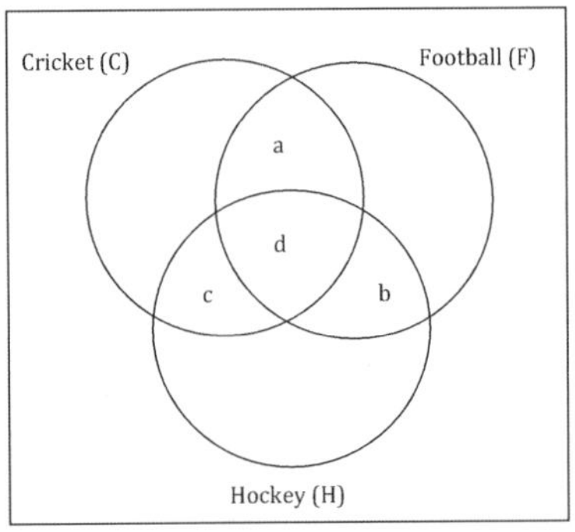

$50 = C + F + H - (a + b + c) + d$

Since $C > F > H >$ (a + b + c)$> d \geq 1$, we assume d = 1, a + b + c = 2, and H = 3.

This follows that F = 4 (minimum)

So, we make the following distribution:

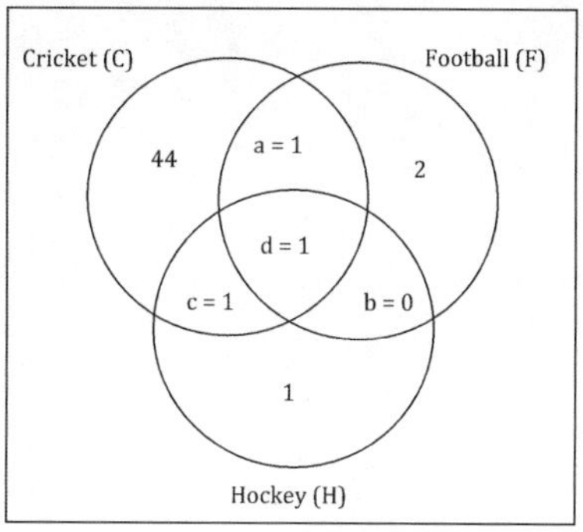

we have C = 50 − 4 − 3 − 2 − 1 = 44, F = 4, H = 3, a = 1, b = 0, c = 1, and d = 1.

Chapter 12

Practice Questions: Word Problems

1. Ann and Bob started traveling from City X at noon. Ann walked to a lake that was between 2 and 3 miles, inclusive, and returned between 1:30 pm and 1:45 pm, inclusive. If Bob walked 3 miles farther from the lake and returned at 4:00 p.m., which of the following statements must be true?

   I. The speed of Bob was greater than that of Ann.

   II. Ann's speed was between 2 and 3 miles per hour.

   III. The maximum speed of Ann is greater than the maximum speed of Bob.

   (A) Only I
   (B) Only II
   (C) Only III
   (D) Both II and III
   (E) I, II and III

2. A man finds a stationary bus at a distance of 500 feet ahead of him and starts running towards it at a speed of 30 feet per second. After 10 seconds, the bus starts moving at a speed of 25 feet per second away from the man. After another 20 seconds, the bus speeds off at a speed of 100 feet per second. If the man kept running towards the bus throughout this time, what was the shortest distance between the man and the bus?

   (A) 50 feet
   (B) 100 feet
   (C) 150 feet
   (D) 200 feet
   (E) 300 feet

3. Danny drove at an average speed of 60 miles per hour for some time and then at an average speed of 120 miles per hour for the rest of the journey. If he made no stops during the trip and his average speed for the entire journey was 100 miles per hour, for what fraction of the total time did he drive at 60 miles per hour?

   (A) $\frac{1}{5}$
   (B) $\frac{1}{3}$
   (C) $\frac{1}{2}$
   (D) $\frac{2}{3}$
   (E) $\frac{4}{5}$

4. A man drove from home at an average speed of 30 miles per hour to a rail station, from where he traveled by train to his office at an average speed of 60 miles per hour. If the entire distance, from his home to the office was 150 miles and the entire trip took 3.5

hours what is the distance from the rail station to his office? Assume there was a loss of time of 30 minutes while boarding the train at the station.

(A) 30 miles

(B) 60 miles

(C) 100 miles

(D) 120 miles

(E) 125 miles

5. A man swims a distance of 12 miles upstream, against the flow of the river, and the same distance downstream, with the flow of the river, to come back to his starting point. If the normal speed of the man in still water, without the effect of the flow of the river, is 4 miles per hour and the total journey took 8 hours, what is the speed of the river current in miles per hour?

   Assume that while going against the flow, the man's speed decreases by an amount equal to the river's speed. Similarly, while going with the flow, the speed of the man increases by an amount equal to the speed of the river.

   (A) 1.5

   (B) 2.0

   (C) 2.5

   (D) 3.0

   (E) 3.5

6. Buses A and B leave City X at a gap of 30 minutes, each traveling at a speed of 40 miles per hour. If another bus C, traveling at a speed of 60 miles per hour in the opposite direction to that of buses A and B on a parallel road, approaches them, what is the time difference between when Bus C crosses Bus A and then Bus B?

   (A) 12 minutes

   (B) 15 minutes

   (C) 20 minutes

   (D) 24 minutes

   (E) 30 minutes

7. If a person had walked at 9 miles per hour instead of 4 miles per hour, he would have covered 7.5 miles more in the same time. What distance did the person actually cover?

   (A) 1.5 miles

   (B) 3.0 miles

   (C) 4.5 miles

   (D) 6.0 miles

   (E) 13.5 miles

8. X and Y are two stations, 320 miles apart. Train P, running at 65 miles per hour, starts at 11:00 am from X and travels towards Y. After traveling for 4 hours, it stops for 30 minutes at Station Z, a station between X and Y. At 2:00 pm, another train, Q, running at 50 miles per hour, starts from Y and travels towards X. At what time do the trains meet?

(A) 3:10 pm

(B) 3:12 pm

(C) 3:20 pm

(D) 3:30 pm

(E) 3:39 pm

9. Buses ply between two cities, X and Y, traveling at the same uniform speed of 40 miles per hour. A man cycling along a straight road connecting the two cities notices that, every 10 minutes, a bus overtakes him, and every 5 minutes, he meets a bus coming from the opposite direction. If the cyclist maintains a constant speed and all buses have a constant time interval between them, what is the speed of the cyclist, in miles per hour?

(A) 6.7

(B) 12.0

(C) 13.3

(D) 15.0

(E) 16.6

10. Traveling at a speed of 2 miles per hour above the normal speed of s miles per hour, a man can cover a distance of d miles in 1 hour less time. Which of the following denotes the correct relationship between s and d ?

(A) $d = \frac{(s-1)^2}{2}$

(B) $d = \frac{s^2-2}{2}$

(C) $d = \frac{s^2+2}{2}$

(D) $d = \frac{s^2-2s}{2}$

(E) $d = \frac{s^2+2s}{2}$

11. Ten equally efficient men can complete a job in 10 days, whereas it takes 15 equally efficient women to complete it in 6 days. If 15 equally efficient men and 9 equally efficient women undertake to complete the job, how many days will it take for them to complete it?

(A) 3

(B) 4

(C) $6\frac{3}{4}$

(D) 12

(E) $18\frac{3}{5}$

12. Thirty-two equally efficient men and 32 equally efficient boys can together complete a certain job in 3 days. Six equally efficient men and 24 equally efficient boys can together complete the same work in 8 days. In how many days can 16 equally efficient men and 4 equally efficient boys complete the same work?

(A) 3

(B) 4

(C) 8

(D) 12

(E) 16

13. Eight equally efficient boys and 12 equally efficient men can complete a project of making 100 clay toys in 9 days. If each boy takes twice the time taken by a man to complete the project, in how many days will 12 such equally efficient men complete the same project?

(A) $4\frac{1}{2}$

(B) $6\frac{3}{4}$

(C) 8

(D) 12

(E) 16

14. Ann starts a job and does $\frac{7}{10}$ of the total work all by herself in 14 days. Joe joins her, and they both complete the remaining work after another 2 days. How many days would it take Joe to complete the entire work all by himself?

(A) 4

(B) 8

(C) 9

(D) 10

(E) 12

15. A boat sprung a leak in the middle of a journey. The crew realized the same after some time; by then some water had already entered. It was calculated that 5 crew members would take 6 hours to remove all the water that had entered the boat. However, if a 6th crew member joined them, the task would be accomplished in 4 hours. How long would it take to remove all the water from the boat if there were 8 crew members?

Assume that water keeps entering the leak at a constant rate and that all crew members are equally efficient. Also, once all the water is removed, the crew members will block the leak to prevent any further water from entering the boat.

(A) 2.4 hours
(B) 2.5 hours
(C) 3.2 hours
(D) 3.5 hours
(E) 3.6 hours

16. A certain number of equally efficient men can complete a job in 40 days. If there were 5 more men instead, the work could be completed in 10 days less than before. How many men were there in the beginning?

(A) 12
(B) 15
(C) 18
(D) 20
(E) 24

17. John does half as much work in one-sixth of the time that Pete takes to do the same work. If together they take 10 hours to complete the same work, how many hours would Pete take to complete the same work alone?

(A) 15
(B) 20
(C) 30
(D) 36
(E) 40

18. Amy and Bob can complete a piece of work in 45 days and 40 days, respectively. They began to work together, but Amy left after a few days. If Bob completed the remaining work in 23 days, after how many days did Amy leave?

(A) 7
(B) 8
(C) 9
(D) 10
(E) 12

19. A contractor employed 30 equally efficient men with the aim of constructing a building in 30 days. However, 10 days later, these men could complete only 25 percent of the work. How many more equally efficient men should the contractor employ to complete the work on time?

(A) 12
(B) 15

(C) 20

(D) 25

(E) 45

20. Amy and Bob can complete a work in 30 days and 20 days, respectively. Amy and Bob decide to work together. However, Bob left the work 3 days before it was completed. For how many days did the entire work last?

(A) 9

(B) 10

(C) $10\frac{4}{5}$

(D) 12

(E) $13\frac{4}{5}$

21. A sum of money is invested at simple interest, partly at 4% and remaining at 7% annual rates of interest. After 2 years, the total interest obtained was \$2,100. If the total investment is \$18,000, what was the sum of money invested at a 4% annual rate of interest?

(A) \$5,500

(B) \$6,000

(C) \$7,000

(D) \$10,500

(E) \$11,000

22. A man invested two equal sums of money in two banks at simple interest, one offering an annual rate of 10% and the other a rate of 20%. If the difference between the interests earned after 2 years is between \$120 and \$140, exclusive, which of the following could be the difference between the amounts earned for the same amounts of money, invested at the same rates of interest as above but at compound interest?

(A) \$130

(B) \$135

(C) \$138

(D) \$154

(E) \$162

23. A sum of money $\$p$, invested at compound interest with a $r\%$ annual rate of interest, and another sum of money $\$q$, invested at simple interest with a $(2r)\%$ annual rate of interest, accumulate the same interest after 2 years. Which of the following gives the correct relationship between p and q ?

(A) $q = p\left(1 + \frac{r}{100}\right)$

(B) $q = \frac{p}{r}\left(1 + \frac{r}{100}\right)^2$

(C) $q = \frac{p}{4}\left(1 + \frac{r}{100}\right)^2$

(D) $q = 4p\left(1 + \frac{r}{100}\right)$

(E) $q = \frac{p}{4}\left(2 + \frac{r}{100}\right)$

24. The difference between the interest accumulated if a sum of money is invested at 10% simple interest for 2 years and the same sum of money invested at 25% compound interest for 2 years is \$87. What is the sum of money invested?

(A) \$160

(B) \$170

(C) \$200

(D) \$220

(E) \$240

25. A man borrowed a certain sum of money from a friend and guaranteed to pay back the entire amount along with the interest accumulated in two equal annual payments of \$200 starting the next year. What is the sum of money the man had borrowed if interest is calculated at an annual 20% rate under compound interest?

(A) \$400.00

(B) \$397.20

(C) \$359.10

(D) \$332.40

(E) \$305.56

26. A sum of money \$$p$, invested at a 6% annual rate of interest under simple interest, fetches \$120 more interest than another sum of money, one-third of the previous sum, invested under simple interest at twice the rate of interest for two more years. If $p > 2,000$ and time t is an integer less than 7, which of the following could be the value of p ?

(A) \$2,400

(B) \$3,000

(C) \$3,600

(D) \$4,000

(E) \$4,500

27. A sum of money invested at a 20% annual rate of compound interest for 3 years amounts to the same value as the same sum of money invested at r% annual rate of compound interest for 2 years. What is the value of r ? Assume $\sqrt{1.2} = 1.1$.

(A) 24%

(B) 30%

(C) 32%

(D) 38%

(E) 40%

28. Amy invested half of her savings in simple interest and the rest in compound interest, both at the same annual rate of interest. If, after 2 years, she earned interest of $180 from the simple interest account and $189 from the compound interest account, what was the value of the total investment made by Amy?

(A) $600

(B) $750

(C) $900

(D) $1,200

(E) $1,800

29. A sum of money is invested at a fixed annual rate of interest under compound interest. If the amount at the beginning of the third year is $1,200 and the amount at the beginning of the fourth year is $1,440, what is the amount invested, rounded to the nearest ten?

(A) $640

(B) $690

(C) $800

(D) $830

(E) $1,040

30. An investment of $2,000 was made at a certain annual rate of interest under compound interest. If, after 6 years, the investment becomes $8,000, what is the rate of interest? Assume that $\sqrt[3]{2} = 1.25$.

(A) 20%

(B) 25%

(C) 33%

(D) 40%

(E) 50%

31. In a group of 12 boys, if two boys, aged 15 and 18, are replaced by three new boys, the average age of the group drops by 1 year. If the average age, in years, of the boys originally present in the group is an integer value between 12 and 15, inclusive, and the average of the three new boys is an integer, what is the average age of the three new boys?

(A) 9

(B) 10

(C) 11

(D) 12

(E) 13

32. What is the per-decade simple average rate of increase in population of a country if the population increased by 20% in the first decade, 25% in the second decade, and 10% in the third decade?

(A) 20.0%

(B) 21.7%

(C) 45.0%

(D) 56.4%

(E) 60.0%

33. The average weights of students in four teams, A, B, C, and D, are 30 kg, 35 kg, 40 kg, and 50 kg, respectively. If there are 40 students in all the teams taken together and their average weight is 40 kg, what is the maximum number of students in team C? There must be at least two students on each team.

(A) 30

(B) 31

(C) 32

(D) 33

(E) 34

34. The average age of a family of 7 members is 25 years. If the youngest member is 4 years old, what was the average age of the family, in years, when the youngest member was born?

(A) 18.5

(B) 21.0

(C) 23.6

(D) 24.5

(E) 27.0

35. The average price of a chair, a table, and a cot is $70; the average price of a table, a cot and a bookshelf is $85. If the price of the book-shelf is $70, what is the price of the chair?

(A) $25

(B) $30

(C) $35

(D) \$40

(E) \$45

36. In a group of friends, one-forth of them, working as managers and having an average salary of \$3,000, get a 50% increase in their salaries. The remaining friends, working as supervisors and having an average salary of \$1,800, get a 20% increase in their salaries. What was the average salary of the entire group after their salaries were increased?

(A) \$2,540

(B) \$2,678

(C) \$2,745

(D) \$3,240

(E) \$3,330

37. In a class, $\frac{3}{4}$ of the students, including $\frac{2}{3}$ of the girls have glasses. If $\frac{3}{5}$ of all students were girls, what percent of the students not having glasses were girls?

(A) 20%

(B) 33%

(C) 35%

(D) 40%

(E) 80%

38. In a club, 60 more than $\frac{1}{2}$ of all the members joined a swimming class and $\frac{1}{3}$ of those who joined the swimming class also joined basketball class. If $\frac{1}{3}$ of all the members in the club joined basketball class, how many members are in the club?

(A) 120

(B) 150

(C) 240

(D) 300

(E) 360

39. Three same sized cakes and four same sized pizzas were ordered by n number of boys, including John. They ate all cakes and pizzas. If each boy except John had the equal share of the cakes and the equal share of the pizzas, while John had double the share of the cakes that each of the other boys had and had half the share of the pizzas that each of the other boys had, which of the following expressions represents the amount of the cakes and the pizzas John ate?

(A) $\frac{6}{2n-1}+\frac{4}{n+1}$

(B) $\frac{3}{2n-1}+\frac{8}{n+1}$

(C) $\frac{6}{2n-1}+\frac{8}{n+1}$

(D) $\frac{3}{2n-1} + \frac{2}{n+1}$

(E) $\frac{6}{n+1} + \frac{4}{2n-1}$

40. Four members at a gym worked out for a month. The amount of time the four members worked out was in the ratio 3 : 4 : 5 : 6. If one of the four members worked out for 60 hours, which of the following cannot be the total number of hours the four members worked out?

(A) 180

(B) 216

(C) 270

(D) 300

(E) 360

41. In a class with 50 students, two groups, A and B, are formed with students in the ratio 2 : 3, respectively. The average weight of all the students is 40 kg. The average weight of the students in group A is 5 kg less than that of the other group. If Joe shifts from Group B to Group A, the average weights of the groups become equal. What is Joe's weight?

(A) 37 kg

(B) 42 kg

(C) 75 kg

(D) 80 kg

(E) 100 kg

42. Eight friends raised some funds among themselves to purchase a camera. Seven of them contributed \$120 each, while the other contributed \$21 more than the average of all eight friends. If they raised an amount just sufficient to purchase the camera, what was the price of the camera?

(A) \$645

(B) \$690

(C) \$778

(D) \$864

(E) \$984

43. In a company, the average salary of senior managers is \$4,500 higher than that of junior managers. If there are 80 junior managers and 100 senior managers, the average salary of all junior and senior managers together is how much higher than the average salary of the junior managers?

(A) \$2,000

(B) \$2,250

(C) \$2,500

(D) \$2,750

(E) \$3,000

44. A store sells laptops and mobiles. Each laptop is priced at \$1,200, and each mobile is priced at \$900. On a particular Tuesday morning, there were a total of 20 items, and the average price of them was \$1,080. After selling a few laptops (no mobiles were sold), the average price of the remaining items was \$1,000 just before the store was shut for the day. If no items were added to his stock, how many laptops were sold on Tuesday?

(A) 4

(B) 8

(C) 9

(D) 10

(E) 12

45. One train left Station X and traveled towards its destination at a speed of 65 miles per hour. Sometime later, another train going at a speed of 105 miles per hour left the same station but traveled in the opposite direction of the first train. After the first train had traveled for 14 hours, the two trains were 1,960 miles apart. How long after the first train did the second train start?

(A) 4 hours

(B) 4 hours 20 minutes

(C) 5 hours

(D) 6 hours 10 minutes

(E) 6 hours 30 minutes

46. A man traveling from one city to another traveled the first one-third of the distance at an average speed of 50 miles per hour, the remaining at a speed of 20 miles per hour, and took 4 hours and 48 minutes for the entire journey. What is the distance (in miles) between the two cities?

(A) 40

(B) 80

(C) 120

(D) 160

(E) 180

47. A train left city LA for city NY at 1:00 pm. At 5:00 pm, another train, traveling at 20 miles per hour faster than the first train, left a city that is 20 miles behind LA, along the same line joining LA and NY, and traveled towards NY. The second train caught up with the first train at 9:00 pm. What is the speed (in miles per hour) of the second train?

(A) 15

(B) 20

(C) 25

(D) 35

(E) 40

48. A man traveled distances in the ratio of 6 : 4 : 5 at speeds in the ratio of 2 : 3 : 5, respectively. If the total distance he traveled was 150 miles and the total time taken was 8 hours, what would have been the time, in hours, taken by the man had he traveled the entire distance at the greatest of the three speeds?

(A) 3.5

(B) 4.0

(C) 4.5

(D) 6.0

(E) 7.5

49. During a journey, a boy traveled at x miles per hour for the first d miles. Thereafter, he increased his speed by 20% for the remaining part of the journey. He realized that the remaining distance was 50% greater than the distance he had already covered. Had the boy covered the entire distance at x miles per hour, what would be the ratio of the total time taken to cover the entire distance for the original scenario and for the revised scenario?

(A) 3 : 5

(B) 4 : 7

(C) 13 : 25

(D) 9 : 10

(E) 1 : 1

50. A man drove a car and traveled 360 miles. Had the speed of the car been 4 miles per hour less, the time taken to travel the same distance would have been 3 hours longer than what it took initially. What was the actual speed, in miles per hour, of the car?

(A) 15

(B) 18

(C) 20

(D) 24

(E) 30

51. A man traveled x miles at an average speed of 5 miles per hour. He then went on to cover another 10 miles at an average speed of 10 miles per hour. If $1 \le x \le 10$, which of the following could be the percent by which his total travel time was greater than it would have been had he traveled at a constant rate of 10 miles per hour?

(A) 8%

(B) 23%

(C) 54%

(D) 67%

(E) 75%

52. A and B are assigned to complete a project. If A worked alone, it would take him 15 hours to finish the work. However, if B had worked alone, he would have required 18 hours to complete the project. If A works for the first 6 hours, and then both A and B work on the project for another 3 hours, how long, in hours, would it take for B to complete the remaining work?

(A) $4\frac{1}{5}$

(B) $5\frac{1}{3}$

(C) $5\frac{3}{5}$

(D) $6\frac{1}{2}$

(E) $6\frac{5}{6}$

53. Pumps A, B, and C are connected to a tank. A and B, working simultaneously, can fill the tank in 12 hours. A and C, working simultaneously, can fill the tank in 15 hours, while B and C, working simultaneously, can fill the tank in 20 hours. In how many hours would A, B, and C, working simultaneously, fill the tank?

(A) 5

(B) 8

(C) 10

(D) 15

(E) $15\frac{2}{3}$

54. A and B can together complete a work in 24 days, while C alone can complete the same work in 28 days. A first works with C for 4 days, following which he leaves, and B continues with C for another 10 days. Finally, B completes the remaining work in another 4 days. In how many days would B alone complete the entire work?

(A) 120

(B) 90

(C) 60

(D) 40

(E) 30

55. A and B start walking from a point P towards a point Q. The ratio of the speeds of A and B is 7 : 1. After reaching point Q, A immediately returns towards P and meets B, who is still walking. If the distance from P to Q is 200 meters, what is the distance, in meters, of the meeting point from point P?

(A) 33

(B) 50

(C) 67

(D) 120

(E) 150

56. Bob starts at a speed of 60 miles per hour from point A and goes towards point B, 150 miles away. At the same time, Chad starts from point B and goes towards point A at a speed of 30 miles per hour. After Chad has traveled 30 miles, he makes a 15-minute stop and then resumes his journey at his normal speed. At what distance, in miles, from point B do Bob and Chad meet?

(A) 30

(B) 45

(C) 60

(D) 90

(E) 105

57. A can complete a work in x days, and B can complete the same work in kx days, where k is a positive integer. If the number of days taken by A and B together to complete the work is an integer less than 5, how many possible values of the ordered set of (k, x) exist?

(A) Three

(B) Four

(C) Five

(D) Eight

(E) Nine

58. A and B start from points P and Q at speeds of 20 miles per hour and 15 miles per hour, respectively. A travels towards Q, while B travels towards P, and they meet at a point M on the way. Had B traveled at 18 miles per hour, A and B would have met at a point N, which is 6 miles farther from M. What is the distance, in miles, from Q to M?

(A) 25

(B) 30

(C) 57

(D) 76

(E) 133

59. Running on a 15-mile circular loop in the same direction, A ran at a constant rate of 12 miles per hour, and B ran at a constant rate of 10 miles per hour. If they began running from the same point on the loop at the same time, after how many hours did the integer number of laps completed by A equal exactly one more than the integer number of laps completed by B?

(A) 4.5

(B) 5.0

(C) 6.0

(D) 7.5

(E) 15.0

60. In a test, Jane solved 20 problem-solving questions and 10 data-sufficiency questions in the stipulated time. Jane always takes twice as much time to solve a data-sufficiency question as a problem-solving question. If she wants to complete 30 problem-solving questions and 20 data-sufficiency questions in the same stipulated time, by what percent should she reduce the time she takes to solve a question? Assume that all questions in the same category take the same amount of time.

(A) 33.3%

(B) 42.9%

(C) 50.0%

(D) 66.7%

(E) 75.0%

61. Brenda takes 45 strides per minute during morning walks (each stride is 2 feet long). However, while going to school, her speed is 50% higher, with each stride being 2.5 feet long. How many more strides does she take per minute while going to school than the number of strides per minute she takes during her morning walks?

(A) 9

(B) 18

(C) 36

(D) 45

(E) 54

62. With five identical servers working at a constant rate, an internet service provider processes 45,000 search requests per hour. How many identical additional servers does the service provider need to process at least 216,000 search requests in 3 hours? Assume that each search request takes the same amount of time.

(A) 1

(B) 3

(C) 4

(D) 5

(E) 8

63. Jamaican Blue Mountain coffee beans cost \$19.00 a pound, and Colombian coffee beans cost \$4.00 a pound. How many pounds of Jamaican Blue Mountain beans must be added to 8 pounds of Colombian beans to produce a blend that, if sold at \$8.40 a pound, results in a 20 percent profit?

(A) 1.5

(B) 2.0

(C) 2.5

(D) 3.0

(E) 4.0

64. During a trip, Francine traveled x percent of the total distance at an average speed of 40 miles per hour and the rest at an average speed of 60 miles per hour. In terms of x, what was Francine's average speed for the entire trip?

(A) $\frac{180 - x}{2}$

(B) $\frac{x + 60}{4}$

(C) $\frac{300 - x}{5}$

(D) $\frac{600}{115 - x}$

(E) $\frac{12,000}{x + 200}$

65. The average (arithmetic mean) cost per book for the 12 books on a certain table is k dollars. A book that costs 42 dollars replaces a book that costs 18 dollars. In terms of k, what will be the average cost per book, in dollars, for the books on the table?

(A) $k + 2$

(B) $k - 2$

(C) $12 + \frac{24}{k}$

(D) $12 - \frac{24}{k}$

(E) $12k - 6$

66. The average (arithmetic mean) score on a test taken by 10 students was x. If the average score of 5 of them was 8, what was the average score, in terms of x, for the remaining 5 students who took the test?

(A) $2x - 8$

(B) $x - 4$

(C) $8 - 2x$

(D) $16 - x$

(E) $8 - \frac{x}{2}$

67. Leo can buy a certain computer for x dollars in State A, excluding sales tax, where the sales tax is t percent, and he can buy the same computer for y dollars in State B, excluding sales tax, where the sales tax is T percent. If $xt > yT$, by what percent is the total cost (inclusive of sales tax) of the computer greater in State A than that in State B?

(A) $\$\left[(x - y) - \frac{(yt - xT)}{100}\right]$

(B) $\$\left[\frac{(100x - 100y + xt - yT)}{100}\right]$

(C) $\$\left[\frac{(x - y - xt + yT)}{100}\right]$

(D) $\$\left[\frac{(xt - yT)}{100}\right]$

(E) $\$\left[(x - y) - \frac{(xt - yT)}{100}\right]$

68. For telephone calls between two particular cities, a telephone company charges \$0.40 per minute if the calls are placed between 5:00 am and 9:00 pm and \$0.25 per minute if the calls are placed between 9:00 pm and 5:00 am. If the charge for a call between the two cities placed at 8:00 pm was \$30.00, how much would a call of the same duration have cost if it had been placed at 11:00 pm?

(A) \$5.00

(B) \$6.25

(C) \$12.50

(D) \$16.00

(E) \$21.00

69. The manager of a theater noted that for every 16 admission tickets sold, the theater sells 3 bags of popcorn at \$5.25 each, 4 sodas at \$4.50 each, and 2 candy bars at \$3.00 each. What is the approximate average (arithmetic mean) amount of all these snack sales per ticket sold?

(A) \$2.00

(B) \$2.10

(C) \$2.50

(D) \$3.50

(E) \$4.40

70. A man travels 720 miles in 8 hours, partly by air and partly by train. Had he traveled all the way by air, he would have saved $\frac{4}{5}$ of the time he was on the train and arrived at his destination 4 hours early. What is the distance, in miles, traveled by the man by air?

(A) 480
(B) 520
(C) 540
(D) 600
(E) 660

71. A man covered one-third of the distance to his destination at 20 miles per hour, one-half of the remaining distance at 12 miles per hour, and the remaining distance at 40 miles per hour. What was his average speed, in miles per hour, for the whole journey?

(A) 12.1
(B) 16.0
(C) 18.9
(D) 22.5
(E) 30.7

72. Jane lives x floors above the ground floor of a high-rise building. It takes her 30 seconds per floor to walk down the steps and 2 seconds per floor to ride the elevator. If it takes Jane the same amount of time to walk down the steps to the ground floor as it does to wait for the elevator for 7 minutes and ride down, then on which floor does Jane live?

(A) 4
(B) 7
(C) 14
(D) 15
(E) 16

73. Mary passed a certain gas station on a highway while traveling west at a constant speed of 50 miles per hour. Then, 15 minutes later, Paul passed the same gas station while traveling west at a constant speed of 60 miles per hour. If both drivers maintained their speeds and remained on the highway for at least 2 hours, how long, in minutes, after Paul passed the gas station did he catch up with Mary?

(A) 30
(B) 45
(C) 60
(D) 75
(E) 85

74. A man, while traveling in a car from City X to City Y to pick up his son, overtakes a bus traveling in the same direction at 10:00 a.m. On reaching City Y at 2:00 pm, the man picks up his son and turns back towards City X without any loss of time. If, on the way back, the man meets the same bus that he had overtaken earlier at 3:00 pm, when would the bus reach City Y? Assume that the man and the bus travel throughout at uniform speeds.

(A) 3:40 pm
(B) 3:50 pm
(C) 4:20 pm
(D) 4:30 pm
(E) 4:40 pm

75. Two hours after Jack started walking from Point P to Point Q, Bob started walking along the same road in the same direction as Jack's. The distance between P and Q is 25 miles. If Jack's walking rate was 3 miles per hour and Bob's was 5 miles per hour, how many miles before Point Q did Bob overtake Jack?

(A) 9
(B) 10
(C) 12
(D) 15
(E) 16

76. Pumping alone at their respective constant rates, one inlet pipe fills an empty tank to $\frac{1}{2}$ of its capacity in 3 hours, and a second inlet pipe fills the same empty tank to $\frac{2}{3}$ of its capacity in 6 hours. How many hours will it take both pipes, pumping simultaneously at their respective constant rates, to fill the empty tank to capacity?

(A) 3.25
(B) 3.60
(C) 4.20
(D) 4.40
(E) 5.52

77. Running at their respective constant rates, Machine X takes 2 days longer to produce w widgets than Machine Y. At these rates, if the two machines together produce $\frac{5}{4}w$ widgets in 3 days, how many days would it take Machine X alone to produce $2w$ widgets?

(A) 4
(B) 6
(C) 8
(D) 10
(E) 12

78. A water sample has 10 percent impurities present. If after each purification process, the impurity level reduces by 60 percent of the present value, what is the minimum number of times that the water sample needs to be purified in order to reduce the impurities present to at most 1 percent?

(A) Two

(B) Three

(C) Four

(D) Five

(E) Six

79. On a certain day, Tim invested \$1,000 for 2 years at 10 percent annual interest, compounded annually, and Lana invested \$2,000 for 2 years at 5 percent annual interest, compounded annually. The interest earned by Tim's investment was how much greater than the interest earned by Lana's investment?

(A) \$5

(B) \$15

(C) \$50

(D) \$100

(E) \$105

80. For telephone calls between two particular cities, a telephone company charges \$0.40 per minute if the calls are placed between 5:00 am and 9:00 pm and \$0.25 per minute if the calls are placed between 9:00 pm and 5:00 am. If the charge for a call between the two cities placed at 1:00 pm was \$10.00, how much would a call of the same duration have cost if it had been placed at 11:00 pm?

(A) \$3.75

(B) \$6.25

(C) \$9.85

(D) \$10.00

(E) \$16.00

81. 4-digit numbers are formed using the digits 3, 4, 8, and 9. If the repetition of digits is allowed, how many of these numbers would be even?

(A) 56

(B) 85

(C) 110

(D) 128

(E) 170

82. How many different license numbers can be allotted using 3 letters and 3 digits, if all the letters and all the digits must be together? Repetition of the letters and the digits is allowed.

(A) $C_3^{26} \times C_3^{10}$

(B) $2 \times C_3^{26} \times C_3^{10}$

(C) $26^3 \times 10^2$

(D) $2 \times 26^3 \times 10^3$

(E) $26^3 \times 10^3$

83. How many 5-digit numbers can be formed containing exactly one 3? Other digits can be repeated.

(A) $4 \times C_4^9$

(B) $5 \times C_4^9$

(C) $4 \times 8 \times 9^4$

(D) $4 \times 9^4 + 9^3$

(E) $9^4 + 4 \times 8 \times 9^3$

84. There are 8 different locks, with exactly one key for each lock. All keys are mixed up. What is maximum number of trials needed to determine which key belongs to which lock?

(A) 14

(B) 28

(C) 56

(D) 64

(E) 128

85. Captain of a team can select only n players out of $(2n + 1)$ players. If the total number of ways, he can select n players is 126, what is the value of n?

(A) 2

(B) 3

(C) 4

(D) 5

(E) 6

86. In a polygon of 8 equal sides, how many different triangles can be drawn using the vertices of the polygon as the vertices of the triangles?

(A) 24

(B) 28

(C) 56

(D) 64

(E) 72

87. A puzzle can be solved by choosing one cube from the puzzle box1 containing 10 cubes, and two cubes from the puzzle box2 containing 5 cubes. Out of how many ways can the puzzle be solved?

(A) 20
(B) 64
(C) 100
(D) 128
(E) 200

88. How many numbers less than 1,000 can be formed out of digits 3, 4, 5, and 6? The repetition of digits is allowed.

(A) 16
(B) 64
(C) 84
(D) 128
(E) 340

89. How many 2-member teams can be formed from 5 individuals provided 2 specific individuals together must not be in the team?

(A) 3
(B) 6
(C) 8
(D) 9
(E) 10

90. In how many ways can 2 boys and 2 girls be chosen from a class comprising 10 boys and 12 girls?

(A) 111
(B) 240
(C) 480
(D) 2,970
(E) 11,880

91. How many 5-letter words can be formed if the first and the last letters are vowels?

(A) $2 \times 5^2 \times 26^3$
(B) $5^2 \times 26^3$
(C) $2 \times 5^2 \times 21^3$
(D) $5^2 \times 21^3$
(E) $5^3 \times 21^2$

92. A criminal speaks either the truth or the lie for each of three questions. To know the fact, he must speak the truth for each question. In how many ways can be hide the truth?

(A) 2
(B) 3
(C) 7
(D) 8
(E) 26

93. There are 5 differently colored cubes. In how many ways can a child arrange them in a row taking only 3 cubes at a time, such that the arrangements make unique patterns?

(A) 6
(B) 10
(C) 15
(D) 60
(E) 120

94. How many distinct words with or without meaning can be formed using all the letters of the word APPEALING?

(A) $\frac{9!}{2! \times 2!}$
(B) $\frac{9!}{2}$
(C) 7!
(D) 9!
(E) $9! \times 2! \times 2!$

95. In the question given above, how many words will have L and G together (in any order)?

(A) $\frac{8!}{2! \times 2!}$
(B) $\frac{8!}{2!}$
(C) 8!
(D) $8! \times 2!$
(E) $\frac{9!}{2! \times 2!}$

96. In the question given above, how many words are possible when all the vowels occur together?

(A) 72
(B) 144
(C) 1,728
(D) 3,456
(E) 4,320

97. A school wishes to select 11 students for an annual parade out of 15 students with similar height. In how many ways can the school make a selection, if the order of selection does not matter?

(A) 165

(B) 1,365

(C) 11!

(D) $1,365 \times (11!)$

(E) $(11!)^2$

98. A UN peacekeeping force wants to form a delegation of 5 generals. It wants to choose 2 out of 5 generals from the UK and 3 out of 6 generals from France. In how many ways can the selection be done?

(A) 100

(B) 200

(C) 462

(D) $100 \times (5!)$

(E) 11!

99. A casting director wishes to cast 5 actors out of 13 shortlisted TV actors for his movies. Out of 13 actors, 5 actors act in soaps, and 8 actors act in reality TV shows. In how many ways can the director select the actors such that there are 2 actors from soaps and 3 actors from reality TV shows?

(A) 66

(B) 110

(C) 356

(D) 560

(E) 6,720

100. A soccer selection panel has to select 15 probable players out of 17 players, such that the captain is always selected and an injured player is not selected. In how many ways can the selection be done?

(A) 15

(B) 120

(C) 136

(D) 560

(E) 680

101. How many 3-element subsets of a set {A, B, C, D, E} are possible such that they do not contain the subset of elements A, C, & D?

(A) 5

(B) 6
(C) 9
(D) 50
(E) 54

102. There are 10 male and 8 female dancers in a club. If the club is to form two groups, each having 1 male and 1 female dancer, in how many ways can it be done?

(A) 143
(B) 160
(C) 1,260
(D) 2,520
(E) 5,040

103. A cricket team of 11 players is to be chosen from 8 batsmen, 6 bowlers and 2 wicket-keepers. In how many ways can the team be chosen if there must be at least 4 batsmen, at least 4 bowlers and exactly 1 wicket-keeper?

(A) 140
(B) 672
(C) 840
(D) 1,652
(E) 6,504

104. In how many ways can a person choose between 4 types of salads and 6 types of toppings for a pizza?

(A) 10
(B) 24
(C) 48
(D) 96
(E) 1,024

105. How many 7-digit telephone numbers starting with 2431 are possible, so that the string of the last three digits neither starts nor ends with '0'?

(A) 648
(B) 729
(C) 810
(D) 900
(E) 1,000

106. In how many ways can 3 men and 2 women be selected from 4 men and 3 women?

(A) 5

(B) 6

(C) 12

(D) 15

(E) 20

107. How many three-digit numbers without the repetition of digits can be formed using the digits 0, 2, 3, and 5 so that they are even?

(A) 6

(B) 10

(C) 15

(D) 20

(E) 40

108. John has four friends. In how many ways can he invite one or more of his friends to a party?

(A) 8

(B) 15

(C) 16

(D) 31

(E) 63

109. In how many ways can the letters of the word INDIANA be arranged to form meaningful or meaningless words?

(A) 630

(B) 700

(C) 729

(D) 800

(E) 810

110. A 5-member team is to be formed for a debate competition and the speaker for the team is to be decided from the chosen members. If there are 8 members from whom the team is to be chosen, in how many ways can it be formed and the speaker is decided?

(A) 100

(B) 200

(C) 280

(D) 300

(E) 360

111. In how many ways can 4 boys and 4 girls be seated in a line so that a particular boy and a particular girl do not sit together?

(A) $6 \times 7!$

(B) $8! - 7!$

(C) $12 \times 7!$

(D) $3 \times 8!$

(E) $48 \times 7!$

112. A school examination consists of 5 questions, each having two parts. In how many ways can a student attempt at least one part of any of the questions?

(A) 624

(B) 625

(C) 1,023

(D) 1,024

(E) 4,095

113. Given that

A = The number of ways 11 players can be selected out of 25 players, and

B = The number of ways 14 players can be selected out of 25 players

Which of the following statements is correct?

(A) $A = \frac{14}{11} \times B$

(B) $A = \frac{11}{14} \times B$

(C) $A = 3 \times B$

(D) $A = \frac{1}{3} \times B$

(E) $A = B$

114. Given that

A = The number of ways 4 marbles can be placed in three containers, and

B = The number of ways 3 marbles can be placed in four containers

Which of the following statements is correct?

(A) $A = \frac{3^4}{4^3} \times B$

(B) $A = \frac{4^3}{3^4} \times B$

(C) $A = \frac{3}{4} \times B$

(D) $A = \frac{4}{3} \times B$

(E) $A = B$

115. How many 4-digit numbers less than 2,000 can be formed using the digits '0,' '1' and '2,' given that the repetition of digits is allowed?

(A) 9

(B) 12

(C) 18

(D) 27

(E) 81

116. In how many ways can a 3-member team be selected from 3 men and 2 women so that at least one woman is always in the team?

(A) 3

(B) 6

(C) 9

(D) 18

(E) 27

117. In how many ways can 3 boys and 3 girls be seated in a line so that the boys and girls sit in alternate positions?

(A) 3

(B) 6

(C) 12

(D) 36

(E) 72

118. In how many ways can 4 boys and 3 girls be seated on a circular table so that a particular boy and a particular girl sit together?

(A) 120

(B) 240

(C) 600

(D) 720

(E) 1,440

119. In how many ways can one mathematics book from 6 different mathematics books and one physics book from 3 identical physics books be selected?

(A) 1

(B) 3

(C) 6

(D) 9

(E) 18

120. How many spy-codes are possible taking all the symbols □ ⊵ ⊗ ⊘ ◊ together such that no spy-code should together have a □ at the beginning and a ◊ at the end?

(A) 20

(B) 60

(C) 84

(D) 114

(E) 120

121. Each participant in a certain study was assigned a sequence of 3 different letters from the set {A, B, C, D, E, F, G, H}. If no sequence was assigned to more than one participant and if 36 of the possible sequences were not assigned, what was the number of participants in the study? (Note: the sequence A, B, E is different from the sequence E, B, A).

(A) 20

(B) 92

(C) 300

(D) 372

(E) 476

122. A box contains 30 balls, of which 10 are black, 10 are green, and 10 are yellow. If two balls are drawn from the box at random, what is the probability that both balls will be green?

(A) 3/29

(B) 1/9

(C) 1/3

(D) 2/3

(E) 1

123. An unbiased coin is tossed thrice. What is the probability of getting at least one Head?

(A) $\frac{1}{8}$

(B) $\frac{3}{8}$

(C) $\frac{1}{2}$

(D) $\frac{7}{8}$

(E) 1

124. A class of 36 students has an equal number of boys and girls. Four students are randomly selected to form a cultural committee. If the first three selected students are boys, what is the probability that the 4th student is also a boy?

(A) 1/33

(B) 1/15

(C) 5/12

(D) 5/11

(E) 12/13

125. A delegation of 9 diplomats comprises 3 Japanese and 6 Chinese. If 3 diplomats are to be chosen at random from the delegation, what is the probability that at least 2 diplomats are Japanese?

(A) 1/12

(B) 3/14

(C) 19/84

(D) 1/3

(E) 1/2

126. John, Peter, and Harry are given an assignment to collectively solve a problem. The probabilities of John, Peter and Harry to solve the problem individually are 4/5, 2/3, and 3/4, respectively. What is the probability that the problem will not be solved?

(A) 1/60

(B) 13/60

(C) 2/3

(D) 47/60

(E) 59/60

127. John and Harry are among the five promising soccer players in the school. If 3 out of 5 players are to be chosen at random, what is the probability that both John and Harry would be chosen?

(A) 1/10

(B) 1/5

(C) 3/10

(D) 2/5

(E) 3/5

128. Three actors are to be chosen out of five—Jack, Steve, Elad, Suzy, and Ali. What is the probability that Jack and Steve would be chosen, but Suzy would be left out?

(A) $\frac{1}{5}$

(B) $\frac{2}{5}$

(C) $\frac{3}{5}$

(D) $\frac{7}{10}$

(E) $\frac{9}{10}$

129. A box contains p number of silver coins and q number of gold coins. One coin is randomly drawn from the box but not replaced; thereafter, a second coin is randomly drawn. What is the probability that the first coin drawn is silver and the second is gold?

(A) $\left(\frac{p}{p+q}\right) \times \left(\frac{q}{p+q}\right)$

(B) $\left(\frac{p}{p+q}\right) \times \left(\frac{p-1}{p+q-1}\right)$

(C) $\left(\frac{pq}{p+q}\right)$

(D) $\left(\frac{p-1}{p+q}\right) \times \left(\frac{q-1}{p+q}\right)$

(E) $\left(\frac{p}{p+q}\right) \times \left(\frac{q}{p+q-1}\right)$

130. A super mall issues coupons to its customers at the entrance of the mall. The coupons are numbered from 100 to 400, inclusive. Customers who have coupons with the numbers divisible by '2' will win gifts. What is the probability that a random customer wins a gift?

(A) 149/300

(B) 150/301

(C) 151/301

(D) 1/2

(E) 151/300

131. There are two groups—Group A comprises of only 50 married men, while Group B comprises of only 40 married women. Between these groups, there are only 5 husband-wife pairs. If one man and one woman is randomly chosen from each group, what is the probability that they make a husband-wife pair?

(A) 1/400

(B) 1/80

(C) 1/18

(D) 1/9

(E) 9/40

132. A box contains 4 black and a few green balls. If two balls are drawn from the box at random, and the probability that both balls are black is 1/6, how many green balls are in the box?

(A) 3

(B) 4

(C) 5

(D) 6

(E) 9

133. A toolbox contains 10 spanners, out of which 3 are defective. If a mechanic randomly selects 2 spanners at a time, what is the probability that at least one spanner will be defective?

(A) 2/9

(B) 3/10

(C) 7/15

(D) 8/15

(E) 14/15

134. A class comprises of 20 students, out of which 15 are boys and the rest are girls. If a committee of 3 students is to be formed, what is the probability that the committee will have only one boy?

(A) 5/38

(B) 3/10

(C) 35/76

(D) 41/76

(E) 7/10

135. The probability that a man can shoot and hit a target is 1/100. If the man fires 100 rounds, what is the probability that all the 100 shots do not miss the target?

(A) $\left(\frac{1}{10}\right)^{200}$

(B) $\frac{10^{200}-1}{10^{200}}$

(C) $\left(\frac{1}{100}\right)^{2}$

(D) $\left(\frac{99}{100}\right)^{100}$

(E) $\frac{(100^{100} - 99^{100})}{100^{100}}$

136. A gambler bets everyday on soccer games. If the probability of his losing the bet each day is 0.4, what is the probability that he wins on only 2 days out of 3 days he bets?

(A) 0.096

(B) 0.144

(C) 0.16

(D) 0.36

(E) 0.432

137. A bag has 20 blue and 10 red balls. What is the probability of drawing two balls simultaneously so that they are not of the same color?

(A) 0.13

(B) 0.26

(C) 0.36

(D) 0.46

(E) 1

138. Probability that a missile will hit its target is 0.6. If three missiles are fired at the target, what is the probability of hitting the target at least once?

(A) 0.096

(B) 0.6

(C) 0.936

(D) 0.964

(E) 1

139. Three balls are drawn successively without repetition from a bag containing 4 white, 3 red and 3 blue balls. What is the probability that the 3 balls are of different colors?

(A) $\frac{1}{20}$

(B) $\frac{1}{5}$

(C) $\frac{2}{9}$

(D) $\frac{5}{21}$

(E) $\frac{3}{10}$

140. There are two bags, one contains 3 white and 2 black balls, while the other contains 4 white and 5 black balls. What is the probability that if one ball is drawn from each bag, they will be of the same color?

(A) $\frac{11}{45}$

(B) $\frac{22}{45}$

(C) $\frac{1}{2}$

(D) $\frac{3}{4}$

(E) 1

141. A box contains 3 blue and 2 red balls, while another box contains 4 green and 2 white balls. A coin is tossed. If 'heads' appears, a ball is drawn from the first box and if a tail appears, a ball is drawn from the second box. What is the probability of drawing a blue ball?

(A) 0.1

(B) 0.2

(C) 0.3

(D) 0.6

(E) 0.7

142. Three balls are simultaneously drawn from a box containing 5 yellow and 4 green balls. What is the probability that the 3 balls are not all of the same color?

(A) 0.167

(B) 0.33

(C) 0.83

(D) 0.89

(E) 1

143. Given that

- A = Probability of drawing 2 red balls from a bag containing 12 red and 10 blue balls
- B = Probability of drawing 2 blue balls from a bag containing 12 red and 10 blue balls
- C = Probability of drawing one red and one blue ball from a bag containing 12 red and 10 blue balls

Which of the following statement is correct?

(A) $A < B < C$

(B) $A > C > B$

(C) $A > B > C$

(D) $C > A > B$

(E) $A < C < B$

144. Given that

- A = Probability of getting all heads when 3 coins are tossed
- B = Probability of getting all tails when 3 coins are tossed
- C = Probability of getting 2 heads and one tail when 3 coins are tossed

Which of the following statement is correct?

(A) $A = 3B = 6C$

(B) $A = C = 3B$

(C) $A = B = C$

(D) $C = 3A = 3B$

(E) $3C = A = B$

145. What is the probability of selecting x from the set $\{-2, 0, 2, 4, 6\}$ such that $x^2 \leq 16$?

(A) $\frac{1}{5}$

(B) $\frac{2}{5}$

(C) $\frac{3}{5}$

(D) $\frac{4}{5}$

(E) $\frac{5}{6}$

146. What is the probability of obtaining an odd number or a prime number on one throw of a dice?

(A) $\frac{1}{6}$

(B) $\frac{1}{3}$

(C) $\frac{2}{3}$

(D) $\frac{3}{4}$

(E) $\frac{5}{6}$

147. What is the probability of getting a number on the first throw greater than that on the second throw when a dice is thrown twice?

(A) $\frac{1}{6}$

(B) $\frac{5}{12}$

(C) $\frac{7}{12}$

(D) $\frac{2}{3}$

(E) $\frac{5}{6}$

148. What is the probability of selecting two cards successively with replacement from five cards numbered '1' to '5' such that the sum of the numbers is '8'?

(A) $\frac{3}{25}$

(B) $\frac{2}{5}$

(C) $\frac{3}{5}$

(D) $\frac{4}{5}$

(E) $\frac{5}{7}$

149. What is the probability of selecting two numbers, one from Set A {1, 3, 6} and the other from Set B {2, 3, 8} such that the sum of the numbers is 9?

(A) $\frac{1}{2}$

(B) $\frac{2}{9}$

(C) $\frac{1}{9}$

(D) $\frac{1}{10}$

(E) $\frac{1}{12}$

150. How many red balls are there in a bag that contains 24 balls of red, blue and green colors, given that the probability of drawing a blue ball is $\frac{1}{4}$ and the probability of drawing a green ball is $\frac{2}{3}$?

(A) 1

(B) 2

(C) 3

(D) 4

(E) 5

151. A box contains 5 black and some green balls. If two balls are drawn from the box at random, and the probability that both balls are green is 1/6, how many green balls are in the box?

(A) 3
(B) 4
(C) 5
(D) 6
(E) 8

152. Each person who attended a company meeting either took only one glass of juice or only one glass of wine, or one glass of each. Of those who attended the meeting, 62 percent took one glass of wine, and 53 percent did not take juice. If the total number of people who attended the meeting is 100, how many glasses were used to serve the drinks if each glass was to be used only once?

(A) 56
(B) 71
(C) 91
(D) 100
(E) 109

153. Last month, 54 cars in a car workshop had body repairs, and 52 cars were repainted. If 24 cars in the workshop had body repairs but were not repainted last month, how many cars were repainted but did not have body repairs?

(A) 10
(B) 16
(C) 18
(D) 20
(E) 22

154. In a school, there are a total of 66 students who have to study at least one of three subjects: Mathematics, Physics, and Economics. There are 28 students who did not take either Physics or Economics; 15 students, who did not take either Mathematics or Economics; 45 students who did not take up Economics; and 30 students, who did not take Mathematics. The number of students who did not take either Mathematics or Economics, exceeded the number of students who did not take Mathematics or Physics by 2. If 5 students took all three subjects, how many students took both Economics and Mathematics?

(A) 1
(B) 5

(C) 6
(D) 7
(E) 10

155. In a survey of 63 people, 33 subscribed to magazine A, 30 subscribed to magazine B, and 17 subscribed to magazine C. For any two of the magazines, 9 subscribed to both magazines. If 5 people in the survey did not subscribe to any of the three magazines, how many subscribed to all three magazines?

(A) 10
(B) 9
(C) 7
(D) 5
(E) 2

156.

Factor	Percent of prospects
Brand name	66%
Safety	44%
Price	38%

The table above lists three factors prospective car buyers consider when buying a car and the percent distribution of the 800 prospective car buyers who cited these three factors as vital, as per a survey. If 25% of the prospects cited both "Brand name" and "Safety" as vital factors, what is the maximum possible number of prospects who cited "Price," but neither "Brand name" nor "Safety" as vital factors?

(A) 120
(B) 136
(C) 160
(D) 240
(E) 300

157. A survey shows that 63% of people in New York like pasta, whereas 76% like burgers. If x% of the people like pasta and burgers both, which of the following statements is correct?

I. $39 \le x \le 63$

II. The maximum percent of people who like only one between pasta and burgers is 61%.

III. The maximum percent of people who like neither pasta nor burgers is 24%.

(A) Only I
(B) Only II
(C) Both I and II
(D) Both II and III
(E) I, II and III

158. On a trip to a theme park, 30 children take rides on a merry-go-round, a roller coaster, and a toy train. A child can take multiple rides, but at most one ride of any type. There are 11 children who ride the merry-go-round and 14 children who ride the roller coaster. There are 10 children who take two different rides, 2 children who take all three rides, and all children take at least one ride. If each ride costs \$5, what is the total cost of all the rides taken by the 30 children?

(A) \$44
(B) \$124
(C) \$150
(D) \$210
(E) \$220

159. In a class, of the total number of students, 36% failed the English test and 30% failed the mathematics test. What is the sum of the maximum and minimum percentages of students who passed in both English and mathematics?

(A) 34
(B) 54
(C) 66
(D) 98
(E) 120

160. A, B, and C are three TV channels. A survey shows that 30%, 20%, and 85% of the people in a locality watch channels A, B, and C, respectively. If 20% of people watch exactly two of the three channels and 5% of people watch none of the three channels, what percent of people watch all three channels?

(A) 0
(B) 5
(C) 10
(D) 15
(E) 20

Chapter 13

Answer Key: Word Problems

(1) C

(2) B

(3) B

(4) D

(5) B

(6) A

(7) D

(8) B

(9) C

(10) E

(11) B

(12) C

(13) D

(14) D

(15) A

(16) B

(17) E

(18) C

(19) B

(20) E

(21) C

(22) D

(23) E

(24) E

(25) E

(26) B

(27) C

(28) E

(29) D

(30) B

(31) C

(32) B

(33) D

(34) D

(35) A

(36) C

(37) E

(38) A

(39) E

(40) D

(41) E

(42) E

(43) C

(44) B

(45) A

(46) C

(47) D

(48) C

(49) D

(50) D

(51) B

(52) A

(53) C

(54) E

(55) B

(56) B

(57) D

(58) C

(59) B

(60) B

(61) A

(62) B

(63) B

(64) E

(65) A

(66) A

(67) E

(68) E

(69) C

(70) C

(71) C

(72) D

(73) D

(74) E

(75) B

(76) B

(77) E

(78) B

(79) A

(80) B

(81) D

(82) D

(83) E

(84) B

(85) C

(86) C

(87) C

(88) C

(89) D

(90) D

(91) B

(92) C

(93) D

(94) B

(95) B

(96) E

(97) B

(98) B

(99) D

(100) A

(101) C

(102) D

(103) D

(104) B

(105) C

(106) C

(107) B

(108) B

(109) A

(110) C

(111) A

(112) C

(113) E

(114) A

(115) D

(116) C

(117) E

(118) B

(119) C

(120) D

(121) C

(122) A

(123) D

(124) D

(125) C

(126) A

(127) C

(128) A

(129) E

(130) C

(131) A

(132) C

(133) D

(134) A

(135) E

(136) E

(137) D

(138) C

(139) B

(140) B

(141) C

(142) C

(143) D

(144) D

(145) A

(146) C

(147) B

(148) A

(149) B

(150) B

(151) B

(152) E

(153) E

(154) C

(155) D

(156) A

(157) E

(158) E

(159) D

(160) C

Chapter 14

Solutions: Word Problems Practice Questions

1. Distance traveled by Ann is between $2 \times 2 = 4$ miles and $2 \times 3 = 6$ miles.

   Time taken by Ann is between 1 hour 30 minutes and 1 hour 45 minutes, i.e. $1\frac{1}{2}$ hours and $1\frac{3}{4}$ hours.

   Thus, we have

   $$\text{Maximum speed of Ann} = \frac{\text{Maximum distance}}{\text{Minimum time}} = \frac{6}{1\frac{1}{2}} = 4 \text{ miles per hour}$$

   $$\text{Minimum speed of Ann} = \frac{\text{Minimum distance}}{\text{Maximum time}} = \frac{4}{1\frac{3}{4}} = \frac{16}{7} = 2.28 \text{ miles per hour}$$

   Since Bob traveled 3 miles farther than Ann did, the total distance covered by Bob is $2 \times 3 = 6$ miles greater than that covered by Ann.

   Thus, distance traveled by Bob is between 10 miles and 12 miles.

   Time taken by Bob = 4 hours.

   Thus, we have

   $$\text{Minimum speed of Bob} = \frac{\text{Minimum Distance}}{\text{Time}} = \frac{10}{4} = 2.5 \text{ miles per hour}$$

   $$\text{Maximum speed of Bob} = \frac{\text{Maximum Distance}}{\text{Time}} = \frac{12}{4} = 3 \text{ miles per hour}$$

   Working with the statements:

   - Statement I:
     Thus, there is an overlap in the range of speeds of Ann and Bob.
     Thus, we may have Ann (4 miles per hour) faster than Bob (3 miles per hour) OR Bob (3 miles per hour) faster than Ann (2.28 miles per hour).
     Thus, Statement I is not true.
   - Statement II:
     Ann's minimum speed is 2.28 miles per hour, while her maximum speed is 4 miles per hour.
     Thus, Ann's speed does not lie between 2 and 3 miles per hour.
     Thus, Statement II is not true.
   - Statement III:
     Maximum speed of Ann is 4 miles per hour, which is greater than the maximum speed of Bob, which is 3 miles per hour.
     Thus, Statement III is true.

   The correct answer is option C.

2. Initial distance between the man and the bus = 500 feet.

   Distance covered by the man in the 10 seconds when the bus was stationary

   $= 30 \times 10 = 300$ feet.

   Distance remaining between the man and the bus = $500 - 300 = 200$ feet

   For the next 20 seconds, the bus was also moving at 25 feet per second.

   Thus, the relative speed between the man and the bus

= 30 - 25 = 5 feet per second

Since the man is still faster than the bus, the man will gain on the bus.

Distance the man gained on the bus

= 5 × 20 = 100 feet.

Thus, distance remaining between the man and the bus = 200 - 100 = 100 feet.

Since, after this, the bus speeds away at a much higher speed (100 ft/second), the distance between the man and the bus would increase.

Thus, the shortest distance between the man and the bus = 100 feet.

The correct answer is option B.

3. Let Danny travel at 60 miles per hour for x hours and at 120 miles per hour for y hours.

Total distance traveled by Danny $= (60x + 120y)$ miles

Total time taken by Danny $= (x + y)$ hours

Since his average speed is 100 miles per hour, we have

$\frac{60x + 120y}{x + y} = 100$

$=> 60x + 120y = 100x + 100y$

$=> \frac{x}{y} = \frac{1}{2}$

Thus, the fraction of the time for which Danny traveled at 60 miles per hour $= \frac{1}{1 + 2} = \frac{1}{3}$

The correct answer is option B.

Alternate approach:

You can deal with this question using a single variable.

Let Danny travel at 60 miles per hour for 1 hours and at 120 miles per hour for p hours.

Total distance traveled by Danny $= (60 + 120p)$ miles

Total time taken by Danny $= (1 + p)$ hours

Since his average speed is 100 miles per hour, we have

$\frac{60 + 120p}{1 + p} = 100$

$=> 60 + 120p = 100 + 100p$

$=> 20p = 40 => p = 2$

Thus, fraction of the time for which Danny traveled at 60 miles per hour $= \frac{1}{1 + p} = \frac{1}{1 + 2} = \frac{1}{3}$

4. Let the man traveled distance d miles by car and the remaining $(150 - d)$ miles by train.

Time taken by car $= \frac{d}{30}$ hours.

Time taken by train = $\frac{150 - d}{60}$ hours.

Time taken while boarding = $\frac{1}{2}$ hours.

Thus, total time = $\left(\frac{d}{30} + \frac{150 - d}{60} + \frac{1}{2}\right)$ hours = $\left(\frac{180 + d}{60}\right)$ hours.

Thus, we have

$\frac{180 + d}{60} = 3.5$

=> $d = 30$

Thus, the required distance = 150 - 30 = 120 miles.

The correct answer is option D.

5. Let the speed of the current be r miles per hour.

Thus, speed of the man while going upstream = $(4 - r)$ miles per hour

Speed of the man while going downstream = $(4 + r)$ miles per hour

Thus, time taken to cover 12 miles upstream = $\left(\frac{12}{4 - r}\right)$ hours

Time taken to cover 12 miles downstream = $\left(\frac{12}{4 + r}\right)$ hours

Since the total time for the trip was 8 hours, we have

$\frac{12}{4 - r} + \frac{12}{4 + r} = 8$

Solving the above equation is difficult, so it is best to use the options to determine the value of r.

We observe that, for $r = 2$, we have $4 - r = 2 \Rightarrow \frac{12}{4 - r} = 6$, and also,

$4 + r = 6 \Rightarrow \frac{12}{4 + r} = 2$.

Thus, 6 + 2 = 8 - Satisfies

The correct answer is option B.

6. Since the buses A and B leave at a gap of 30 minutes, i.e. $\frac{1}{2}$ an hour and they travel at 40 miles per hour, the distance between the buses

$= 40 \times \frac{1}{2} = 20$ miles

Let bus C meets one of the other two buses at some time.

The second bus is 20 miles away from bus C.

Since both the other bus and bus C travel in opposite directions, their relative speed

= 40 + 60 = 100 miles per hour

Thus, time taken by bus C to meet the other bus

$= \frac{20}{100} = \frac{1}{5}$ hour

$= \frac{1}{5} \times 60 = 12$ minutes

Thus, the time difference is 12 minutes.

The correct answer is option A.

7. Let the distance covered by that person be d miles.

Time taken while walking at 4 miles per hour = $\left(\frac{d}{4}\right)$ hours

Distance he covers while walking at 9 miles per hour = $(d + 7.5)$ miles

Thus, time taken = $\left(\frac{d + 7.5}{9}\right)$ hours

Since the time is same in both the cases, we have

$\frac{d}{4} = \frac{d + 7.5}{9}$

=> $9d = 4d + 30$

=> $d = 6$

The correct answer is option D.

Alternate approach:

Difference in speeds = 9 – 4 = 5 miles per hour

Difference in distances = 7.5 miles

The difference in distances is due to his higher speed.

Thus, the time for which he traveled = $\frac{7.5}{5} = 1.5$ hours

Thus, actual distance covered = $1.5 \times 4 = 6$ miles.

8. Distance traveled by Train P in 4 hours, i.e. by 3:00 pm = $4 \times 65 = 260$ miles

Distance traveled by Train Q in 1 hour, i.e. by 3:00 pm = 50 miles

Since the distance between the stations is 320 miles, the distance between the trains at 3:00 pm

= 320 – (260 + 50) = 10 miles

We know that train P is stationary for the next 30 minutes.

Distance covered by Train Q in 30 minutes, i.e. $\frac{1}{2}$ an hour = $50 \times \frac{1}{2} = 25$ miles.

Since 25 miles is greater than the distance between the trains (10 miles), the trains would meet at a time when Train P was still stationary.

Thus, time taken by Train Q to meet Train P

$= \frac{10}{50}$ hour $= \frac{1}{5}$ hour

$= \frac{1}{5} \times 60$ minutes = 12 minutes

Thus, time of meeting is 3:12 pm

The correct answer is option B.

9. Let the speed of the man be x miles per hour.

Speed of each bus = 40 miles per hour.

Let the time interval between any two buses be t hours.

(This means that a person, standing at any point on the road will observe buses coming from either direction at a constant time difference, i.e. if he sees one bus at 12:00 going from X to Y, he will observe the next bus, going from Y to X, t hours after 12, the next bus going from X to Y, $2t$ hours after 12, and so on.)

Let us consider the buses moving in the same direction as the cyclist:

Since the time interval between the two buses is t hours, the distance between the two buses = $40t$ miles

The relative speed of the cyclist and the second bus, both moving in the same direction

= $(40 - x)$ miles per hour

Thus, time taken by the second bus to overtake the cyclist

$= \left(\frac{40t}{40 - x}\right)$ hours

Thus, we have

$$\frac{40t}{40 - x} = 10$$

$\Rightarrow 40t = 10\,(40 - x) \ldots$(i)

Let us consider the buses moving in the opposite direction to the cyclist:

Since the time interval between the two buses is t hours, the distance between the two buses is also $40t$ miles

The relative speed of the cyclist and the second bus, both moving in the same direction

= $(40 + x)$ miles per hour

Thus, time taken by the second bus to overtake the cyclist

$= \left(\frac{40t}{40 + x}\right)$ hours

Thus, we have

$$\frac{40t}{40 + x} = 5$$

$\Rightarrow 40t = 5\,(40 + x) \ldots$(ii)

Thus, from (i) and (ii), we have

$10\,(40 - x) = 5\,(40 + x)$

$\Rightarrow 80 - 2x = 40 + x$

$\Rightarrow x = \frac{40}{3} = 13.3$

The correct answer is option C.

10. Traveling at normal speed s, time taken to cover distance d

$= \frac{d}{s}$ hours

New speed = $(s + 2)$ miles per hour

Thus, time taken

$= \frac{d}{s+2}$ hours

Since the time difference is 1 hour, we have

$\frac{d}{s} - \frac{d}{s+2} = 1$

$=> \frac{d(s+2-s)}{s(s+2)} = 1$

$=> 2d = s^2 + 2s$

$=> d = \frac{s^2 + 2s}{2}$

The correct answer is option E.

11. 10 men can complete a job in 10 days.

Thus, to complete the work in 1 day, number of men required = $10 \times 10 = 100$

15 women can complete the work in 6 days.

Thus, to complete the work in 1 day, number of women required = $15 \times 6 = 90$

Thus, we have

100 men ≡ 90 women

$=> 15$ men $\equiv \frac{90}{100} \times 15 = 13.5$ women

Thus, 15 men and 9 women together ≡ (13.5 + 9) women = 22.5 women

We know that 90 women can complete the work in 1 day.

Thus, time taken by 22.5 women to complete the work = $\frac{1 \times 90}{22.5} = 4$ days.

The correct answer is option B.

12. 32 men and 32 boys can complete a work in 3 days.

Thus, to complete the work in 1 day, number of people required would be thrice the number

= 32×3 men and 32×3 boys

= 96 men and 96 boys

6 men and 24 boys can complete the work in 8 days.

Thus, to complete the work in 1 day, number of people required would be eight times the number

= 6×8 men and 24×8 boys

= 48 men and 192 boys

Thus, as the amount of work done is same, we have

96 men + 96 boys ≡ 48 men + 192 boys

=> 48 men ≡ 96 boys

=> 1 man ≡ 2 boys

=> 6 men and 24 boys together ≡ $(6 \times 2 + 24)$ boys = 36 boys

Thus, 36 boys can complete the work in 8 days.

Again, 16 men and 4 boys together ≡ $(16 \times 2 + 4)$ boys = 36 boys

Thus, 16 men and 4 boys together can complete the work in 8 days.

The correct answer is option C.

13. Since each boy takes twice the time taken by a man, 8 boys are equivalent to $\frac{8}{2} = 4$ men.

Thus, 8 boys and 12 men ≡ $(4 + 12)$ men = 16 men

Thus, we have

16 men can complete a project in 9 days

=> Time taken by 12 men to complete the project = $\frac{16 \times 9}{12} = 12$ days

The correct answer is option D.

14. We know that Ann completes $\frac{7}{10}$ of the work in 14 days.

Ann and Joe work for 2 days to complete the remaining $\left(1 - \frac{7}{10}\right) = \frac{3}{10}$ of the work.

Fraction of the total work done by Ann in 1 day = $\left(\frac{7}{10} \times \frac{1}{14}\right) = \frac{1}{20}$

Thus, fraction of the total work done by Ann in 2 days = $2 \times \frac{1}{20} = \frac{1}{10}$

Thus, fraction of the total work done by Joe in 2 days = $\frac{3}{10} - \frac{1}{10} = \frac{1}{5}$

Thus, time taken by Joe to complete the entire work = $2 \div \left(\frac{1}{5}\right) = 10$ days

The correct answer is option D.

15. Let the amount of water already in the boat be x liters.

Let water enter the boat at a uniform rate of w liters per hour.

Also, let each crew member remove r liters of water per hour.

Thus, we have

- After 6 hours:
  Total water that would have been in the boat if not removed = $(x + 6w)$ liters
  Total water removed by 5 crew members = $5 \times 6 \times r = 30r$ liters
  => $x + 6w = 30r$ ...(i)
- After 4 hours:
  Total water that would have been in the boat if not removed = $(x + 4w)$ liters
  Total water removed by 6 crew members = $6 \times 4 \times r = 24r$ liters
  => $x + 4w = 24r$ ...(ii)

Subtracting (ii) from (i):

$2w = 6r \Rightarrow w = 3r \ldots$ (iii)

$\Rightarrow x + 6(3r) = 30r \Rightarrow x = 12r \ldots$ (iv)

Let the time taken by 8 crew members be t hours.

Thus, we have

Total water that would have been in the boat if not removed = $(x + tw)$ liters

Total water removed by the 8 crew members = $8 \times t \times r = 8tr$ liters

Thus, we have

$x + tw = 8tr$

From (iii) and (iv):

$\Rightarrow 12r + 3tr = 8tr$

$\Rightarrow 12r = 5tr$

$\Rightarrow t = \frac{12}{5} = 2.4$ hours

The correct answer is option A.

16. Let the number of men present initially be x.

Thus, x men can complete the work in 40 days

$\Rightarrow$ 1 man can complete the work in $40x$ days ... (i)

Also, $(x + 5)$ men can complete the work in $(40 - 10) = 30$ days

$\Rightarrow$ 1 man can complete the work in $30(x + 5)$ days ... (ii)

Thus, as the amount of work completed is same, from (i) and (ii):

$40x = 30(x + 5)$

$\Rightarrow x = 15$

The correct answer is option B.

17. Let Pete complete x units of work in 1 hour.

Thus, John completes $\frac{x}{2}$ units of work in $\frac{1}{6}$ hour.

$\Rightarrow$ John completes $\left(\frac{x}{2} \times 6\right) = 3x$ units of work in 1 hour.

Thus, Pete and John together complete $(x + 3x) = 4x$ units of work in 1 hour.

Thus, the amount of work they complete in 10 hours = $4x \times 10 = 40x$ units.

Since Pete alone completes x units in 1 hour, time he takes to complete $40x$ units of work

$= \frac{40x}{x} = 40$ hours

The correct answer is option E.

18. We know that Amy and Bob can complete a piece of work in 45 days and 40 days, respectively.

Part of work completed by Amy in 1 day = $\frac{1}{45}$

Part of work completed by Bob in 1 day = $\frac{1}{40}$

Let Amy leave after t days.

Thus, part of work completed by Amy and Bob in t days = $\left(\frac{1}{45} + \frac{1}{40}\right) \times t = \frac{17t}{360}$

The remaining work was completed by Bob in 23 days.

Part of work completed by Bob in 23 days = $\frac{23}{40}$

Thus, we have

$\frac{23}{40} + \frac{17t}{360} = 1$

=> $207 + 17t = 360$

=> $17t = 153$

=> $t = 9$

The correct answer is option C.

19. We know that 30 men completed 25% of the work in 10 days.

Thus, 30 men can complete the remaining 75% of work in $\left(10 \times \frac{75}{25}\right) = 30$ days.

Thus, to complete the remaining work in 1 day, number of men required = $30 \times 30 = 900$

Since the total work has to be completed in 30 days, the remaining work needs to be completed in the remaining $(30 - 10) = 20$ days.

Thus, to complete the remaining work in 20 days, number of men required = $\frac{900}{20} = 45$

Thus, number of additional men required = $45 - 30 = 15$

The correct answer is option B.

20. Since Bob left 3 days before completion, Amy had worked alone for the last 3 days.

We know that Amy can complete the entire work in 30 days.

Thus, fraction of work she can completed in one day = $\frac{1}{30}$

=> Fraction of work she completed in 3 days = $3 \times \frac{1}{30} = \frac{1}{10}$

Thus, the remaining $\left(1 - \frac{1}{10}\right) = \frac{9}{10}$ of the work was completed by Amy and Bob together.

Since Bob can complete the work in 20 days, part of the work he does in 1 day = $\frac{1}{20}$

Thus, fraction of work completed by Amy and Bob in 1 day = $\left(\frac{1}{30} + \frac{1}{20}\right) = \frac{1}{12}$

Thus, time taken to complete the remaining $\frac{9}{10}$ part of the work by Amy and Bob together

$= \frac{9}{10} \div \frac{1}{12}$

$= \frac{9}{10} \times 12 = 10.8$

$= 10\frac{4}{5}$ days

Thus, the duration of the entire work $= 10\frac{4}{5} + 3 = 13\frac{4}{5}$ days

The correct answer is option E.

21. Let the sum of money invested at 4% rate of interest be $\$x$

Thus, the sum of money invested at 7% rate of interest = $\$\,(18{,}000 - x)$

Thus, interest on $\$x$ at 4% rate of interest in 2 years

$= \$\left(\frac{x \times 2 \times 4}{100}\right) = \$\left(\frac{8x}{100}\right)$

Also, interest on $\$\,(18{,}000 - x)$ at 7% rate of interest in 2 years

$= \$\left(\frac{(18{,}000 - x) \times 2 \times 7}{100}\right) = \$\left(\frac{14\,(18{,}000 - x)}{100}\right)$

Since total interest is \$2,100, we have

$\frac{8x}{100} + \frac{14\,(18{,}000 - x)}{100} = 2{,}100$

$\Rightarrow 14 \times 18{,}000 - 6x = 210{,}000$

$\Rightarrow 6x = 252{,}000 - 210{,}000$

$\Rightarrow x = 7{,}000$

The correct answer is option C.

22. Let the sums of money invested at 10% and 20% rates of interest be $\$x$ each.

Since the difference between the interests earned after three years is between \$120 and \$140, we have

$120 < \frac{x \times 20 \times 2}{100} - \frac{x \times 10 \times 2}{100} < 140$

$\Rightarrow 120 < \frac{20x}{100} < 140$

$\Rightarrow 600 < x < 700$

We need to find the difference between the amounts earned after 2 years at compound interest at the same rates as above.

Thus, the required difference

$= x\left(1 + \frac{20}{100}\right)^2 - x\left(1 + \frac{10}{100}\right)^2 = x\left\{\left(1 + \frac{20}{100}\right)^2 - \left(1 + \frac{10}{100}\right)^2\right\}$

$= x\left\{\left(1 + \frac{20}{100}\right) - \left(1 + \frac{10}{100}\right)\right\}\left\{\left(1 + \frac{20}{100}\right) + \left(1 + \frac{10}{100}\right)\right\}$

$= x\left(\frac{10}{100}\right)\left(2 + \frac{30}{100}\right) = x\,(0.1 \times 2.3) = 0.23x$

We know that $600 < x < 700$

Thus, the required difference is

$600 \times 0.23 < 0.23x < 700 \times 0.23$

$=> 138 < 0.23x < 161$

Thus, the required difference lies between \$138 and \$161.

Only Option D lies in the above range.

The correct answer is option D.

23. Interest accumulated on $\$p$ at $r\%$ rate under compound interest after period of two years

$$= \$ \left\{p\left(1 + \frac{r}{100}\right)^2 - p\right\}$$

$$= \$p \left\{\frac{r}{100}\left(2 + \frac{r}{100}\right)\right\}$$

Interest accumulated on $\$q$ at $(2r)\%$ rate under simple interest after period of two years

$$= \$ \left(\frac{q \times 2r \times 2}{100}\right)$$

$$= \$ \left(\frac{4qr}{100}\right)$$

Since the interests are equal, we have

$$p \left\{\frac{r}{100}\left(2 + \frac{r}{100}\right)\right\} = \frac{4qr}{100}$$

$$=> p\left(2 + \frac{r}{100}\right) = 4q$$

$$=> q = \frac{p}{4}\left(2 + \frac{r}{100}\right)$$

The correct answer is option E.

24. Let the sums of money invested in each case be $\$x$

The interest accumulated at 10% simple interest for 2 years

$$= \$ \left(\frac{x \times 10 \times 2}{100}\right) = \$ \left(\frac{x}{5}\right)$$

The interest accumulated at 25% compound interest for 2 years

$$= \$ \left\{x\left(1 + \frac{25}{100}\right)^2 - x\right\} = \$x \left\{\left(1 + \frac{1}{4}\right)^2 - 1\right\}$$

$$= \$x \left\{\frac{1}{4}\left(2 + \frac{1}{4}\right)\right\} = \$ \left(\frac{9x}{16}\right)$$

Note: $25\% = \frac{25}{100} = \frac{1}{4}$

Since the difference between the interests is \$87, we have

$$\frac{9x}{16} - \frac{x}{5} = 87$$

$$=> \frac{45x - 16x}{80} = 87 => \frac{29x}{80} = 87$$

$=> x = \$240$

The correct answer is option E.

25. Let the sum of money borrowed be $\$100x$.

Thus, the amount after 1 year = $\$\left(100x + \frac{100x \times 20 \times 1}{100}\right) = \$120x$

Amount repaid at that point of time = \$200

Thus, amount due = $\$(120x - 200)$

Interest accumulated on the above in 1 more year

$= \$\left\{\frac{(120x - 200) \times 20 \times 1}{100}\right\} = \$(24x - 40)$

Thus, the amount due

$= \$\{(120x - 200) + (24x - 40)\}$

Since this is cleared after paying \$200, we have

$(120x - 200) + (24x - 40) = 200$

$=> 144x = 440$

$=> 100x = \frac{440}{144} \times 100 \approx 3 \times 100 = 300$

Note: $144 \times 3 = 432 \approx 440 => \frac{440}{144} \approx 3$ (the value is slightly greater than 3)

Thus, the actual answer should be slightly greater than \$300.

The correct answer is option E.

26. Let the time for which the first sum of money is invested be t years.

Thus, the time for which the second sum of money is invested = $(t + 2)$ years.

Interest accumulated on $\$p$ for t years at 6% rate of interest

$= \$\left(\frac{p \times 6 \times t}{100}\right) = \$\left(\frac{6pt}{100}\right)$

In the second scenario, the sum of money invested is $\$\left(\frac{p}{3}\right)$ and the rate of interest is $2 \times 6 = 12\%$

Interest accumulated on $\$\left(\frac{p}{3}\right)$ for $(t + 2)$ years at 12% rate of interest

$$= \$\left(\frac{\left(\frac{p}{3}\right) \times 12 \times (t + 2)}{100}\right) = \$\left(\frac{4p(t + 2)}{100}\right)$$

Since the difference between the two interests is \$120, we have

$\frac{6pt}{100} - \frac{4p(t + 2)}{100} = 120 => 2pt - 8p = 12,000$

$=> p(t - 4) = 6,000 => t = \frac{6,000}{p} + 4$

Since t is an integer, p must be a factor of 6,000.

Also, since $t < 7$, we have

$$\frac{6,000}{p} + 4 < 7 => \frac{6,000}{p} < 3$$

$$=> p > \frac{6,000}{3} => p > 2,000$$

Thus, greater than 2,000 and a factor of 6,000 can only be 3,000 and 6,000.

Among the options, 3,000 is the correct answer.

Thus, we have $p = 3,000$

The correct answer is option B.

27. Let the sum of money invested in either scenario be $\$x$

The amount for $\$x$ at 20% rate for 3 years = $\$x\left(1 + \frac{20}{100}\right)^3$

The amount for $\$x$ at $r\%$ rate for 2 years = $\$x\left(1 + \frac{r}{100}\right)^2$

Since the amount in both the scenarios are the same, we have

$$x\left(1 + \frac{20}{100}\right)^3 = x\left(1 + \frac{r}{100}\right)^2$$

$$=> \left(1 + \frac{r}{100}\right)^2 = \left(1 + \frac{20}{100}\right)^3 = \frac{6}{5} \times \left(\frac{6}{5}\right)^2 = 1.2 \times \left(\frac{6}{5}\right)^2$$

Taking square root on both sides:

$$\left(1 + \frac{r}{100}\right) = \sqrt{1.2} \times \left(\frac{6}{5}\right) = 1.1 \times 1.2 = 1.32$$

$$=> \frac{r}{100} = 0.32$$

$$=> r = 32$$

The correct answer is option C.

28. Let the rates of interest in each account be $r\%$

Let her investments for each account be $\$x$

Interest earned on $\$x$ at $r\%$ simple interest after 2 years = $\$\left(\frac{2xr}{100}\right)$

Thus, we have

$$\frac{2xr}{100} = 180 => \frac{xr}{100} = 90 \ldots \text{(i)}$$

Interest earned on $\$x$ at $r\%$ compound interest after 2 years = $\$x\left(\left(1 + \frac{r}{100}\right)^2 - 1\right)$

Thus, we have

$$x\left\{\left(1 + \frac{r}{100}\right)^2 - 1\right\} = 189$$

$$=> x\left(\frac{r}{100}\right)\left(2 + \frac{r}{100}\right) = 189$$

$=> \frac{xr}{100}\left(2 + \frac{r}{100}\right) = 189$

Thus, using (i), we have

$=> 90\left(2 + \frac{r}{100}\right) = 189$

$=> 2 + \frac{r}{100} = \frac{189}{90} = 2.1$

$=> \frac{r}{100} = 0.1$

$=> r = 10$

Thus, from (i):

$\frac{x \times 10}{100} = 90$

$=> x = 900$

Thus, Amy's total investment = $\$\,(x + x) = \$1,800$

The correct answer is option E.

Alternate approach:

Simple interest after 2 years is $180

=> Simple interest every year is $90

Since compound interest after 2 years is $189 and compound interest for the first year is the same as the simple interest every year, we have

Compound interest for the second year is $189 – $90 = $99

Thus, there is a 10% increase in the compound interest from the first year to the second year.

Thus, rate of interest is 10%

Since simple interest every year is $90, we have

10% of the investment under simple interest is $90

Thus, the investment under simple interest is $(90 × 10) = $900

Since investment under simple interest and compound interest is the same, we have

Total investment = $900 + $900 = $1,800

29. Let the amount invested be $\$x$

Let the rate of interest be $r\%$

Thus, the amount in the beginning of the third year, i.e. after 2 years

$= \$x\left(1 + \frac{r}{100}\right)^2 = \$1,200 \ldots \text{(i)}$

The amount in the beginning of the fourth year, i.e. after 3 years

$= \$x\left(1 + \frac{r}{100}\right)^3 = \$1,440 \ldots \text{(ii)}$

Dividing (ii) by (i):

$$\frac{x\left(1+\frac{r}{100}\right)^3}{x\left(1+\frac{r}{100}\right)^2} = \frac{1,440}{1,200}$$

$$=> \left(1+\frac{r}{100}\right) = \frac{6}{5}$$

$$=> \frac{r}{100} = \frac{1}{5} => r = 20\%$$

Thus, from (i):

$$x\left(1+\frac{20}{100}\right)^2 = 1,200$$

$$=> x = 1,200 \times \left(\frac{5}{6}\right)^2 = 1,200 \times \frac{5}{6} \times \frac{5}{6} = \frac{2,500}{3} = 833.3$$

$=> x = 830$ (After rounding to the nearest tens)

The correct answer is option D.

30. Let the rate of interest be $r\%$

Thus, the amount after 6 years = $\$2,000\left(1+\frac{r}{100}\right)^6$

Thus, we have

$$2,000\left(1+\frac{r}{100}\right)^6 = 8,000$$

$$=> \left(1+\frac{r}{100}\right)^6 = 4$$

Taking square root on both sides:

$$\left(1+\frac{r}{100}\right)^3 = 2$$

$$=> 1+\frac{r}{100} = \sqrt[3]{2} = 1.25$$

$$=> r = 25$$

The correct answer is option B.

31. After replacing two boys, aged 15 years and 18 years, by three other boys, the average of the group reduces by 1 year.

Let the initial average age of the group be x years.

Thus, the total age of the 12 boys = $12x$ years.

The total age of the group of 10 boys after the two boys are removed from the group

= $12x - 15 - 18 = (12x - 33)$ years

Let the average age of the three new boys be y years.

Thus, total age of the three boys = $3y$ years

Thus, after these 3 boys join the group, the total age of the group of 13 boys

= $(12x - 33 + 3y)$ years

Since the average age of the group reduces by 1 year, we have

$$\frac{12x - 33 + 3y}{13} = x - 1$$

$$=> 12x - 33 + 3y = 13x - 13$$

$$=> 3y = x + 20$$

$$=> y = \frac{x + 20}{3} \ldots \text{(i)}$$

We know that the initial average age of the boys in the group lies between 12 and 15 years, inclusive.

Thus, we have $12 \leq x \leq 15$

Thus, from (i), the only integer value of y occurs if $x = 13$

Thus, the value of $y = \frac{13 + 20}{3} = 11$

The correct answer is option C.

32. Let the population at the beginning of the first decade be 100.

Thus, the population after the first decade, after a 20% increase = 120% of 100 = 120

The population after the second decade, after a 25% increase = 125% of 120 = 150

The population after the third decade, after a 10% increase = 110% of 150 = 165

Thus, the overall percent increase in population after 3 decades

$$= \frac{165 - 100}{100} \times 100 = 65\%$$

Thus, the simple average rate of increase in population per decade

$$= \frac{65}{3}\% = 21.7\%$$

The correct answer is option B.

33. Let the number of students in the teams A, B, C, and D be a, b, c and d.

Since there are 40 students, we have

$$a + b + c + d = 40 \ldots \text{(i)}$$

We know that the teams, A, B, C and D, have average weights of 30 kg, 35 kg, 40 kg and 50 kg, respectively.

Since the average of all students taken together is 40 years, we have

$$\frac{30a + 35b + 40c + 50d}{a + b + c + d} = 40$$

$$=> 30a + 35b + 40c + 50d = 40a + 40b + 40c + 40d$$

$$=> 2a + b = 2d$$

$$=> b = 2(d - a) \ldots \text{(ii)}$$

We know that each team must have at least two students.

Also, we need to maximize the value of c, thus, we need to minimize the values of a, b and d.

Since $a = b = d = 2$ does not satisfy (ii), we increase the value of d to 3, keeping $a = 2$.

$=> b = 2\,(3 - 2) = 2$

Thus, the smallest possible values of a, b and d are:

$a = 2$, $b = 2$ and $d = 3$

Thus, from (i), we have

$c = 40 - (a + b + d) = 40 - 7 = 33$

The correct answer is option D.

34. Total age of the 7 members having average age of 25 years = $7 \times 25 = 175$ years

We know that the youngest member is 4 years old.

Thus, when the youngest member was born, i.e. 4 years ago, the age of each member would reduce by 4 years.

Thus, the total age would reduce by $7 \times 4 = 28$ years.

Thus, the total age of the $7 - 1 = 6$ members of the family

$= 175 - 28 = 147$ years

Thus, average age of the members of the family

$= \frac{147}{6} = 24.5$ years

The correct answer is option D.

35. Let the price of the chair, table, cot and bookshelf be \$$a$, \$$b$, \$$c$ and \$$d$, respectively.

Since the average price of a chair, a table, and a cot is \$70, we have

$\frac{a + b + c}{3} = 70$

$=> a + b + c = 210 \ldots$ (i)

Since the average price of the table, the cot and a bookshelf is \$85, we have

$\frac{b + c + d}{3} = 85$

$=> b + c + d = 255 \ldots$ (ii)

Subtracting (i) from (ii):

$d - a = 45$

Since $d = 70$, we have

$a = d - 45 = 25$

The correct answer is option A.

36. We know that for $\frac{1}{4}$ of the people, the salary increase was 50% on a base salary of \$3,000 and for the remaining $\frac{3}{4}$ of the people, the salary increase was 20% on a base salary of \$1,800.

Let the number of friends be $4n$.

After increase in salaries, we have

Average salary of n friends having 50% increase in salary = $\$\left(\frac{150 \times 3,000}{100}\right) = \$4,500.$

Average salary of $3n$ friends having 20% increase in salary = $\$\left(\frac{120 \times 1,800}{100}\right) = \$2,160.$

Thus, average salary of the entire group

$= \frac{n \times 4,500 + 3n \times 2,160}{4n} = \$2,745$

Note: Here, we could have done without the use of n (since n cancels out eventually) and instead we could have worked assuming only 4 friends.

The correct answer is option C.

Alternate approach:

Let there be 4 friends.

=> Number of managers = $\frac{1}{4} \times 4 = 1$

=> Number of supervisors = $\frac{3}{4} \times 4 = 3$

Thus, increased average salary of managers = $1 \times (3,000 + 50\% \text{ of } 3,000) = \$4,500$

Similarly, increased average salary of supervisors = $3 \times (1,800 + 20\% \text{ of } 1,800) = \$2,160$

Thus, average salary of the entire group

$= \frac{1 \times 4,500 + 3 \times 2160}{4} = \$2,745$

37. Since the problem asks us a percent value, we can choose a suitable initial value for the ease of calculations, since the initial value does not affect the final answer.

Since the ratios used are $\frac{3}{4}, \frac{2}{3}, \& \frac{3}{5}$, we choose a value of the total students, which is divisible by 4, 3 and 5, say 60.

Thus, let the total number of students = 60

Number of students having glasses = $\frac{3}{4} \times 60 = 45$

Number of students not having glasses = $60 - 45 = 15$

Number of girls = $\frac{3}{5} \times 60 = 36$

Number of girls having glasses = $\frac{2}{3} \times 36 = 24$

Number of girls not having glasses = $36 - 24 = 12$

Thus, the percent of the students not having glasses, who were girls

$= \frac{\text{Number of girls not having glasses}}{\text{Number of students not having glasses}} \times 100$

$= \frac{12}{15} \times 100\%$

$= 80\%$

The correct answer is option E.

38. Let the number of members in the club = x.

Thus, the number of members who joined a swimming class = $\left(60 + \frac{x}{2}\right)$.

Thus, the number of members who joined the swimming class also joined basketball class = $\frac{1}{3}\left(60 + \frac{x}{2}\right)$.

Since $\frac{1}{3}$ of all the members joined basketball class, we have

$\frac{1}{3}\left(60 + \frac{x}{2}\right) = \frac{x}{3}$

$=> 20 + \frac{x}{6} = \frac{x}{3}$

$=> \frac{x}{3} - \frac{x}{6} = 20$

$=> \frac{x}{6} = 20$

$=> x = 120$

The correct answer is option A.

39. Since John ate double the share of cakes that each of the other $(n - 1)$ boys ate, let's distribute the three cakes among $(n + 1)$ boys.

The fraction $\left(\frac{3}{n+1}\right)$ represents the share of each boy, while $\left(2 \times \frac{3}{n+1} = \frac{6}{n+1}\right)$ represents the share of John.

Again since John ate half the share of pizzas that each of the other $(n - 1)$ boys ate, let's distribute four pizzas among $\left(n - 1 + \frac{1}{2} = n - \frac{1}{2}\right)$ boys.

In the expression $\left[(n-1)+\frac{1}{2}\right]$, we first excluded John by applying $(n-1)$ and then added him by applying $\frac{1}{2}$.

The fraction $\left(\frac{4}{n-\frac{1}{2}}\right)$ represents the share of each boy, while $\left(\frac{1}{2}\times\frac{4}{n-\frac{1}{2}}=\frac{2\times2}{2n-1}=\frac{4}{2n-1}\right)$ represents the share of John.

Thus, John had $\left(\frac{6}{n+1}+\frac{4}{2n-1}\right)$ share of the cakes and the pizzas.

The correct answer is option E.

Alternate approach 1:

Say the share of cakes of each of the $(n-1)$ boys is x, thus, John's share of cakes = $2x$.

=> $x(n-1)+2x=3$

=> $nx+x=3$

=> $x(n+1)=3$

=> $x=\dfrac{3}{n+1}$

=> John's share of cakes = $2x=\dfrac{6}{n+1}$

Similarly, say the share of pizzas of each of the $(n-1)$ boys is y, thus, John's share of pizzas = $\dfrac{y}{2}$.

=> $y(n-1)+\dfrac{y}{2}=4$

=> $ny-\dfrac{y}{2}=4$

=> $y\left(n-\dfrac{1}{2}\right)=4$

=> $y=\dfrac{4}{n-\dfrac{1}{2}}$

=> $y=\dfrac{8}{2n-1}$

=> John's share of pizzas = $\dfrac{y}{2}=\dfrac{4}{2n-1}$

John's share of cakes and pizzas = $\dfrac{6}{n+1}+\dfrac{4}{2n-1}$

Alternate approach 2:

This question can be solved by assuming a suitable value of n.

Say there are 10 number of boys, incl. John.

Since John had double the share of cake, let's cut the cakes in $9 + 2 = 11$ pieces.

John's share of the cakes $= 2 \times \frac{3}{11} = \frac{6}{11}$

Again, since John had half the share of pizza, let's slice the pizzas in $9 + \frac{1}{2} = \frac{19}{2}$ pieces.

John's share of the pizzas $= \frac{1}{2} \times \frac{4}{\frac{19}{2}} = \frac{4}{19}$

Only option E qualifies, the correct answer.

40. The ratio of the time durations = 3 : 4 : 5 : 6

If the first person worked for 60 hours:

The modified ratio would be

= **3 × 20** : 4 × 20 : 5 × 20 : 6 × 20

= 60 : 80 : 100 : 120

Thus, total time = 60 + 80 + 100 + 120 = 360 hours - Option E is possible

If the second person worked for 60 hours:

The modified ratio would be

= 3 × 15 : **4 × 15** : 5 × 15 : 6 × 15

= 45 : 60 : 75 : 90

Thus, total time = 45 + 60 + 75 + 90 = 270 hours - Option C is possible

Having considered that the first member spent 60 hours at the gym, the total number of possible hours was 360 hours, while having considered that the second member spent 60 hours in the gym, the total number of possible hours was 270 hours (less than 360 hours). It is evident that when we consider that the third member and the fourth member spent 60 hours in the gym, the total number of possible hours in each case would be less than 270 hours. Thus, it is for sure than option D (300 hours) value which is greater than 270 hours, is not a possible value; thus, the correct answer.

However, for the sake of understanding, we consider the scenario when the third member and the fourth member spent 60 hours at the gym.

If the third person worked for 60 hours:

The modified ratio would be

$= 3 \times 12 : 4 \times 12 : \mathbf{5 \times 12} : 6 \times 12$

$= 36 : 48 : 60 : 72$

Thus, total time = 36 + 48 + 60 + 72 = 216 hours - Option B is possible

If the fourth person worked for 60 hours:

The modified ratio would be

$= 3 \times 10 : 4 \times 10 : 5 \times 10 : \mathbf{6 \times 10}$

$= 30 : 40 : 50 : 60$

Thus, total time = 30 + 40 + 50 + 60 = 180 hours - Option A is possible

Thus, option D is not possible.

The correct answer is option D.

41. Ratio of number of students in groups A and B = 2 : 3

Number of students in Group A $= 50\left(\frac{2}{2+3}\right) = 20$

Number of students in Group B $= 50\left(\frac{3}{2+3}\right) = 30$

Let the average weight of students in groups A and B be a kg and b kg, respectively.

Since the average weight of all students is 40 kg, we have

$$\frac{20a + 30b}{20 + 30} = 40$$

$=> 2a + 3b = 200 \ldots$ (i)

Also, the average weight of the students in Group A is 5 kg less than that of the other group.

Thus, we have

$a = b - 5 \ldots$ (ii)

Thus, from (i) and (ii), we have

$2(b - 5) + 3b = 200 \Rightarrow b = 42 \Rightarrow a = 37$

When Joe moves from Group B to Group A, the average weight of the students in either group becomes the same. Since the average weight of all students (50) is 40 kg, after Joe shifts, the average weight of each group must become equal to the overall average, i.e. 40 kg.

Working with Group A:

Total weight of 20 students in Group A having an average weight of 37 kg

$= 20 \times 37 = 740$ kg

After Joe shifts to Group A, number of students in Group A = 21

Thus, the total weight of 21 students having an average weight of 40 kg = $21 \times 40 = 840$ kg

Thus, Joe's weight = 840 – 740 = 100 kg

The correct answer is option E.

42. Let the average contribution of all 8 friends be $\$x$

Contribution from 7 friends = $\$(120 \times 7) = \$840$

Contribution of the last friend = $\$(x + 21)$

Thus, we have

$840 + (x + 21) = 8x$

$\Rightarrow 7x = 861$

$\Rightarrow x = 123$

Thus, cost of the camera = $\$(8 \times 123) = \$984$

The correct answer is option E.

43. Let the average salary of junior managers be $\$x$

Thus, the average salary of senior managers = $\$(x + 4,500)$

We know that there are 80 junior managers and 100 senior managers, i.e. total 180 employees.

Thus, combined salary of all junior managers = $\$(80x)$

Also, combined salary of all senior managers = $\$(100(x + 4500))$

Thus, the combined salary of all junior and senior managers = $\$\{80x + 100(x + 4,500)\}$

Thus, average salary of all junior and senior managers = $\$\left\{\frac{80x + 100(x + 4,500)}{180}\right\}$

Thus, required difference

$$= \$\left\{\frac{80x + 100(x + 4500)}{180} - x\right\} = \$\left\{\frac{80x + 100(x + 4,500) - 180x}{180}\right\}$$

$= \$\left(\frac{4,500 \times 100}{180}\right) = \$2,500$

The correct answer is option C.

Alternate approach:

We can assume any value for the average salary of the junior managers. The actual value chosen is irrelevant since, depending on this value, all other values would increase or decrease by the same amount, resulting in the required 'DIFFERENCE' remaining the same.

Thus, to simplify calculations, let us assume the average salary of junior managers is '0.'

Thus, the average salary of senior managers = $4,500

We know that there are 80 junior managers and 100 senior managers, i.e., a total of 180 employees.

Thus, the combined salary of all junior managers is '0'.

Also, the combined salary of all senior managers = $\$ (100 \times 4,500)$

Thus, the combined salary of all junior and senior managers = $450,000

Thus, the average salary of all junior and senior managers = $\$\left(\frac{450,000}{180}\right) = \$2,500$

Thus, the required difference = $2,500

44. The prices of each laptop and each mobile are $1,200 and $900, respectively.

Total price of all items on Tuesday morning = $\$(1,080 \times 20) = \$21,600$

We know that only laptops were sold on Tuesday.

Let the number of laptops sold be x.

Thus, reduction in total price of all items = $\$ (1,200x)$

Thus, total price of all items = $\$ (21,600 - 1,200x)$

Number of items left = $(20 - x)$

Since the final average price was $1,000, we have

$1,000\,(20 - x) = 21,600 - 1,200x$

$=> 200x = 1,600$

$=> x = 8$

The correct answer is option B.

45. Distance traveled by the first train at 65 miles per hour in 14 hours = $65 \times 14 = 910$ miles

Since the distance between the two trains is 1960 miles and they travel in opposite directions, the sum of the distances covered by the two trains equals the distance between them.

Thus, the distance traveled by the second train = 1,960 - 910 = 1,050 miles.

Since the speed of the second train is 105 miles per hour, time taken to travel 1,050 miles $= \frac{1,050}{105} = 10$ hours

Thus, the second train started 14 - 10 = 4 hours after the first train started.

The correct answer is option A.

46. Let the total distance be d miles.

Thus, his speed for $\frac{d}{3}$ miles is 50 miles per hour and his speed for the remaining $\frac{2d}{3}$ miles is 20 miles per hour.

Since the total time taken is 4 hours and 48 minutes, i.e. $4\frac{48}{60} = 4\frac{4}{5}$ hours, we have

$$\frac{\frac{d}{3}}{50} + \frac{\frac{2d}{3}}{20} = 4\frac{4}{5}$$

$$=> \frac{d}{150} + \frac{d}{30} = \frac{24}{5} => \frac{6d}{150} = \frac{24}{5}$$

$$=> d = 120$$

The correct answer is option C.

47. Let the speed of the first train be x miles per hour.

Thus, the speed of the second train = $(x + 20)$ miles per hour.

In the time from 1:00 pm to 9:00 pm, i.e. 8 hours, the distance traveled by the first train

= $8x$ milestraveled

In the time from 5:00 pm to 9:00 pm, i.e. 4 hours, the distance traveled by the second train

= $4(x + 20)$ miles

Since the second train started 20 miles behind LA and caught up with the first train at 9:00 pm, the distance traveled by the second train is 20 miles greater than the distance traveled by the first train.

Thus, we have

$$4(x + 20) = 8x + 20$$

$$=> 4x = 60$$

$$=> x = 15$$

Thus, the speed of the second train = $x + 20 = 35$ miles per hour

The correct answer is option D.

48. Ratio of distances = 6 : 4 : 5

Ratio of speeds = 2 : 3 : 5

Thus, ratio of time taken = $\frac{6}{2} : \frac{4}{3} : \frac{5}{5} = 3 : \frac{4}{3} : 1$

Thus, let the actual times taken be $3x$, $\frac{4}{3}x$ and x hours, respectively.

Since the total time taken is 8 hours, we have

$$3x + \frac{4}{3}x + x = 8 => \frac{16}{3}x = 8$$

$$=> x = \frac{3}{2}$$

The distances, in miles, traveled are

$\frac{6}{6+4+5} \times 150 = 60; \frac{4}{6+4+5} \times 150 = 40; \frac{5}{6+4+5} \times 150 = 50$

The corresponding times, in hours, taken are

$3x = \frac{9}{2}; \frac{4}{3}x = 2; x = \frac{3}{2}$

Thus, the corresponding speeds, in miles per hour, are

$\frac{60}{\frac{9}{2}} = \frac{40}{3} = 13.33; \frac{40}{2} = 20; \frac{50}{\frac{3}{2}} = \frac{100}{3} = 33.33$

Thus, the maximum speed is $\frac{100}{3}$ miles per hour.

Thus, the time taken if the entire distance is covered at this speed

$= \frac{150}{\frac{100}{3}} = 4.5$ hours

The correct answer is option C.

Alternate approach:

Distance	6	4	5	
Original speed	2	3	5	
Original time (Distance/speed)	3	$\frac{4}{3}$	1	Total time = $3 + \frac{4}{3} + 1 = \mathbf{\frac{16}{3}}$
				Taking maximum speed (5)
New speed	5	5	5	
New time	$\frac{6}{5}$	$\frac{4}{5}$	1	Total time = $\frac{6}{5} + \frac{4}{5} + 1 = \mathbf{3}$

Thus, if the original time is $\frac{16}{3}$, the new time is 3.

=> If the original time is 8 hours, the new time

$= \frac{3}{\left(\frac{16}{3}\right)} \times 8 = \frac{9}{16} \times 8$

$= \frac{9}{2} = 4.5$ hours

49. Speed over the first d miles = x miles per hour

Thus, time taken = $\left(\frac{d}{x}\right)$ hours

Speed over the remaining 150% of $d = \frac{3d}{2}$ miles is 120% of $x = \frac{6x}{5}$ miles per hour

Thus, time taken = $\frac{\left(\frac{3d}{2}\right)}{\left(\frac{6x}{5}\right)} = \frac{5}{4}\left(\frac{d}{x}\right)$ hours

Thus, total time = $\left(\frac{d}{x}\right) + \frac{5}{4}\left(\frac{d}{x}\right) = \frac{9}{4}\left(\frac{d}{x}\right)$ hours . . . (i)

Total distance traveled = $d + \frac{3d}{2} = \frac{5d}{2}$ miles

Thus, if the boy traveled at x miles per hour, time taken = $\frac{5}{2}\left(\frac{d}{x}\right)$ hours ... (ii)

Thus, required ratio of the time taken

$$= \frac{9}{4}\left(\frac{d}{x}\right) : \frac{5}{2}\left(\frac{d}{x}\right) = 9 : 10$$

The correct answer is option D.

Alternate approach:

We can assume any suitable value of x and d since the final answer does not depend on those values.

Let $x = 50$ miles per hour and $d = 100$ miles.

Thus, time taken over the first d miles = $\frac{100}{50} = 2$ hours

Remaining distance = 150% of 100 = 150 miles

Speed over the remaining distance = 120% of 50 = 60 miles per hour

Thus, time taken for the remaining distance = $\frac{150}{60} = 2.5$ hours

Thus, total time = 2 + 2.5 = 4.5 hours

Total distance = 100 + 150 = 250 miles

Had the boy traveled the entire distance at 50 miles per hour, time taken = $\frac{250}{50} = 5$ hours

Thus, required ratio = 4.5 : 5 = 9 : 10

50. Let the actual speed of the car be x miles per hour

Distance traveled = 360 miles

Time taken to cover the distance = $\left(\frac{360}{x}\right)$ hours

New speed of the car = $(x - 4)$ miles per hour

Thus, time taken to cover 360 miles = $\left(\frac{360}{x-4}\right)$ hours

Since the time taken in the second case was 3 hours greater than in the first case, we have

$$\frac{360}{x-4} - \frac{360}{x} = 3$$

$$=> \frac{360\{x - (x-4)\}}{(x-4)x} = 3$$

$$=> x^2 - 4x - 480 = 0$$

$$=> (x - 24)(x + 20) = 0$$

$=> x = 24$ (Since the speed must be positive, $x \neq -20$)

The correct answer is option D.

51. Time taken to cover x miles at 5 miles per hour = $\left(\frac{x}{5}\right)$ hours

Time taken to cover another 10 miles at 10 miles per hour = 1 hour

Thus, total time taken = $\left(\frac{x}{5} + 1\right)$ hours

Total distance traveled = $(x + 10)$ miles

Thus, time taken had he traveled throughout at 10 miles per hour

$= \left(\frac{x + 10}{10}\right) = \left(\frac{x}{10} + 1\right)$ hours

Thus, the required percent difference

$$= \frac{\left(\frac{x}{5} + 1\right) - \left(\frac{x}{10} + 1\right)}{\left(\frac{x}{10} + 1\right)} \times 100\%$$

$$= 100\left(\frac{x}{x + 10}\right)\%$$

Thus, we have

- If $x = 1$ (smallest value)

  Required percent difference = $100 \times \left(\frac{1}{11}\right) = 9.09\%$
- If $x = 10$ (largest value)

  Required percent difference = $100 \times \left(\frac{10}{20}\right) = 50\%$

Thus, the required percent difference must lie between 9.09% and 50%.

Only option B satisfies the above condition.

The correct answer is option B.

Alternate approach:

Let us assume that $x = 10$

So in the original scenario, the man traveled 10 km @ 5 km/hr and another 10 km @ 10 km/hr speeds.

Thus, total time traveled = $\frac{10}{5} + \frac{10}{10} = 2 + 1 = 3$ hrs

Total time for the revised scenario = $\frac{20}{10} = 2$ hrs

Thus, the required percent difference

$$= \frac{3 - 2}{2} \times 100\% = 50\%$$

Since $1 \le x \le 10$, the answers which are less or equal to 50% may be correct. But there are two qualifying options: A (8%) and B (23%).

Need we solve the question with another approach to eliminate one of the options? The answer is NO! There is no need to do so since this is a PS question and only one option can be correct. If option A (8%) is correct, it is for sure that option B (23%) is also correct; however, if option B (23%) is correct, it is NOT certain that option A (8%) is also correct. Thus, the best bet is option B. It can be intuited that 8% would qualify if $x < 1$.

52. Part of the work done by A in 1 hour = $\frac{1}{15}$

Part of the work done by B in 1 hour = $\frac{1}{18}$

Part of the work completed by A alone in 6 hours = $\frac{1}{15} \times 6 = \frac{2}{5}$

Part of the work completed by A and B together in next 3 hours = $\left(\frac{1}{15} + \frac{1}{18}\right) \times 3 = \frac{11}{30}$

Thus, part of the work remaining = $1 - \left(\frac{2}{5} + \frac{11}{30}\right) = \frac{7}{30}$

Thus, time taken by B to complete the remaining work alone

$= \frac{\left(\frac{7}{30}\right)}{\left(\frac{1}{18}\right)} = \frac{7}{30} \times 18 = 4\frac{1}{5}$ hours

The correct answer is option A.

53. Since A and B take 12 hours to fill the tank, part of the tank filled by (A + B) in 1 hour = $\frac{1}{12}$

Since A and C take 15 hours to fill the tank, part of the tank filled by (A + C) in 1 hour = $\frac{1}{15}$

Since B and C take 20 hours to fill the tank, part of the tank filled by (B + C) in 1 hour = $\frac{1}{20}$

Adding the above three results, we have

Part of the tank filled by 2 (A + B + C) in 1 hour = $\frac{1}{12} + \frac{1}{15} + \frac{1}{20} = \frac{1}{5}$

Thus, part of the tank filled by (A + B + C) in 1 hour = = $\frac{1}{2} \times \frac{1}{5} = \frac{1}{10}$

Thus, time taken to complete the work with A, B and C working together = $\frac{1}{\left(\frac{1}{10}\right)} = 10$ hours

The correct answer is option C.

54. We know that

Time taken by A and B together to complete a work = 24 days

Time taken by C alone to complete the same work = 28 days

The pattern in which A, B and C worked to complete the work is:

A and C worked for 4 days; B and C worked for 10 days; and B worked for 4 days

=> A worked for 4 days, B worked for 14 days and C worked for 14 days

=> A and B worked for 4 days, C worked for 14 days and B worked for 14 – 4 = 10 days

Part of work completed by A and B in 4 days = $\frac{4}{24} = \frac{1}{6}$

Part of work completed by C in 14 days = $\frac{14}{28} = \frac{1}{2}$

Thus, part of work completed by B in 10 days = $1 - \left(\frac{1}{6} + \frac{1}{2}\right) = \frac{1}{3}$

Thus, time taken by B to complete the entire work = $\frac{10}{\left(\frac{1}{3}\right)} = 30$ days

The correct answer is option E.

55. We can represent the situation using a diagram, as shown below:

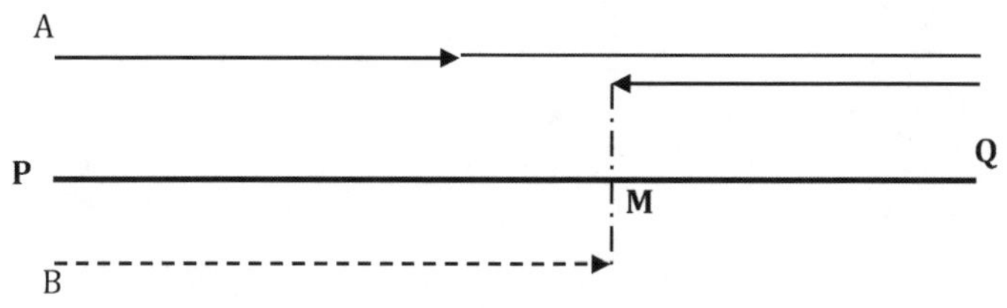

Assuming M to be the meeting point, and distance from P to M be d meters, we have

Distance from P to Q = 200 meters

=> Distance from Q to M = $(200 - d)$ meters

Thus, distance traveled by A = (Distance from P to Q) + (Distance from Q to M) = 200 + $(200 - d)$

= $(400 - d)$ meters

Distance traveled by B = Distance from P to M

= d meters

Since both A and B travel for the same time, the ratio of the distances they cover is equal to the ratio of their speeds, given as 7 : 1.

Thus, we have

$$\frac{400 - d}{d} = \frac{7}{1}$$

=> $d = 50$

The correct answer is option B.

Alternate approach:

Both A and B travel for the same amount of time. So if, B travels x distance, then A will travel $7x$ distance as the ratio of their speeds is 7 : 1.

Total distance traveled by both of them together = $x + 7x = 8x$

According to the diagram, we see both of them together traveled $2 \times 200 = 400$ meters.

Thus, $8x = 400 \Rightarrow x = 50$

56. We know that Bob and Chad started at the same time. However, Chad stopped for 15 minutes on the way. Hence, Chad would have covered a distance equivalent to 15 minutes had he not made the stop.

However, this scenario is exactly the same as if Chad had started 15 minutes after Bob did, hence covering a distance equivalent to 15 minutes had he started with Bob.

Thus, in the first 15 minutes, distance traveled by Bob at 60 miles per hour

$= 60 \times \frac{15}{60} = 15$ miles

Distance remaining between Bob and Chad when Chad started his journey

$= 150 - 15 = 135$ miles

Since Bob and Chad move in opposite directions at speeds of 60 miles per hour and 30 miles per hour, we have

Relative speed $= 60 + 30 = 90$ miles per hour

Thus, time taken for Bob and Chad to meet

$= \frac{\text{Distance}}{\text{Relative speed}} = \frac{135}{90}$

$= 1.5$ hours

In 1.5 hours, distance covered by Chad $= 1.5 \times 30 = 45$ miles

Thus, they meet at a distance of 45 miles from point B.

The correct answer is option B.

57. We know that

Time taken by A to complete the work $= x$ days

Time taken by B to complete the work $= kx$ days

Thus, part of the work completed by A and B together in 1 day

$= \frac{1}{x} + \frac{1}{kx} = \frac{k+1}{kx}$

Thus, time taken by A and B to complete the work $= \left(\frac{kx}{k+1}\right)$ days

We know that $\left(\frac{kx}{k+1}\right)$ is an integer.

Thus, $(k + 1)$ is a factor of kx

However, k and $(k + 1)$ are consecutive integers, and hence, are co-prime to one another. Thus, $(k + 1)$ or a part of $(k + 1)$ cannot be a factor of k.

Thus, $(k + 1)$ must be a factor of x . . . (i)

We also know that:

$\frac{kx}{k+1} < 5$

$=> \frac{kx}{k+1} \leq 4 \ldots$ (ii)

Thus, from (i) and (ii), the possible scenarios are:

- $k = 1 => k + 1 = 2$
  $=>$ 2 is a factor of x, i.e. x is an even number
  $=>$ Possible values of x are: 2, 4, 6 or 8
  $=>$ Possible values of (k, x) are $(1, 2)$, $(1, 4)$, $(1, 6)$, $(1, 8)$
  Thus, there are four possible values of (k, x)

- $k = 2 \Rightarrow k + 1 = 3$
  => 3 is a factor of x
  => Possible values of x are 3 or 6
  => Possible values of $(k,\ x)$ are: $(2,\ 3)$, $(2,\ 6)$
  Thus, there are two possible values of $(k,\ x)$
- $k = 3 \Rightarrow k + 1 = 4$
  => 4 is a factor of x
  => Possible value of x is 4
  => Possible value of $(k,\ x)$ is $(3,\ 4)$
  Thus, there is one possible value of $(k,\ x)$
- $k = 4 \Rightarrow k + 1 = 5$
  => 5 is a factor of x
  => Possible value of x is 5
  => Possible value of $(k,\ x)$ is $(4,\ 5)$
  Thus, there is one possible value of $(k,\ x)$

For any higher value of k, the inequality given in (ii) is not satisfied.

Thus, the number of possible values of the ordered set of $(k,\ x)$ are:

$4 + 2 + 1 + 1 = 8$

The correct answer is option D.

58. We can represent the situation using a diagram, as shown below:

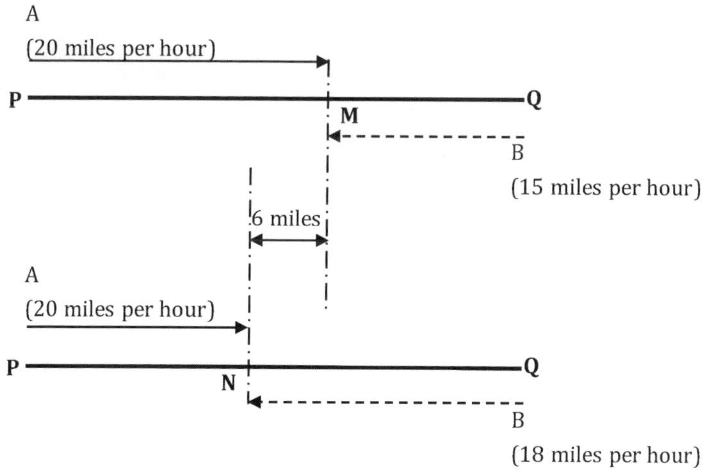

Since in either case, A and B start simultaneously, the ratio of their distances is equal to the ratio of their speeds.

Thus, we have

- First case: $\frac{\text{PM}}{\text{QM}} = \frac{20}{15} = \frac{4}{3}$

  $\Rightarrow \text{PM} = \frac{4}{3}\text{QM} \ldots \text{(i)}$

- Second case: $\frac{PN}{QN} = \frac{20}{18} = \frac{10}{9}$

  => $\frac{PM - 6}{QM + 6} = \frac{10}{9}$

Thus, from (i), we have

$$\frac{\frac{4}{3}QM - 6}{QM + 6} = \frac{10}{9} => \frac{4QM - 18}{3QM + 18} = \frac{10}{9}$$

=> $36QM - 162 = 30QM + 180$

=> $QM = 57$

The correct answer is option C.

Alternate approach:

Say PM = x and MQ = y

=> PN = $x - 6$ and NQ = $y + 6$

In the original scenario:

Time taken by A to reach point M = Time taken by B to reach point M

=> $\frac{x}{20} = \frac{y}{15}$ => $x = \frac{4y}{3}$

In the revised scenario:

Time taken by A to reach point N = Time taken by B to reach point N

=> $\frac{x - 6}{20} = \frac{y + 6}{18}$ => $\frac{x - 6}{10} = \frac{y + 6}{9}$

=> $9(x - 6) = 10(y + 6)$

=> $9x - 54 = 10y + 60$

=> $9x = 10y + 114$

=> $9 \times \frac{4y}{3} = 10y + 114$

=> $12y = 10y + 114$

=> $2y = 114$

=> $y = 57$ = QM

59. Let the required time be t hours.

Thus, distance covered by A, running at 12 miles per hour, in t hours = $12t$ miles

Distance covered by B, running at 10 miles per hour, in t hours = $10t$ miles

We know that A completed one lap more than that completed by B.

Thus, the distance covered by A is 15 miles more than that covered by B after t hours.

Thus, we have

$12t - 10t = 15$

=> $t = 7.5$

Note:

In 7.5 hours, distance covered by A = $7.5 \times 12 = 90$ miles = $\frac{90}{15}$ = 6 laps (an integer value)

In 7.5 hours, distance covered by B = $7.5 \times 10 = 75$ miles = $\frac{75}{15}$ = 5 laps (an integer value)

The difference between the numbers of laps covered by A and B = 6 - 5 = 1

The correct answer is option B.

Alternate approach:

Since A and B are running in the same direction, their relative speed

= 12 − 10 = 2 miles/hr.

Given that A leads B by one lap or 15 miles, we have

Time = $\frac{\text{Distance}}{\text{Speed}} = \frac{15}{2}$ = 7.5 hours.

60. We know that Jane always takes twice as much time to solve a data-sufficiency question as a problem-solving question.

Thus, we have

20 problem-solving and 10 data-sufficiency ≡ 20 problem-solving and 20 problem-solving

≡ 40 problem-solving questions

Similarly, we have

30 problem-solving and 20 data-sufficiency ≡ 30 problem-solving and 40 problem-solving

≡ 70 problem-solving questions

Thus, Jane had solved 40 problem-solving questions in the stipulated time and she wants to solve 70 problem-solving questions in the same time.

Let the time taken be t hours.

Thus, time taken per question initially = $\frac{t}{40}$ hours.

Time taken per question finally = $\frac{t}{70}$ hours.

Thus, percent reduction in time = $\frac{\frac{t}{40} - \frac{t}{70}}{\frac{t}{40}} \times 100$

$= \left(1 - \frac{40}{70}\right) \times 100 = \frac{3}{7} \times 100\%$

= 42.9%

The correct answer is option B.

Alternatively, we can assume that the time taken be 280 hours. (LCM of 40 and 70)

Thus, time taken per question initially = $\frac{280}{40}$ = 7 hours

Time taken per question finally = $\frac{280}{70}$ = 4 hours

Thus, percent reduction in time

$= \frac{7-4}{7} \times 100\%$

$= \frac{3}{7} \times 100\%$

$= 42.9\%$

61. We know that during morning walks, Brenda takes 45 strides, each 2 feet long, per minute.

Thus, distance she covers per minute = $45 \times 2 = 90$ feet.

While going to school, Brenda's speed is 50% higher

Thus, the distance she covers per minute while going to school = 150% of 90 = 135 feet.

While going to school, her each stride is 2.5 feet long.

Thus, the number of strides she takes per minute = $\frac{135}{2.5} = \frac{135 \times 2}{5} = 54$

Thus, the number of more strides she takes while going to school than during her morning walks

= 54 - 45 = 9

The correct answer is option A.

62. We know that

5 servers can process 45,000 search requests per hour

=> 1 server in 1 hour can process $\frac{45{,}000}{5} = 9{,}000$ search requests ... (i)

The service provider needs to process 216,000 search requests in 3 hours.

Thus, the number of search requests required to be processed in 1 hour $= \frac{216{,}000}{3} = 72{,}000$

Thus, from (i), we observe that the required number of search requests processed in 1 hour is $\frac{72{,}000}{9{,}000} = 8$ times the number of search requests that can be processed by 1 server in 1 hour.

Thus, the number of servers required is 8.

However, there are already 5 servers present.

Thus, the number of additional servers required = 8 - 5 = 3.

The correct answer is option B.

63. The percent profit made by selling the final mixture at \$8.40 a pound = 20%.

Thus, we have

$(100 + 20)\%$ of Cost price of the final mixture per pound = \$8.40

=> Cost price of the final mixture per pound = $\$\left(\frac{100}{120} \times 8.40\right) = \$7.$

Let the quantity of Jamaican Blue Mountain beans used be x pounds.

Thus, on mixing x pounds of Jamaican Blue Mountain beans priced at \$19.00 per pound and 8 pounds of Colombian coffee beans priced at \$4.00 per pound, we get a mixture which costs \$7 per pound.

Thus, we have

$$\frac{x \times 19 + 8 \times 4}{x + 8} = 7$$

$$=> 19x + 32 = 7(x + 8)$$

$$=> 19x + 32 = 7x + 56$$

$$=> 12x = 24$$

$$=> x = 2$$

The correct answer is option B.

64. Since we need the expression for average speed, we can assume any suitable value of the total distance since the distance does not affect the value of the average speed.

Let the total distance be 100.

Distance traveled at 40 miles per hour

$= x\%$ of 100

$= x$

Thus, time (in hours) taken to cover the above distance

$$= \frac{x}{40}$$

Distance traveled at 60 miles per hour

$= (100 - x)$

Thus, time (in hours) taken to cover the above distance

$$= \frac{(100 - x)}{60}$$

Thus, average speed

$$= \frac{(\text{Total distance})}{(\text{Total time})}$$

$$= \frac{100}{\frac{x}{40} + \frac{(100 - x)}{60}}$$

$$= \frac{120 \times 100}{3x + 2(100 - x)}$$

$$= \frac{12,000}{x + 200}$$

The correct answer is option E.

Alternate approach:

Let us plug in a value for x, calculate the average speed and identify the correct answer by substituting that value of x in each option.

Let the total distance be 100.

Let $x = 40$ (40 is the best choice since in that case, the distances would be easily divisible by the corresponding values of the speeds).

Distance traveled at 40 miles per hour

$= 40\%$ of $100 = 40$

Thus, time (in hours) taken to cover the above distance

$= \frac{40}{40} = 1$

Distance traveled at 60 miles per hour

$= (100 - 40) = 60$

Thus, time (in hours) taken to cover the above distance

$= \frac{60}{60} = 1$

Thus, average speed

$= \frac{(\text{Total distance})}{(\text{Total time})} = \frac{100}{1+1} = 50$

Substituting $x = 40$ in each option, we get:

Option A: $\frac{180 - x}{2} = \frac{140}{2} = 70$ - Does not satisfy

Option B: $\frac{x + 60}{4} = \frac{100}{4} = 25$ - Does not satisfy

Option C: $\frac{300 - x}{5} = \frac{260}{5} = 52$ - Does not satisfy

Option D: $\frac{600}{115 - x} = \frac{600}{75} = 80$ - Does not satisfy

Option E: $\frac{12,000}{x + 200} = \frac{12,000}{240} = 50$ - Satisfies

65. The average cost per book for 12 books = $\$k$

Thus, the total cost of the 12 books = $\$12k$

Cost of the book removed = \$18

Cost of the book added = \$42

Thus, cost of the books on the table

$= \$\,(12k - 18 + 42)$

$= \$\,(12k + 24)$

Number of books on the table $= 12 - 1 + 1 = 12$

Thus, average cost per book on the table

$= \frac{\text{Total cost of all books}}{\text{Number of books}}$

$= \$\left(\frac{12k + 24}{12}\right)$

$= \$\,(k + 2)$

The correct answer is option A.

Alternate approach:

After one book costing \$18 was removed and one book costing \$42 was added, the total cost of all the books increased by \$ (42 − 18) = \$24

Since there are 12 books, the average cost increased by $\$\frac{24}{12} = \$2$

Since the initial average was $\$k$, the new average = $\$ (k + 2)$

66. The average score on a test taken by 10 students = x

Thus, the total score of the 10 students = $10x$

The average score of 5 students = 8

Thus, the total score of the 5 students = $8 \times 5 = 40$

Thus, the total score of the remaining 5 students = $10x - 40$

Thus, the average score of the remaining 5 students = $\frac{10x - 40}{5} = 2x - 8$

The correct answer is option A.

67. Sales tax applicable in State A

$= \$(t\% \text{ of } x)$

$= \$\left(\frac{xt}{100}\right)$

Thus, total cost of the computer in State A

$= \$\left(x + \frac{xt}{100}\right)$

$= \$\frac{x(100 + t)}{100} \ldots \text{(i)}$

Sales tax applicable in State B

$= \$(T\% \text{ of } y)$

$= \$\left(\frac{yT}{100}\right)$

Thus, total cost of the computer in State A

$= \$\left(y + \frac{yT}{100}\right)$

$= \$\frac{y(100 + T)}{100} \ldots \text{(ii)}$

Since $xt > PT$, we know that the total cost of the computer is more in State A.

The amount by which the total cost of computer is more in State A than that in State B = $\$\left(\frac{x(100 + t)}{100} - \frac{y(100 + T)}{100}\right)$

$= \$\left[(x - y) - \frac{(xt - yT)}{100}\right]$

The correct answer is option E.

68. Calls between 5:00 am and 9:00 pm are charged at $0.40 per minute.

Calls between 9:00 pm and 5:00 am are charged at $0.25 per minute.

A call at 8:00 pm, i.e. between 5:00 am and 9:00 pm, is charged $30.00

The call charges would be $0.40 per minute from 8:00 pm to 9:00 pm, i.e. for 60 minutes, after which, the call charges would change to $0.25 per minute.

Assuming that the call does not extend till a time when the call rates are different, the duration of the call comes to be

$= \frac{30}{0.40} = 75$ minutes

Since the duration is more than 60 minutes, a different call charge would be applicable for the last part of the call.

Charge for the first 60 minutes

$= \$\,(60 \times 0.40)$

$= \$24.00$

Thus, the charge on a call for the last part, billed at $0.25 per minute

$= \$(30.00 - 24.00) = \$6.00$

Thus, the duration of the last part of the call

$= \frac{6}{0.25} = 24$ minutes

Thus, the total duration of the call = (60 + 24) = 84 minutes

We need to calculate the charge on a call of the same duration, made at 11:00 pm, i.e. between 9:00 pm and 5:00 am.

Since the duration is 84 minutes, the entire call would be at the same charge of $0.25 per minute.

Thus, the charge on the call

$= \$\,(84 \times 0.25)$

$= \$21.00$

The correct answer is option E.

Alternate approach:

We know that the first 60 minutes of the call is charged at $0.40 per minute giving a total charge of $\$\,(60 \times 0.40) = \$24.00$

The remaining amount billed = $\$\,(30.00 - 24.00) = \$6.00$

For the new call, the entire duration would be charged at $0.25 per minute.

Thus, the initial 60 minutes which had been charged at $0.40 per minute, would be now charged at $0.25 per minute, giving a charge = $\$\,(60 \times 0.25) = \$15.00$

However, the remaining part of the call would still be charged at $0.25 per minute, as before, resulting in the charge of $6.00

Thus, the total charge = $\$\,(15.00 + 6.00) = \$21.00$

69. Total sales of popcorn (3 bags at \$5.25 each)

$= \$(5.25 \times 3)$

$= \$15.75$

Total sales of soda (4 at \$4.50 each)

$= \$(4.50 \times 4)$

$= \$18.00$

Total sales of candy bars (2 at \$3.00 each)

$= \$(3.00 \times 2)$

$= \$6.00$

Thus, total sales of all snacks (for every 16 admission tickets)

$= \$(15.75 + 18.00 + 6.00)$

$= \$39.75$

Thus, the average amount of snacks sold per ticket

$= \$\left(\frac{39.75}{16}\right)$

$=\approx \$\left(\frac{40}{16}\right)$

$= \$\left(\frac{5}{2}\right)$

$= \$2.50$

(Note: The actual answer would be slightly less than \$2.50 (it actually is \$2.48) since we approximated \$39.75 to \$40. However, we only want an approximate answer, and, \$2.48 when approximated can be written as \$2.50)

The correct answer is option C.

70. Let the time taken by the man while traveling by air and while traveling by train be t hours and T hours, respectively.

Since the total time is 8 hours, we have

$T + t = 8 \ldots$ (i)

The man would have saved $\frac{4}{5}$ of the time he was in train had he traveled all the way by air.

=> Time saved $= \frac{4T}{5}$

Again, we know that he would reach 4 hours early had he traveled entirely by air.

Thus, the time saved is 4 hours

$=> \frac{4T}{5} = 4$

$=> T = 5$

$=> t = 8 - 5 = 3 \ldots$ using equation (i)

Had the man traveled entirely by air, he would have taken 8 - 4 = 4 hours.

Thus, speed of the man by air = $\frac{720}{4}$ = 180 miles per hour.

Thus, the distance, in miles, originally traveled by the man by air

$= 180 \times t$

$= 180 \times 3 = 540$

The correct answer is option C.

71. Since we need to find the average speed, we may assume any suitable value of the distance.

Since we need to take $\frac{1}{3}$ of the distance, and also divide a part of the distance by 12, 20 and 40 (since speeds are given as 20, 12 and 40 miles per hour), it is best to assume the distance to be a multiple of 120 (LCM of 20, 12 and 40).

To be on the safer side, let us assume the distance to be 360 miles.

Distance traveled at 20 miles per hour = $\frac{1}{3} \times 360 = 120$ miles.

Thus, time taken = $\frac{120}{20} = 6$ hours ... (i)

Distance left to be covered = 360 - 120 = 240 miles.

Distance traveled at 12 miles per hour = $\frac{1}{2}$ × Distance left to be covered = $\frac{1}{2} \times 240 = 120$ miles.

Thus, time taken = $\frac{120}{12} = 10$ hours ... (ii)

Distance left to be covered = 240 - 120 = 120 miles.

This distance is traveled at 40 miles per hour.

Thus, time taken = $\frac{120}{40} = 3$ hours ... (ii)

Thus, from (i), (ii) and (iii), we have

Total time taken = 6 + 10 + 3 = 19 hours.

Thus, average speed

$= \frac{\text{Total distance}}{\text{Total time}}$

$= \frac{360}{19}$

$\frac{360}{19} > \frac{360}{20} = 18$

$\frac{360}{19} < \frac{361}{19} = 19$

Thus, the actual value of the average speed will be between 18 and 19.

The correct answer is option C.

72. Number of floors Jane needs to cover = x.

Thus, time taken to walk down the steps for all x floors taking 30 seconds per floor

$= 30x$ seconds

Time taken to wait for the elevator and ride for x floors taking 2 seconds per floor

$= 7$ minutes $+ 2x$ seconds

$= (420 + 2x)$ seconds

Thus, we have

$30x = 420 + 2x$

$\Rightarrow x = \frac{420}{28} = 15$

The correct answer is option D.

73. Distance covered by Mary in 15 minutes traveling at 50 miles per hour

$= 50 \times \frac{15}{60}$

$= 12.5$ miles

Thus, when Paul crossed the gas station, the distance between Mary and Paul = 12.5 miles.

Since both Mary and Paul are traveling in the same direction, their relative speed

= Speed of Paul – Speed of Mary

$= 60 - 50 = 10$ miles per hour.

Thus, time taken by Paul to catch up with Mary

$= \frac{\text{Distance between them}}{\text{Relative speed}}$

$= \frac{12.5}{10}$

$= 1.25$ hours

$= 1.25 \times 60$ minutes

$= 75$ minutes

The correct answer is option D.

74. At 10:00 am, the man and the bus were at the same point (since the man overtook the bus), say point Z.

The man reached City Y at 2:00 pm.

Thus, time taken by the man to reach City Y from point Z

= Time difference between 2:00 pm and 10:00 am

= 4 hours

On the way back the man meets the bus at 3:00 pm at point W.

Thus, the time taken by the bus to travel from point Z to point W

= Time difference between 3:00 pm and 10:00 am

= 5 hours

The above situation can be represented in the diagram below:

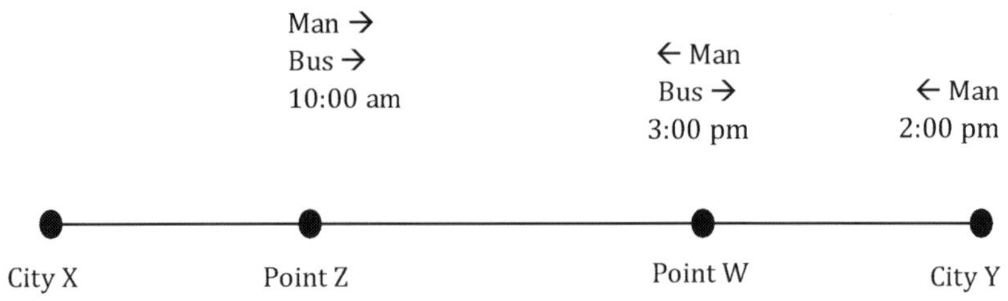

Thus, from the diagram above, we can say that:

The distance ZY is equivalent to 4 hours travel of the man ... (i)

The distance WY is equivalent to 1 hour travel of the man ... (ii)

From (i) and (ii):

The distance WZ is equivalent to $(4 - 1) = 3$ hours travel of the man ... (iii)

The distance WZ is equivalent to 5 hours travel of the bus ... (iv)

Thus, from (iii) and (iv), we have

3 hours travel of the man $\equiv$ 5 hours travel of the bus

=> 1 hour travel of the man (dist. WY) $\equiv \frac{5}{3}$ hour travel of the bus (dist. WY)

Thus, time taken by the bus to travel from point W to City Y

$= \frac{5}{3}$ hours $= 1\frac{2}{3}$ hours

= 1 hour 40 minutes

Thus, the bus reached City Y 1 hour 40 minutes after 3:00 pm

= 4:40 pm

The correct answer is option E.

75. Speed of Jack = 3 miles per hour.

Distance covered by Jack (before Bob started walking) in 2 hours $= 3 \times 2 = 6$ miles.

Thus, total distance between Jack and Bob = 6 miles.

Speed of Bob = 5 miles per hour.

Relative speed (in same direction) of Jack and Bob $= 5 - 3 = 2$ miles per hour.

Thus, time taken for them to meet

$= \frac{\text{Distance}}{\text{Relative speed}}$

$= \frac{6}{2}$

= 3 hours

Distance covered by Bob in these 3 hours $= 3 \times 5 = 15$ miles.

Thus, the number of miles before Q where Bob overtook Jack

$= 25 - 15$

$= 10$

The correct answer is option B.

76. Time taken by the first inlet pipe to fill the tank to $\frac{1}{2}$ of capacity = 3 hours.

=> Fraction of the tank filled by the first inlet pipe in 1 hour = $\frac{1}{2} \times \frac{1}{3} = \frac{1}{6}$

Time taken by the second inlet pipe to fill the tank to $\frac{2}{3}$ of capacity = 6 hours.

=> Fraction of the tank filled by the second inlet pipe in 1 hour = $\frac{2}{3} \times \frac{1}{6} = \frac{1}{9}$

Thus, fraction of the tank filled in 1 hour by both pipes working together

$= \frac{1}{6} + \frac{1}{9}$

$= \frac{5}{18}$

=> Time taken by both pipes to fill the tank to capacity

$= \frac{1}{\left(\frac{5}{18}\right)} = \frac{18}{5}$

$= 3.6$ hours

The correct answer is option B.

77. Let Machine Y takes a days to produce w widgets.

Thus, Machine X takes $(a + 2)$ days to produce w widgets.

Thus, total widgets produced by machines X and Y in 1 day

$= \frac{w}{a} + \frac{w}{a + 2}$

$= w\left(\frac{1}{a} + \frac{1}{a + 2}\right)$

$= \frac{w\,(2a + 2)}{a\,(a + 2)}$

Thus, total widgets produced by machines X and Y in 3 days

$= \frac{3w\,(2a + 2)}{a\,(a + 2)}$

Thus, we have

$\frac{3w\,(2a + 2)}{a\,(a + 2)} = \frac{5}{4}w$

$=> \frac{3\,(2a + 2)}{a\,(a + 2)} = \frac{5}{4}$

$=> 24a + 24 = 5a^2 + 10a$

$=> 5a^2 - 14a - 24 = 0$

$=> 5a^2 - 20a + 6a - 24 = 0$

$=> (a - 4)\,(5a + 6) = 0$

Since a is positive, we have

$a = 4$

Thus, time taken by Machine X to produce w widgets = $(a + 2) = 6$ days

Thus, time taken by Machine X to produce $2w$ widgets = $2 \times 6 = 12$ days

The correct answer is option E.

78. Let the quantity of the sample be x.

Thus, amount of impurity present

= 10% of x

$= \frac{10}{100} \times x$

$= \frac{x}{10}$

After each purification process, the amount of impurity reduces by 60%.

Let the required number of times the purification has to be done = n.

Thus, the final amount of impurity

$= \frac{x}{10} \times \left(1 - \frac{60}{100}\right)^n$

Thus, we have

$\frac{x}{10} \times \left(1 - \frac{60}{100}\right)^n \leq 1\%$ of x

$=> \frac{x}{10} \times \left(1 - \frac{60}{100}\right)^n \leq \frac{x}{100}$

$=> \left(\frac{2}{5}\right)^n \leq \frac{1}{10}$

$=> \left(\frac{2 \times 2}{5 \times 2}\right)^n \leq \frac{1}{10}$

$=> \left(\frac{4}{10}\right)^n \leq \frac{1}{10}$

$=> 4^n \leq 10^{(n-1)}$

As per the answer options, for $n = 2$, $4^2 = 16 \nleq 10^1$ - Does not satisfy

For $n = 3$, $4^3 \leq 10^2 => 64 \leq 100$ - Satisfies

The correct answer is option B.

79. Under compound interest, if a sum of money P is invested at r% annual rate of interest for t years, the amount A is given by:

$A = P\left(1 + \frac{r}{100}\right)^t$

Thus, the interest accumulated

$= A - P$

$= P\left(1 + \frac{r}{100}\right)^t - P$

For Tim:

$P = \$1,000,\ r = 10\%,\ t = 2$ years

$=> \text{Interest} = \$\left(1,000\left(1 + \frac{10}{100}\right)^2 - 1,000\right)$

$= \$\left(1,000\left(\frac{11}{10} \times \frac{11}{10}\right) - 1,000\right)$

$= \$(10 \times 121 - 1,000)$

$= \$(1,210 - 1,000)$

$= \$210$

For Lana:

$P = \$2,000,\ r = 5\%,\ t = 2$ years

$=> \text{Interest} = \$\left(2,000\left(1 + \frac{5}{100}\right)^2 - 2,000\right)$

$= \$\left(2,000\left(\frac{21}{20} \times \frac{21}{20}\right) - 2,000\right)$

$= \$(5 \times 441 - 2,000)$

$= \$(2,205 - 2,000)$

$= \$205$

Thus, the different in total interest accumulated after two years

$= \$\ (210 - 205)$

$= \$5$

The correct answer is option A.

80. Calls between 5:00 am and 9:00 pm are charged at \$0.40 per minute.

Calls between 9:00 pm and 5:00 am are charged at \$0.25 per minute.

A call at 1:00 pm, i.e. between 5:00 am and 9:00 pm, is charged \$10.00

The duration of the call $= \frac{10}{0.40} = 25$ minutes

We find that the duration of the call does not extend till a time when the call rates are different.

Thus, the charge on a call of the same duration, made at 11:00 pm, i.e. between 9:00 pm and 5:00 am with charge \$0.25 per minute.

$= \$\ (0.25 \times 25)$

$= \$6.25$

Alternate approach:

We can say that the rate of charge for the calls between 9:00 pm and 5:00 am is $\frac{25}{40} = \frac{5}{8}$ of the rate of charge for the calls between 5:00 am and 9:00 pm.

Thus, the charge on a call of the same duration, made at 11:00 pm, i.e. between 9:00 pm and 5:00 am

$= \$\left(\frac{5}{8} \times 10\right) = \$6.25$

The correct answer is option B.

81. For a 4-digit number to be even, its units digit must be either 4 or 8.

The number of ways, units place can be filled = 2 (either 4 or 8)

The number of ways, tens place can be filled = 4; given that the repetition of digits is allowed.

The number of ways, hundreds place can be filled = 4

The number of ways, thousands place can be filled = 4

So, the total number of ways = $2 \times 4 \times 4 \times 4 = 128$ ways.

The correct answer is option D.

82. The number of ways, each letter can be chosen = 26
So, the total number of ways of choosing 26 letters for three places = 26^3.

Similarly, number of ways, each digit can be chosen = 10
So, the total number of ways of choosing 10 digits for three places = 10^3.

These two blocks of the digits and the letters can be arranged in 2 ways (At the beginning or at the end).

So, the total number of ways = $2 \times 26^3 \times 10^3$ ways.

The correct answer is option D.

83. '3' can be placed in the following ways in 5-digit numbers: 3XXXX, X3XXX, XX3XX, XXX3X, and XXXX3. Since '3' has to appear only once, we have 9 choices for each vacant place out of digits 0-9 (excluding the digit '3').

But wait! A 5-digit number cannot begin with a digit '0,' so the 5-digit numbers that do not begin with '3' will have only 8 number of ways to its fill ten-thousandth place.

For numbers X3XXX, XX3XX, XXX3X, and XXXX3, the total number of ways = $4 \times 8 \times 9^3$

For number 3XXXX, the total number of ways = 9^4

Hence, the total number of ways = $9^4 + 4 \times 8 \times 9^3$

The correct answer is option E.

84. We can start with the first key. After the maximum 7 number of trials, we can conclude that the last key belongs to the 8th lock. Similarly, for the 2nd key, there would be 6 trails. So, in total trying with all keys, we will need $7 + 6 + 5 + 4 + 3 + 2 + 1 = 28$ trials.

The correct answer is option B.

85. The number of ways n players can be selected out of $(2n + 1)$ players:

$$C_n^{2n+1} = \frac{(2n+1)!}{n! \times (2n + 1 - n)!}$$
$$= \frac{(2n+1)!}{n! \times (n+1)!} = 126 \text{ (Given)}$$

The better approach to get the value of n would be to plug in values from the options. We recommend that you follow a strategy on which option to try first while plugging in?

In the GMAT exam, if the option values are arranged in ascending order, follow B & D strategy while plugging in.

B & D strategy

(1) Plug in option B's value first. If the right-hand side (RHS) value matches the left-hand side (LHS) value, option B is the correct answer; otherwise,

(2) If the value derived (LHS) is greater than the desired (RHS) value, option A could be the correct answer. Now plug in option A's value and check if the right-hand side (RHS) value matches the left-hand side (LHS) value.

(3) If the value derived (LHS) is less than the desired (RHS) value, options A and B cannot be the correct answers. Try option D instead of option C.

(4) If, for option D, the right-hand side (RHS) value matches the left-hand side (LHS) value, option D is the correct answer; otherwise,

(5) If the value derived (LHS) is greater than the desired (RHS) value, option C is the correct answer.

(6) If the value derived (LHS) is less than the desired (RHS) value, option E is the correct answer.

Let us try with option B value, $n = 3$, we get $\frac{(2n+1)!}{n! \times (n+1)!} = \frac{7!}{3! \times 4!} = 35$. Since the derived value (35) is less than the desired value (126), hence, option B or A cannot be the correct answers.

Let us try with option D value, $n = 5$, we get $\frac{(2n+1)!}{n! \times (n+1)!} = \frac{11!}{5! \times 6!} = 462$. Since the derived value (462) is more than the desired value (126), hence, option C is the correct answer. We can mark the answer without even trying option C. For the sake of your curiosity, we present to you the value.

At $n = 4$, we have $\frac{(2n+1)!}{n! \times (n+1)!} = \frac{9!}{4! \times 5!} = 126$ (the desired value)

The correct answer is option C.

86. A triangle can be drawn taking any 3 vertices of a polygon.

So, the number of triangles = $C_3^8 = \frac{8 \times 7 \times 6}{1 \times 2 \times 3} = 56$

The correct answer is option C.

A regular Octagon (8-sided polygon) will not have 3 or more number of co-linear vertices, hence, there is no need to think of deducting a few number of dummy triangles formed out of co-linear points.

What if the question had been:

How many triangles are possible out of 8 points lying in a plane, out of which 4 points are co-linear?

The answer would not be 56. There would be a few dummy triangles formed out of 4 co-linear points, and we must deduct those from 56.

Number of dummy triangles = $C_3^4 = C_1^4 = 4$

So, the total number of actual triangles = $56 - 4 = 52$ triangles

87. The number of ways of choosing 1 cube from Puzzle box1 = $C_1^{10} = 10$

The number of ways of choosing 2 cubes from Puzzle box2 = $C_2^5 = \frac{5 \times 4}{1 \times 2} = 10$

Total number of ways = $10 \times 10 = 100$

The correct answer is option C.

88. 'Numbers less than 1,000' means all 1-digit, 2-digit and 3-digit numbers (1 to 999).

The number of ways of making 1-digit numbers out of the given four digits 3, 4, 5, and 6 = 4.

The number of ways of making 2-digit numbers out of the given four digits 3, 4, 5, and 6 = $4 \times 4 = 16$. (Remember that the repetition of digits is allowed. Questions do not restrict us to use the given digits once.)

Similarly, the number of ways of making 3-digit numbers out of the given four digits 3, 4, 5, and 6 = $4 \times 4 \times 4 = 64$.

Total 'numbers less than 1,000' = 4 + 16 + 64 = 84.

The correct answer is option C.

89. The number of ways of choosing any 2 members from 5 individuals = $C_2^5 = \frac{5 \times 4}{1 \times 2} = 10$

Since 2 specific individuals **together** must not be in the team, the number of ways 2 specific individuals can be chosen = 1.

Hence, the total number of ways = 10 − 1 = 9.

The correct answer is option D.

Remember that the question asks that the 2 specific individuals **together** must not be in the team, however, one of them alone can be a part of the team.

Had you excluded both the individuals while calculating the number of ways, you could have wrongly calculated the answer as $C_2^3 = C_1^3 = 3$.

90. The number of ways of choosing 2 boys out of 10 boys = $C_2^{10} = \frac{10 \times 9}{1 \times 2} = 45$

The number of ways of choosing 2 girls out of 12 girls = $C_2^{12} = \frac{12 \times 11}{1 \times 2} = 66$

Hence, the total number of ways = $45 \times 66 = 2,970$.

The correct answer is option D.

91. The number of ways the first place of 5-letter words can be filled (Only vowels) = 5.

The number of ways the second place of 5-letter word can be filled = 26. Remember that the second, third and fourth place can be filled with any letter. **Not only consonants!**

The number of ways the third place of 5-letter words can be filled = 26.

The number of ways the fourth place of 5-letter words can be filled = 26.

The number of ways the fifth place of 5-letter words can be filled (Only vowels) = 5.

Total number of ways = $5^2 \times 26^3$

The correct answer is option B.

92. The number of ways he can answer each question = 2 (Truth/false)
Hence, the number of ways he can answer 3 questions = $2 \times 2 \times 2 = 8$

The number of ways he can speak truth for each question = 1
So, the number of ways he can speak truth for 3 questions = $1 \times 1 \times 1 = 1$

Hence, the number of ways he can hide the truth = $8 - 1 = 7$

The correct answer is option C.

93. The number of ways 3 cubes can be chosen = $C_3^5 = \frac{5 \times 4}{1 \times 2} = 10$

The number of ways the 3 chosen cubes can be arranged to present unique pattern $= 3! = 3 \times 2 \times 1 = 6$

Hence, the total number of ways = $10 \times 6 = 60$

The correct answer is option D.

This question can be solved alternatively.

The number of ways 3 cubes can be chosen and arranged = $P_3^5 = 5 \times 4 \times 3 = 60$

94. We know that if there are total n objects, out of these there are p number of one type of identical objects, q number of another type of identical objects, and r number of another type of identical objects, then the number of ways these objects can be arranged in a row to have unique patterns = $\frac{n!}{p! \times q! \times r!}$

There are 9 letters APPEALING; so $n = 9$, and there are 2 Ps and 2 As, so $p = 2$, and $q = 2$.

Thus, the answer is $\frac{9!}{2! \times 2!}$.

The correct answer is option A.

95. Treat L and G as a single object. Though there are 9 letters, this would imply that there are 8 objects to be arranged in a line. So, the total number of words would be $\frac{8!}{2! \times 2!}$. (2 Ps, and 2 As)

However, L and G could interchange their positions as 'LG' or 'GL' in 2! ways. Thus, the correct answer is $2! \times \left[\frac{8!}{2! \times 2!}\right] = \frac{8!}{2!}$

The correct answer is option B.

96. There are 4 vowels (A, A, E, and I) and 5 consonants (P, P, L, N, and G). Treating the vowels as one object, there would be total 6 objects to be arranged in a line. This can be done in $\frac{6!}{2!}$ ways (2 Ps) and the vowels can be arranged in $\frac{4!}{2!}$ ways (2 As). Thus, the number of ways $= \frac{6!}{2!} \times \frac{4!}{2!} = 4,320$.

The correct answer is option E.

97. The number of ways 11 students of out 15 can be selected

$= C_{11}^{15} = C_{(15-11)}^{15} = C_4^{15} = \frac{15 \times 14 \times 13 \times 12}{1 \times 2 \times 3 \times 4} = 1,365$

Note: $C_r^n = C_{n-r}^n$

It is important that the arrangement of 11 students must be ignored as the question states that the order of selection does not matter.

The correct answer is option B.

98. The number of ways 2 generals from 5 UK generals can be selected $= C_2^5 = \frac{5 \times 4}{1 \times 2} = 10$

The number of ways 3 generals from 6 French generals can be selected $= C_3^6 = \frac{6 \times 5 \times 4}{1 \times 2 \times 3} = 20$

Total number of ways $= 10 \times 20 = 200$

The correct answer is option B.

99. The number of ways 2 actors from soaps can be selected $= C_2^5 = \frac{5 \times 4}{1 \times 2} = 10$

The number of ways 3 actors from reality TV shows can be selected $= C_3^8 = \frac{8 \times 7 \times 6}{1 \times 2 \times 3} = 56$

Total number of ways = $10 \times 56 = 560$

The correct answer is option D.

100. Since the captain is to be chosen, the number of players left to be chosen = 15 - 1 = 14.

Again, since the captain is already chosen and the injured player is not to be chosen, number of players left for selection = 17 – 1 – 1 = 15.

The number of ways 14 players can be selected out of 15 players = $C_{14}^{15} = C_{1}^{15} = 15$.
Note:- $C_r^n = C_{n-r}^n$

The correct answer is option A.

101. The number of ways 3 elements can be selected out of 5 elements = $C_3^5 = \frac{5 \times 4 \times 3}{1 \times 2 \times 3} = 10$.

Since the sub-set {A, C, D} is to be excluded, the required number of sets = 10 – 1 = 9.
Note: in a set, the order or the arrangement of the elements is immaterial.

The correct answer is option C.

102. The number of ways 1 male can be selected for the group1 = 10.
The number of ways 1 female can be selected for the group1 = 8.

Hence, total number of ways, 1 male and 1 female can be selected for the group1 = $10 \times 8 = 80$.

Similarly,

The number of ways 1 male can be selected for the group2 = 9 (since one male is already selected for group1.)
The number of ways 1 female can be selected for the group2 = 7. (One female is already selected for group1.)

Hence, the total number of ways 1 male and 1 female can be selected for the group2 = $9 \times 7 = 63$.

Now, which pair of dancers will join which group is not specified as group numbers are not fixed. So, if we do 80×63 = 5,040 ways, we will get an ordered selection, which is incorrect. Since there are two groups, the answer should be 5,040/2! = 2,520.

The correct answer is option D.

Alternate approach:

We can select 2 males for the two groups in $C_2^{10} = 45$ ways.
We can select 2 females for the two groups in $C_2^8 = 28$ ways.

Let us say the males are M1, M2 and the females are F1, F2. These 4 people can be grouped in 2 ways: (M1F1 and M2F2) or (M1F2 and M2F1).

Thus, the answer is 45 × 28 × 2 = 2,520.

Had you done this way—Total number of ways = $C_2^{10} \times C_2^8 = 1,260$, it would have been wrong since the question asks for 1 male and 1 female for each group, but 1,260 represents selection of 2 males and 2 females. This means 4 dancers for only one group.

103. **Selection 1:** 4 batsmen, 6 bowlers, and 1 wicket-keeper

The number of ways = $C_4^8 \times C_6^6 \times C_1^2 = C_4^8 \times C_0^6 \times C_1^2 = \left(\frac{8 \times 7 \times 6 \times 5}{1 \times 2 \times 3 \times 4}\right) \times 1 \times 2 = 140$
Note:- $C_r^n = C_{n-r}^n$ **and** $C_0^n = 1$

Selection 2: 5 batsmen, 5 bowlers, and 1 wicket-keeper

The number of ways = $C_5^8 \times C_5^6 \times C_1^2 = C_3^8 \times C_1^6 \times C_1^2 = \left(\frac{8 \times 7 \times 6}{1 \times 2 \times 3}\right) \times 6 \times 2 = 672$

Selection 3: 6 batsmen, 4 bowlers, and 1 wicket-keeper

The number of ways = $C_6^8 \times C_4^6 \times C_1^2 = C_2^8 \times C_2^6 \times C_1^2 = \left(\frac{8 \times 7}{1 \times 2}\right) \times \left(\frac{6 \times 5}{1 \times 2}\right) \times 2 = 840$

Hence, the total number of ways = 140 + 672 + 840 = 1,652.

The correct answer is option D.

104. The person can choose one out of 4 salads in 4 ways and one topping out of 6 toppings in 6 ways.

Thus, the total number of ways of selecting a salad AND a topping = 4 × 6 = 24.

The correct answer is option B.

105. The condition: the string of the last three digits should not start nor end in '0'.

Thus, the first of the last three digits can be filled using digits from '1' to '9' i.e. 9 ways (excluding '0').

The second of the last three digits can be filled using digits from '0' to '9' i.e. 10 ways (including '0').

The third of the last three digits can be filled using digits from '1' to '9' i.e. 9 ways (excluding '0').

Thus, the number of such telephone numbers = $9 \times 10 \times 9 = 810$.

The correct answer is option C.

106. We can select 3 men from four in $C_3^4 = 4$ ways.

We can select 2 women from three in $C_2^3 = 3$ ways.

Thus, we can select 3 men AND 2 women in $C_3^4 \times C_2^3 = 4 \times 3 = 12$ ways.

The correct answer is option C.

107. To make a 3-digit number, we need to fill up 3 blank positions using 0, 2, 3, and 5 digits. Since the number should be even, the units digit should be '0' or '2'.

Case I: The units digit is '0': _ _ 0

The first place can be filled using any of '2,' '3,' and '5' in 3 ways.

The second place can be filled in one less way (since repetition is not allowed) i.e. 2 ways.

Thus, we can have $3 \times 2 = 6$ such numbers.

Case II: The units digit is '2': _ _ 2

The first place can be filled using any of '3' or '5' in 2 ways (since '0' cannot be used in the first place).

The second place can be filled in two ways (since 0 can be used) i.e. 2 ways.

Thus, we can have $2 \times 2 = 4$ such numbers.

Thus, total # of such numbers = $6 + 4 = 10$.

The correct answer is option B.

108. John can invite one friend in C_1^4 ways.

He can similarly invite two or three or all four friends in C_2^4, C_3^4 and C_4^4 ways, respectively.

Thus, the number of ways of doing this = $C_1^4 + C_2^4 + C_3^4 + C_4^4 = 4 + 6 + 4 + 1 = 15$.

Alternatively, we can say that John has two options for each friend: either to call him or not to call him.

Thus, total number of choices with John = $2 \times 2 \times 2 \times 2 = 2^4 = 16$

These 16 choices include a case when John invites 'none'. Thus, we need to remove that case.

Hence, total ways = $2^4 - 1 = 15$.

The correct answer is option B.

109. INDIANA consists of 7 letters; of which 2 are Is, 2 are Ns and 2 are As.

Thus, the number of ways in which the letters can be arranged

$$= \frac{n!}{p! \times q! \times r!} = \frac{7!}{2! \times 2! \times 2!} = \frac{7 \times 6 \times 5 \times 4 \times 3 \times 2 \times 1}{8} = 7 \times 6 \times 5 \times 3 = 630.$$

The correct answer is option A.

110. First we need to select five members from the 8 probable members in C_5^8 ways.

Then we need to decide the speaker from the selected 5 members in C_1^5 ways.

Thus, the total number of ways of selecting the team AND deciding the speaker

$$= C_5^8 \times C_1^5 = \frac{8!}{5! \times 3!} \times 5 = \frac{8!}{4! \times 3!} = \frac{8 \times 7 \times 6 \times 5 \times 4!}{4! \times 3!} = 8 \times 7 \times 5 = 280.$$

The correct answer is option C.

111. Let us find the total number of arrangements without any constraint.

Eight people can be seated in 8! ways.

Let us find the number of cases when the particular boy and girl sit together.

We group the two of them as one unit. This results in 7 students to be arranged in a line.

Thus, it can be arranged in 7! ways.

But, for each of the above arrangements, the particular boy and the particular girl can interchange their position in 2! ways.

Thus, the total number of ways = $7! \times 2!$

Thus, the number of ways when the particular boy and the particular girl do not sit together

$= 8! - 7! \times 2! = 7! \times (8 - 2) = 6 \times 7!$

The correct answer is option A.

112. The student has four ways to deal with each question: attempt both parts; attempt only the first part, attempt only the second part; and leave both parts.

Since there are 5 questions each having 4 ways, total number of ways for a student = $4 \times 4 \times 4 \times 4 \times 4 = 4^5$.

Out of these 4^5 ways, there is one way in which he/she does not attempt any part of any question and that we need to ignore.

Thus, the number of possible ways = $4^5 - 1 = 2^{10} - 1 = 1,024 - 1 = 1,023$.

The correct answer is option C.

113. Since the question asks for calculating the number of ways of selecting, we will apply 'Combination' and not 'Permutation'.

We know that, the number of ways of selecting r objects out of n objects = $\boldsymbol{C_r^n} = \dfrac{\boldsymbol{n!}}{\boldsymbol{r! \times (n-r)!}}$

A = # of ways 11 players can be selected out of 25 players = $C_{11}^{25} = \dfrac{25!}{11! \times (14)!}$

B = # of ways 14 players can be selected out of 25 players = $C_{14}^{25} = \dfrac{25!}{14! \times (11)!}$

As we see that A and B are equal in value, there is no need to calculate the values.

The correct answer is option E.

114. Since the question asks for calculating the number of ways of assigning marbles to containers, we will apply 'Permutation,' since there can be different ways of assigning marbles to containers.

We know that the number of ways of assigning r objects in n places = n^r

A = # of ways four marbles can be placed in three containers = 3^4.

B = # of ways three marbles can be placed in four containers = 4^3.

Thus, $A = \frac{3^4}{4^3} \times B$

The correct answer is option A.

115. Since the question asks for calculating the # of numbers formed using digits, we will apply 'Permutation' since for the same digits, many numbers can be formed by changing the positions of the digits.

Since 4-digit numbers are to be formed, we need to fill up four blank positions: _ _ _ _.

The first position can be filled up in one way (only possible digit is '1' since the number should be less than 2,000).

The second, third and fourth positions can be filled up in three ways each (since each of '0,' '1,' and '2' can be used).

Thus, the total # of such numbers = $1 \times 3 \times 3 \times 3 = 27$.

The correct answer is option D.

116. Since the question asks for calculating the number of ways of selecting, we will apply 'Combination' and not 'Permutation'.

We know that the number of ways of selecting r objects out of n objects = $C_r^n = \frac{n!}{r! \times (n-r)!}$

Since at least one woman needs to be a part of the team, we have two cases:

1. Only one woman (from two) and two men (from three): $C_1^2 \times C_2^3 = 2 \times 3 = 6$ ways

2. Both women (from two) and one man (from three): $C_2^2 \times C_1^3 = 1 \times 3 = 3$ ways

Thus, the total of ways = 6 + 3 = 9 ways.

The correct answer is option C.

117. Since the question asks for calculating the seating arrangements, we will apply 'Permutation'.

We know that the # of ways of arranging n objects in a line = $n!$

The boys and girls can sit in alternate positions in either of the two ways:

1. B G B G B G

2. G B G B G B

The three boys and the three girls can be separately arranged in their positions among themselves in 3! = 6 ways each

Thus, the total number of ways for each of the cases = $6 \times 6 = 36$ ways.

Thus, the total number of ways = $36 \times 2 = 72$ ways.

The correct answer is option E.

118. Since the question asks for calculating the seating arrangements, we will apply 'Permutation'.

We know that the # of ways of arranging n objects in a circle = $(n - 1)!$

We group the particular boy and the particular girl together. Thus, we now have 6 things to arrange in a circle: 3 boys, 2 girls and the boy-girl pair.

This can be done in $(6 - 1)! = 5! = 120$ ways.

For each such way, the boy and girl can interchange positions.

Thus, total ways = $120 \times 2 = 240$.

The correct answer is option B.

119. Since the question asks for calculating the number of ways of selecting, we will apply 'Combination' and not 'Permutation'.

We know that, number of ways of selecting r objects out of n objects = $C_r^n = \frac{n!}{r! \times (n-r)!}$

The number of ways of selecting one mathematics book from 6 different books = $C_1^6 = 6$ ways.

The number of ways of selecting one physics book from 3 identical books = 1 way (since there is nothing to choose if the objects are identical; thus, the number of identical objects is immaterial.).

Thus, there are a total of $6 \times 1 = 6$ ways.

The correct answer is option C.

120. There are a total of 5 symbols and we have to take all of them together to form codes, considering that the □ must NOT be at the beginning and the ◊ must NOT be at the end.

The number of ways 5 symbols can be arranged = $P_5^5 = 5! = 120$. We will deduct the number of codes that start the □ with and ends with the ◊; thus, this asks for the arrangement of 3 symbols.

The number of ways 3 symbols can be arranged = $P_3^3 = 3! = 6$.

Thus, the desired number of codes = $120 - 6 = 114$.

The correct answer is option D.

121. Each person is assigned a sequence of three different letters from the set of 8 letters: {A, B, C, D, E, F, G, H}.

Also, the order of the letters in a sequence differentiates one sequence from the other.

The number of ways of selecting 3 letters from the 8 letters = C_3^8

$= \frac{8 \times 7 \times 6}{3 \times 2 \times 1} = 56$

The number of ways the order of the letters can be decided for the same three letters

= The number of ways 3 letters can be arranged in a line

$= 3 \times 2 \times 1 = 6$

Thus, total number of distinct sequences = $56 \times 6 = 336$

Alternately, we have

The total number of sequences formed $= P_3^8$

$= 8 \times 7 \times 6 = 336$

Since 36 of the sequences were not assigned, the number of participants

$= 336 - 36 = 300$

The correct answer is option C.

122. Probability of an event $= \dfrac{\text{Number of favorable events}}{\text{Number of total events}}$;

Here, number of favorable events = number of ways 2 green balls can be selected from 10 balls $= C_2^{10}$;

Number of total events = number of ways any 2 balls can be selected from 30 balls $= C_2^{30}$

Hence, probability of getting both green balls $= \dfrac{C_2^{10}}{C_2^{30}} = \dfrac{\frac{10\times 9}{1\times 2}}{\frac{30\times 29}{1\times 2}} = \dfrac{3}{29}$

The correct answer is option A.

123. Since the coin is unbiased, it means that probabilities of getting a 'Head' or getting a 'Tail' upon a single toss are equal = 1/2.

Traditional approach of solving this question would be to find the probability for each time the coin turns Head and then add them, but that is a lengthy process.

An efficient approach would be to find out the probability of not getting a Head EVEN once, and then deducting it from total probability (= 1); this will assure that we get the Heads—one time, two times, and all three times.

So, probability of getting at least one Head

= 1 − (Probability of getting Tails in all 3 tosses)

$= 1 - \left(\frac{1}{2} \times \frac{1}{2} \times \frac{1}{2}\right) = 1 - \frac{1}{8} = \frac{7}{8}$

The correct answer is option D.

124. Number of students = 36; number of Boys = number of girls = 18

Since the 3 selected students are boys, the number of boys remaining = 15. & total number of students remaining = 33.

So, the probability of selecting 4th boy

$$= \frac{\text{Number of Boys remaining}}{\text{Total no. of students remaining}} = 15/33 = 5/11$$

The correct answer is option D.

Remember that this question is different from a similar question: *...what is the probability that all the students selected are boys?* The difference is that in this question, 4 boys and no girl are to be chosen, while in the earlier one, 3 boys were already chosen, and only the 4th boy was to be chosen.

We will deal with this kind of question in subsequent questions.

125. Number of diplomats = 9; Number of Japanese = 3; & Number of Chinese = 6.

The probability of choosing at least 2 Japanese

= P (Japanese = 2, Chinese = 1) + P (Japanese = 3, Chinese = none)

$$\Rightarrow P\text{ (Japanese = 2, Chinese = 1)} = \frac{C_2^3 \times C_1^6}{C_3^9} = \frac{C_1^3 \times C_1^6}{C_3^9} = \frac{3 \times 6}{\frac{9\times8\times7}{1\times2\times3}} = 3/14; [C_r^n = C_{n-r}^n]$$

$$\Rightarrow P\text{ (Japanese = 3, Chinese = none)} = \frac{C_3^3}{C_3^9} = \frac{1}{\frac{9\times8\times7}{1\times2\times3}} = 1/84; [C_n^n = 1]$$

Hence, the probability of choosing at least 2 Japanese = 3/14 + 1/84 = 19/84

The correct answer is option C.

126. The problem will not be solved in only condition when none among John, Peter, and Harry can solve it. Even if one of them could crack it, the problem will be solved as they have to solve it collectively.

So, probability of not solving the problem = P(John could not solve it) × P(Peter could not solve it) × P(Harry could not solve it)

$$=> \left(1 - \frac{4}{5}\right) \times \left(1 - \frac{2}{3}\right) \times \left(1 - \frac{3}{4}\right) = \frac{1}{5} \times \frac{1}{3} \times \frac{1}{4} = \frac{1}{60}$$

The correct answer is option A.

127. Total number of players = 5;

Number of players apart from John and Harry = 3;

Probability of choosing both John and Harry (2 players) only one among the rest

$$= \frac{C_2^2 \times C_1^3}{C_3^5} = \frac{1 \times 3}{\frac{5\times4\times3}{1\times2\times3}} = 3/10$$

The correct answer is option C.

128. Total Number of actors = 5;

Since Jack and Steve need to be in the selection and Suzy is to be left out, only one selection matters.

Number of actors apart from Jack, Steve, and Suzy = 2;

Probability of choosing 3 actors, incl. Jack and Steve, but not Suzy = $\frac{C_1^2}{C_3^5} = 1/5$

The correct answer is option A.

129. Number of silver coins = p; Number of gold coins = q;
so total umber of coins = $p + q$;

The probability of getting 1st coin a silver and 2nd coin a gold
= P (silver coin) × P (gold coin)

$$\Rightarrow \text{P (silver coin)} = \frac{C_1^p}{C_1^{p+q}} = \frac{p}{p+q}$$

After 1st draw, there would be only $(p + q - 1)$ coins left.

$$\Rightarrow \text{So, P (gold coin)} = \frac{C_1^q}{C_1^{p+q-1}} = \frac{q}{p+q-1}$$

⇒ Hence, the probability of getting 1st coin a silver and 2nd coin a gold

$$= \left(\frac{p}{p+q}\right) \times \left(\frac{q}{p+q-1}\right)$$

The correct answer is option E.

You could have solved this question by picking some values for p and q. Say $p = 2$ and $q = 4$; after solving, plug in the values of p and q in the options, and cross check which option matches the derived value. The derived value is 4/15 and matches with option E.

130. A number is divisible by 2, if it is an even number.

Between 100 and 400, there are $400 - 100 + 1 = 301$ numbers. Out of these, 151 numbers are Even and 150 numbers are Odd.

So, the probability of winning a gift = $\frac{\text{Number of even numbered coupons}}{\text{Total number of coupons}}$

= 151/301

The correct answer is option C.

131. Probability of choosing any paired husband from Group A = 5/50 = 1/10;

The chosen husband will have only one married woman in Group B as his wife. So, the probability of choosing paired wife = 1/40;

So, the probability of choosing a paired husband-wife = (1/10) × (1/40) = 1/400;

The correct answer is option A.

The answer (5/50) × (5/40) = 1/80 is wrong because choosing any husband from Group A and any wife from Group B will not necessarily make the correct husband-wife pair.

132. Say the number of green balls are x, so the total number of balls = $(4 + x)$;

Number of ways both ball drawn will be black = C_2^4;

Total number of ways 2 balls can be drawn = C_2^{4+x};

Hence, the probability of getting both balls being black =

$$\frac{C_2^4}{C_2^{(4+x)}} = \frac{\frac{4\times3}{1\times2}}{\frac{(4+x)\times(3+x)}{1\times2}} = \frac{1}{6} \text{ (given)};$$

This reduces to $\frac{4 \times 3}{(4 + x) \times (3 + x)} = \frac{1}{6}$;

We do not recommend you to solve the quadratic equation to get the value of x, you must plug in the values of x from the options and check if right hand side and left hand side are equal. If yes, the option is the correct answer.

Trying with $x = 5$ will satisfy the equation as $\frac{4 \times 3}{9 \times 8} = \frac{1}{6} \Rightarrow \frac{1}{6} = \frac{1}{6}$.

The correct answer is option C.

133. The better approach would be to find out the probability of getting both non-defective spanners, and then deduct the probability from '1' to get the answer. It will make sure that either one or both spanners are defective.

So, the probability of getting both non-defective spanners

$$= \frac{C_2^7}{C_2^{10}} = \frac{\frac{7 \times 6}{1 \times 2}}{\frac{10 \times 9}{1 \times 2}} = \frac{7}{15}$$

Hence, the probability of getting at least one defective spanner = 1 − 7/15 = 8/15

The correct answer is option D.

134. Number of students = 20; Number of boys = 15; hence, the number of girls = 5;

So, the probability of getting 1 boy and 2 girls

$$= \frac{C_1^{15} \times C_2^5}{C_3^{20}} = \frac{(15/1) \times (5 \times 4/1 \times 2)}{(20 \times 19 \times 18)/(1 \times 2 \times 3)} = \frac{5}{38}$$

The correct answer is option A.

135. The meaning of 'all the 100 shots do not miss the target' is that at least one shot hits the target. It would be very time-consuming to calculate the answer if we follow the traditional approach as this involves handling 100 terms. Let us do it in another way.

Let all the 100 shots miss and then we would deduct the derived value from '1' to get the answer.

Given that:

Probability that a shot hits the target = 1/100

Thus, probability that a shot misses the target = 1 − 1/100 = 99/100

Probability that all the 100 shots miss the target $= \left(\frac{99}{100}\right)^{100}$

We need to determine the probability that ALL shots do not miss.

Required probability = 1 − P(All shots miss the target)

$= 1 - \left(\frac{99}{100}\right)^{100}$

$= \frac{(100^{100} - 99^{100})}{100^{100}}$

The correct answer is option E.

136. Probability of losing the bet each day = 0.4. It means that the probability of winning the bet each day = 1 - 0.4 = 0.6.

Probability (Winning = 2 days)
= (Number of ways of choosing any 2 winning days) × (Probability of winning on 2 days, and losing on the other day)
$= [C_2^3] \times [(0.6) \times (0.6) \times (0.4)] = 0.432;$

The correct answer is option E.

137. Since we do not want balls of the same color, we need to draw one red AND one blue ball.

We can select one red and one blue ball in C_1^{10} and C_1^{20} ways, respectively.

We can select two balls from the 30 balls in C_2^{30} ways.

Thus, the required probability $= \frac{C_1^{10} \times C_1^{20}}{C_2^{30}} = \frac{10 \times 20}{\frac{30 \times 29}{1 \times 2}} = \frac{10 \times 20}{15 \times 39} = \frac{40}{87} =\approx 0.46$

The correct answer is option D.

138. Probability of a missile striking a target = 0.6.
Thus, probability of the missile missing the target = 0.4.

Calculating 'the probability of hitting the target at least once' with the traditional approach would be time-consuming. The only possibility that is NOT desired is: when no missile hits the target, and thereby subtracting not-desired probability form '1', will give us the answer.

For three missiles, the probability that all three miss the target = 0.4 × 0.4 × 0.4 = 0.064.

Thus, probability that at least one missile hits the target
= 1 - (Probability that all three missiles miss the target)
= 1 - 0.064 = 0.936.

The correct answer is option C.

139. Three balls of different colors imply one white, one red and one blue ball.

These may be selected in C_1^4, C_1^3 and C_1^3 ways, respectively.

However, their drawings can be done in 3! = 6 ways.

There are a total of 4 + 3 + 3 = 10 balls.

Ways of selecting three balls successively without replacement = $C_1^{10} \times C_1^9 \times C_1^8$.

Thus, the required probability = $\frac{6 \times C_1^4 \times C_1^3 \times C_1^3}{C_1^{10} \times C_1^9 \times C_1^8} = \frac{6 \times 4 \times 3 \times 3}{10 \times 9 \times 8} = \frac{3}{10}$

The correct answer is option E.

140. We can select balls of the same color if the balls are either both white or both black.

Thus, the necessary condition is:

- The ball from 1st bag is white AND that from 2nd bag is also white

  OR

- The ball from 1st bag is black AND that from 2nd bag is also black

Thus, the required probability = $\frac{C_1^3 \times C_1^4}{C_1^5 \times C_1^9} + \frac{C_1^2 \times C_1^5}{C_1^5 \times C_1^9} = \frac{3 \times 4}{5 \times 9} + \frac{2 \times 5}{5 \times 9} = \frac{12}{45} + \frac{10}{45} = \frac{22}{45}$

The correct answer is option B.

141. A blue ball can be drawn only if a ball is drawn from the first box as the second box does not have blue balls.

Thus, the only condition is: Heads appears AND a blue ball is drawn.

Probability of a Head appearing = $\frac{1}{2}$

Probability of drawing a blue ball = $\frac{C_1^3}{C_1^{(3+2)}} = \frac{C_1^3}{C_1^5} = \frac{3}{5}$

Thus, the required probability = $\frac{1}{2} \times \frac{3}{5} = \frac{3}{10} = 0.3$

The correct answer is option C.

142. The ball would all be of the same color if all three are yellow OR all are green.

Thus, probability $= \frac{C_3^5}{C_3^9} + \frac{C_3^4}{C_3^9} = \left[\frac{\frac{5 \times 4 \times 3}{3!}}{\frac{9 \times 8 \times 7}{3!}}\right] + \left[\frac{\frac{4 \times 3 \times 2}{3!}}{\frac{9 \times 8 \times 7}{3!}}\right] = \frac{10}{84} + \frac{4}{84} = \frac{14}{84} = \frac{1}{6}$

Thus, probability that balls are not all of the same color $= 1 - \frac{1}{6} = \frac{5}{6} = 0.83$

Alternatively, since the balls are not all of the same color, there are only two possible cases:

(1) Two yellow AND one green

OR

(2) Two green AND one yellow

Hence, the required probability $= \frac{C_2^5 \times C_1^4}{C_3^9} + \frac{C_1^5 \times C_2^4}{C_3^9} = \frac{10 \times 4}{84} + \frac{5 \times 6}{84} = \frac{70}{84} = \frac{5}{6}$

The correct answer is option C.

143. Traditional approach:

- $A = \frac{C_2^{12}}{C_2^{22}} = \left[\frac{12 \times 11}{22 \times 21}\right] \times \left[\frac{1 \times 2}{1 \times 2}\right] = \frac{2}{7}$
- $B = \frac{C_2^{10}}{C_2^{22}} = \left[\frac{10 \times 9}{22 \times 21}\right] \times \left[\frac{1 \times 2}{1 \times 2}\right] = \frac{15}{77}$

We can rewrite $A = \frac{2}{7} = \frac{22}{77}$, thus, $A > B$; or options A and E are wrong.

- $C = \frac{C_1^{12} \times C_1^{10}}{C_2^{22}} = \left[\frac{12 \times 10}{22 \times 21}\right] \times 2 = \frac{40}{77}$

So, the answer is $C > A > B$.

The correct answer is option D.

Logical deduction approach:

Since A = Probability of drawing two **red** balls from a bag containing 12 red and 10 blue balls, and B = Probability of drawing two **blue** balls from a bag containing 12 red and 10 blue balls, value of A would be greater than B as there are two more red balls for A compared to blue balls for B; so, A > B.

Again, C = Probability of drawing **one red and one blue** ball from a bag containing 12 red and 10 blue balls, value of C would be greater than A and greater than B as in case of C, there are more chances **one red** and **one blue**, versus **both red** and **both blue**; so, C > A > B.

144. Let us calculate A.

We know that probability of head = $\frac{1}{2}$.

Hence, the required probability of all heads = $\left(\frac{1}{2} \times \frac{1}{2} \times \frac{1}{2}\right) = \frac{1}{8}$

Similarly, B = probability of all tails = $\left(\frac{1}{2} \times \frac{1}{2} \times \frac{1}{2}\right) = \frac{1}{8}$

So, A = B; it means that options A and b are wrong.

Let us calculate C.

C = Probability of getting two heads and one tail = $C_1^3 \times \left(\frac{1}{2} \times \frac{1}{2} \times \frac{1}{2}\right) = \frac{3}{8}$;

$C_1^3 = 3$ is the number of ways in which two heads may appear out of three tosses: [1st & 2nd]; [1st & 3rd]; [2nd & 3rd]

Thus, $C = 3A = 3B$

The correct answer is option D.

145. Since $x^2 \leq 16$, we can see that from the numbers in the given set $-2,\ 0,\ 2$ and 4 satisfy:

- $(-2)^2 = 4 < 16$ (qualifies)
- $0^2 = 0 < 16$ (qualifies)
- $2^2 = 4 < 16$ (qualifies)

- $4^2 = 16$ (qualifies)

- $6^2 = 36 > 16$ (does not qualify)

Thus, there are four favorable values of x out of the five values in the set.

Hence, the required probability = $\frac{4}{5}$

The correct answer is option D.

146. We know that,

Probability of A **or** B = P (A) + P (B) – P(A ∩ B)

There are three odd numbers: 1, 3, & 5 and there are three prime numbers: 2, 3, & 5.

There are two numbers which are both odd as well as prime: 3 & 5.

Thus, the required probability = P (odd) + P (prime) – P (odd and prime)

$= \frac{3}{6} + \frac{3}{6} - \frac{2}{6} = \frac{4}{6} = \frac{2}{3}$

The correct answer is option C.

147. There are a total of $6 \times 6 = 36$ cases when a dice is thrown twice.

Of these, there are six cases when the throws give the same number.

Thus, there are 36 – 6 = 30 cases where the numbers on the two faces are different.

Of these 30 cases, it is obvious that half the cases i.e. 15 would have the number on the first throw greater than the other and similarly 15 more for the other way round.

Thus, the required probability = $\frac{15}{36} = \frac{5}{12}$

The correct answer is option B.

148. There are five cards numbered '1' to '5'.

Thus, two cards (same card can now be used twice) which add up to eight could be (5, 3); (3, 5) or (4, 4).

Thus, the required probability = Probability of getting

- 1st card = 5 and 2nd card = 3

  OR

- 1st card = 3 and 2nd card = 5

  OR

- 1st card = 4 and 2nd card = 4

$= \frac{1}{5} \times \frac{1}{5} + \frac{1}{5} \times \frac{1}{5} \times \frac{1}{5} \times \frac{1}{5} = \frac{3}{25}.$

The correct answer is option A.

149. The possible cases whn one number from A and one from B add up to 9 are (1, 8) and (6, 3).

Thus, there are two favorable cases.

Total number of ways of selecting one number from A and one from B = $3 \times 3 = 9$.

Hence, the required probability = $\frac{2}{9}$

The correct answer is option B.

150. We know that,

$P(A) + P(\overline{A}) = 1$; where $P(\overline{A})$ is the probability of the event A not happening

Probability of picking a red ball = P (Red)

= 1 − [P (Blue] + P (Green)]

$= 1 - \left[\frac{1}{4} + \frac{2}{3}\right] = \frac{1}{12} = \frac{2}{24}$

Hence, the number of red balls

= (Total number of balls) × P (Red)

$= 24 \times \frac{2}{24} = 2$

The correct answer is option B.

151. Say the number of green balls are x, so the total number of balls = $(5 + x)$;

Number of ways both ball drawn will be green = C_2^x;

Total number of ways 2 balls can be drawn = C_2^{5+x};

Hence, the probability of getting both balls being green =

$$\frac{C_2^x}{C_2^{(5+x)}} = \frac{[x \times (x-1)]/[1 \times 2]}{[(5+x) \times (4+x)]/[1 \times 2]} = \frac{1}{6} \text{ (given)};$$

This reduces to $\frac{x(x-1)}{(5+x) \times (4+x)} = \frac{1}{6}$;

We do not recommend you to solve the quadratic equation to get the value of x, you must plug in the values of x from the options and check if right hand side and left hand side are equal. If yes, the option is the correct answer.

Trying with $x = 4$ will satisfy the equation as $\frac{4 \times 3}{9 \times 8} = \frac{1}{6} \Rightarrow \frac{1}{6} = \frac{1}{6}$

The correct answer is option B.

152. Since 53% of the total people did not take juice, we can say that $(100 - 53) = 47\%$ of the total people took juice.

Let the number of people who took both be y.

Let us represent the above information using a Venn-diagram, as shown below:

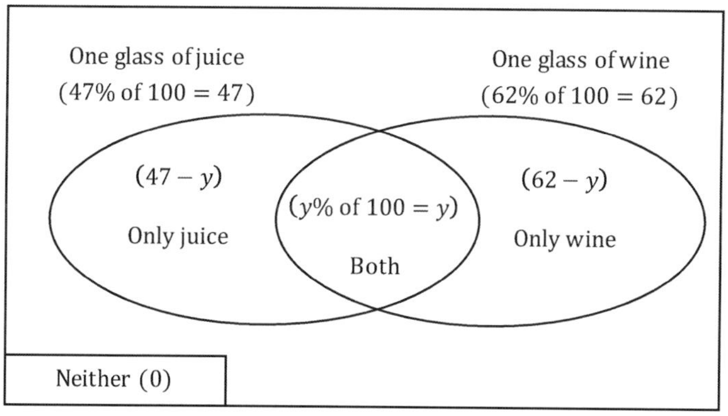

We know that 53% did not take juice

$=> 62 - y = 53$

$=> y = 9$

Thus, the final (solved) Venn-diagram is

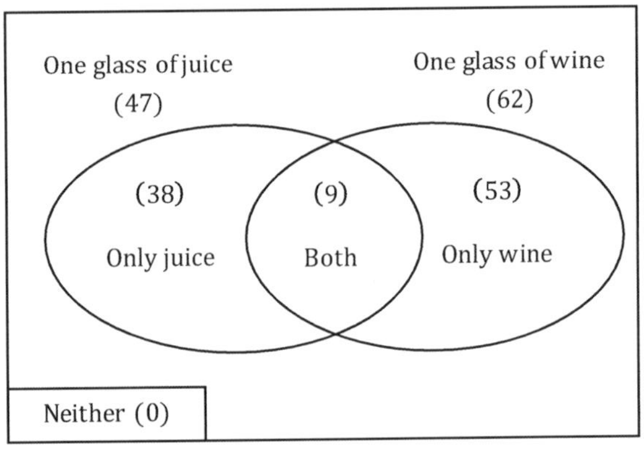

Thus, we have

Number of glasses needed to serve those who took only juice = 38.

Number of glasses needed to serve those who took only wine = 53.

Number of glasses needed to serve those who took both = $9 \times 2 = 18$.

Thus, total number of glasses used = $38 + 53 + 18 = 109$.

Alternate approach:

First we calculate the value of those who took juice, i.e., 47.

Now, we observe that if we simply add the values of the two sets;, i.e., those who took juice and those who took wine, their intersection would be added twice.

Since each glass is to be used only once, the people in the intersection would actually use two glasses each.

Thus, by simply adding 47 and 62, we should get the total number of glasses.

Thus, total number of glasses used = $47 + 62 = 109$.

The correct answer is option E.

153. Let us represent the above information using a Venn-diagram, as shown below:

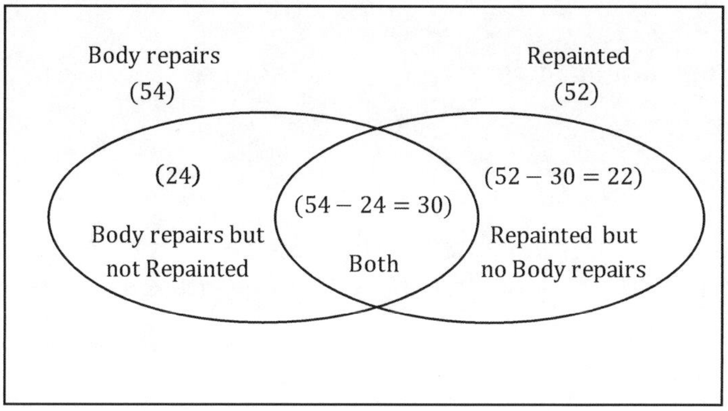

Number of cars having both jobs = 54 - 24 = 30.

Thus, the number of cars that were repainted but did not have body repairs = 52 - 30 = 22

The correct answer is option E.

154. The above information can be represented in a Venn-diagram as shown below:

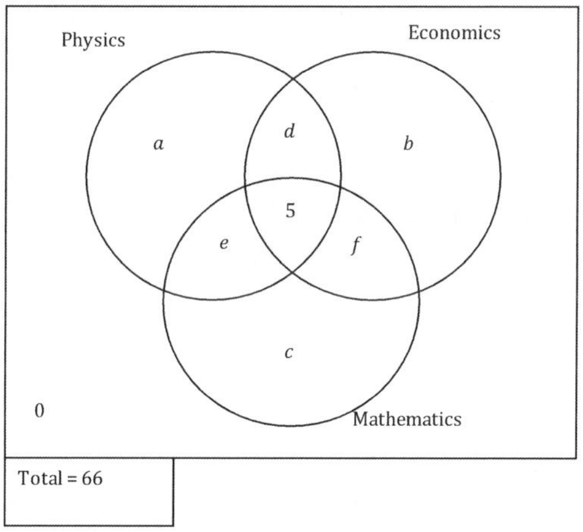

Thus, we have

- 28 students did not take up either Physics or Economics
  => $c = 28 \ldots$(i)

- 15 students did not take up either Mathematics or Economics
  => $a = 15 \ldots$(ii)

- 45 students did not take up Economics
  => $a + c + e = 45$

=> $e = 45 - (a + c) = 45 - (28 + 15)$
=> $e = 2 \ldots$ (iii)

- 30 students did not take up Mathematics
  => $a + d + b = 30$
  => $d + b = 30 - a = 30 - 15$ => $d + b = 15 \ldots$ (iv)

- Number of students who did not take up either Mathematics or Economics, exceeded the number of students who did not take up Mathematics or Physics by 2
  => $a = b + 2$ => $b = a - 2 = 15 - 2$
  => $b = 13 \ldots$ (v)

  Thus, from (iv) and (v), we have
  $d = 15 - b = 15 - 13$
  => $d = 2 \ldots$ (vi)

- Total number of students is 66
  => $a + b + c + d + e + f + 5 = 66$
  => $f = 66 - (a + b + c + d + e + 5) = 66 - (15 + 13 + 28 + 2 + 2 + 5)$
  => $f = 1$

  Thus, the number of students who took up both Economics and Mathematics
  $= f + 5 = 1 + 5$
  $= 6$

The complete Venn-diagram is shown below for reference:

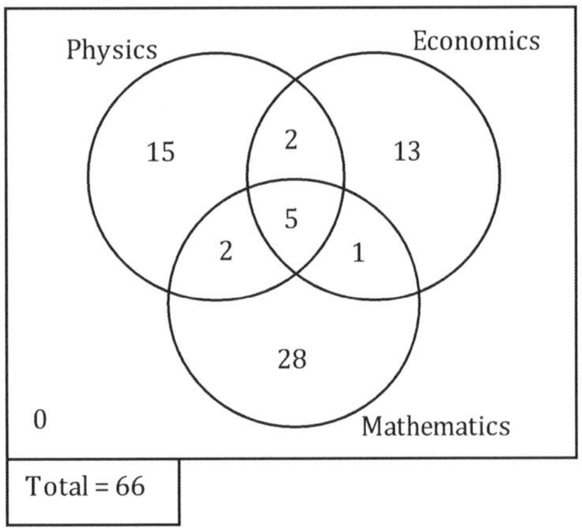

The correct answer is option C.

155. The data can be represented in a Venn-diagram as shown below:

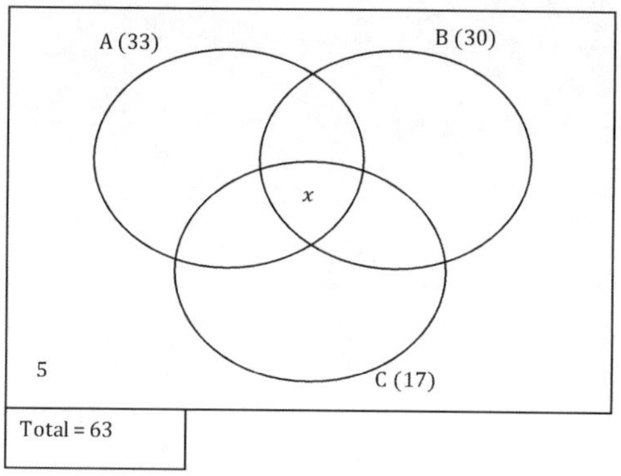

We know that:

A & B = B & C = A & C = 9

Also, we know that:

A or B or C =63 − 5 = 58

Thus, we have

A or B or C = A + B + C− (A & B + B & C + A & C) +A & B & C

$=> 58 = 33 + 30 + 17 - (9 + 9 + 9) + x = 58 = 80 - 27 + x$

$=> x = 5$

Alternate approach:

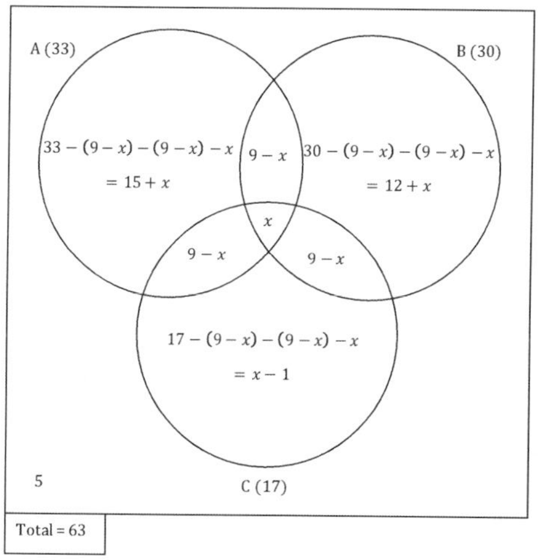

Based on the Venn-diagram shown above, we have

$(15 + x) + (12 + x) + (x - 1) + (9 - x) + (9 - x) + (9 - x) + x + 5 = 63$

$=> 58 + x = 63$

$=> x = 5$

The correct answer is option D.

156.

Factor	Percent of prospects
Brand name	66%
Safety	44%
Price	38%

Percent of prospects citing "Brand name" = 66%

Percent of prospects citing "Safety" = 44%

Percent of prospects citing both of the above = 25%

For two sets A & B:

(A or B) = A + B - (A and B)

Thus, percent of prospects citing "Brand name" or "Safety"

= 66% + 44% - 25%

= 85%

The number of prospects who cited "Price," but neither "Brand Name" nor "Safety," will be the maximum if there is a MINIMUM intersection (overlap) between those who cited "Price" and those who cited "Brand Name or Safety." Also, the percent of prospects who did not cite any of the three should be '0%', as shown in the diagram below:

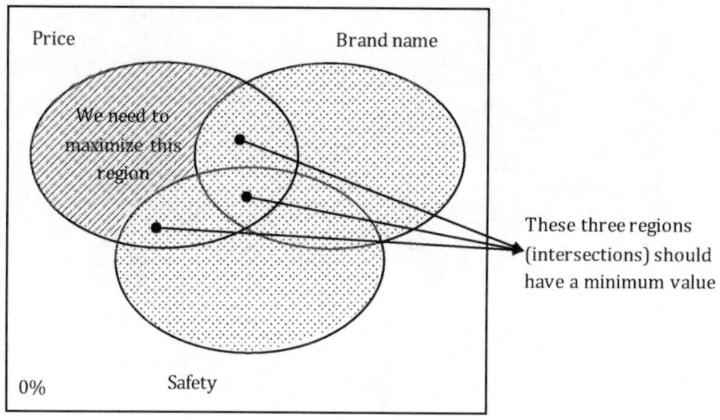

The above diagram may be simplified to:

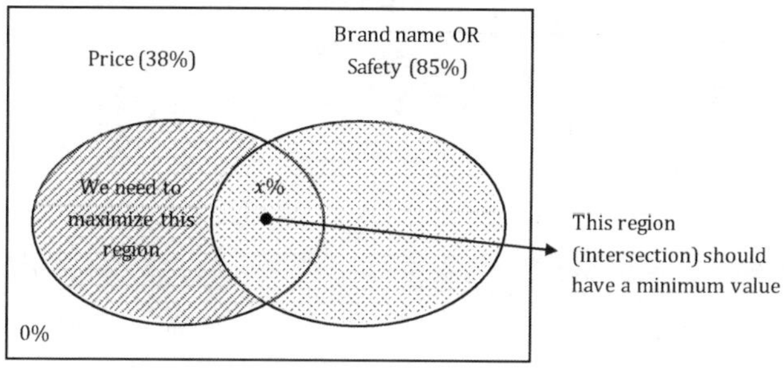

Thus, we have

$38 + 85 - x = 100$

$=> x = 23$

Thus, the required maximum value

$= (38\% - 23\%)$ of 800

$= 15\%$ of 800

$= \frac{15}{100} \times 800$

$= 120$

The correct answer is option A.

157.

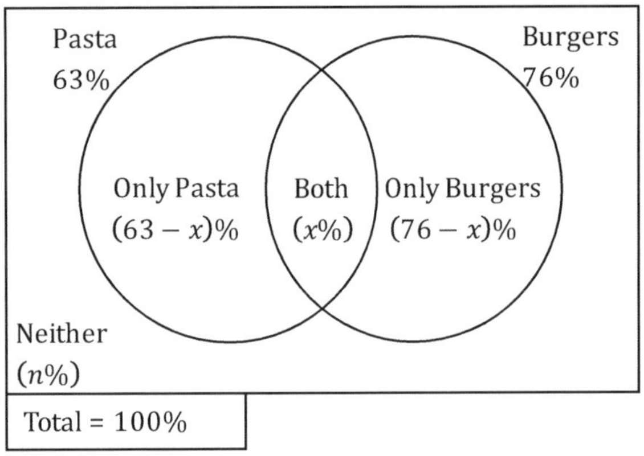

We have

$63 + 76 - x + n = 100$

$=> x = n + 39 \ldots$(i)

Thus, the minimum value of x occurs when $n = 0$

$=> x$ (Minimum) = 39

Again, the maximum value of x occurs when the entire 63% (Pasta) forms a part of 76% (Burgers); thus, the maximum value of the intersection of Pasta and Burgers being 63%.

Thus, we have

$39 \le x \le 63$ - Statement I is correct

Percent of people who like only one among Pasta and Burgers

$= (63 - x) + (76 - x) = 139 - 2x$

The above is a maximum when x is minimum, i.e., 39

Thus, the maximum percent of people who like only one among Pasta and Burgers

$= 139 - 2 \times 39 = 61$ - Statement II is correct

People who do not like either Pasta or Burgers

$= n = x - 39 \ldots$ From (i)

The value of n is maximum when x is maximum, i.e., 63

Thus, the maximum percent of people who do not like either pasta or burgers

$= 63 - 39 = 24$ - Statement III is correct

Thus, all three statements are correct.

The correct answer is option E.

158. The three rides are merry-go-round (M), roller coaster (R) and toy train (T).

From the diagram, we have

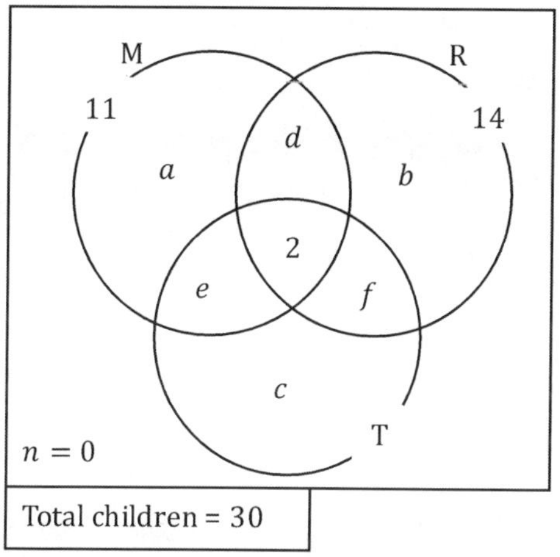

We have

$d + e + f = 10$

We need to determine the total cost of all rides. Thus, we need to know the total number of rides.

$(a + b + c)$ children take one ride each, $(d + e + f)$ children take 2 rides, and 2 children take three rides.

Thus, total number of rides

$= (a + b + c) \times 1 + (d + e + f) \times 2 + 2 \times 3$

$= (a + b + c + d + e + f + 2) + (d + e + f) + 4$

$= 30 + 10 + 4 = 44$

Thus, the total cost of all rides

$= \$(5 \times 44) = \$220$

The correct answer is option E.

159.

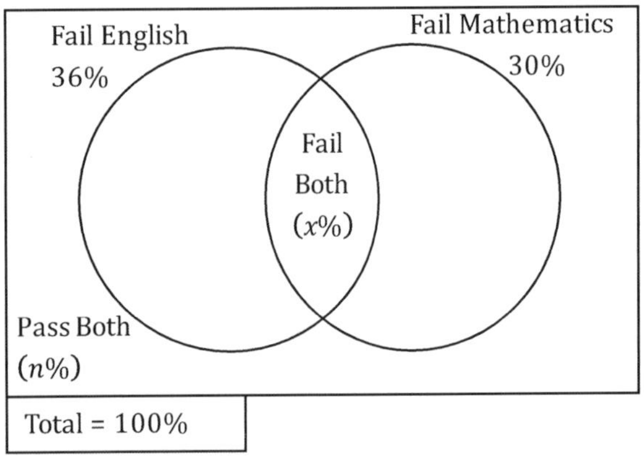

Let us first maximize the value of n.

The value of n is maximized if all of those who fail in Mathematics (30%) forms a part of those who fail in English (36%).

Thus, $n = 100 - 36 = 64\%$ (Maximum)

Let us now minimize the value of n.

The value of n is minimized if there is no overlap between those who fail in Mathematics and those who fail in English.

Thus, $n = 100 - (36 + 30) = 34\%$ (Minimum)

Thus, the required value = 64% + 34% = 98%

The correct answer is option D.

160.

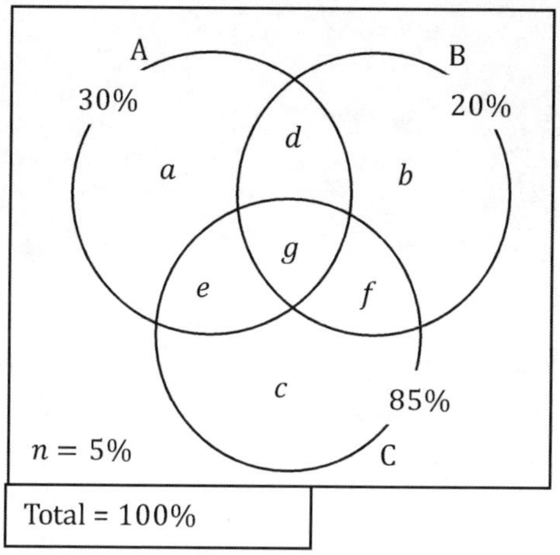

We have $d + e + f = 20$

For three sets A, B, C, we have

A or B or C =(A + B + C)−[(A and B)+(B and C)+(C and A)]+(A and B and C)

$\Rightarrow 100 - 5 = 30 + 20 + 85 - [(d + g) + (f + g) + (e + g)] + g$

$\Rightarrow 95 = 135 - ((d + e + f) + 3g) + g$

$\Rightarrow 95 = 135 - (20 + 3g) + g$

$\Rightarrow g = 10$

The correct answer is option C.

Chapter 15

Statistics

Statistics is the study of the collection, analysis, and interpretation of data. Some of the most commonly used statistical measures are the Mean, Median, Range, and Standard Deviation.

Let us discuss them.

15.1 Arithmetic mean:

- For a set of n numbers $x_1, x_2, x_3, \dots x_n$, the arithmetic mean is calculated as:

$$\overline{x} = \frac{\text{Sum of the } n \text{ numbers}}{n} = \frac{x_1 + x_2 + x_3 + \cdots + x_n}{n}$$

- If $x_1, x_2, x_3, \dots x_n$ be n observations and $f_1, f_2, f_3, \dots f_n$, be their corresponding frequencies, their arithmetic mean is calculated as:

$$\overline{x} = \frac{x_1 \times f_1 + x_2 \times f_2 + \dots x_n \times f_n}{f_1 + f_2 + \cdots + f_n}$$

For ease of calculations, the mean can be easily calculated using the method of 'deviation about the mean', as shown below:

Let the assumed mean be x_a

The observations are modified by subtracting x_a from each of the values as follows:

$x_1 \equiv x_1 - x_a$

$x_2 \equiv x_2 - x_a$

...

...

$x_n \equiv x_n - x_a$

The means are now calculated as shown:

- For a set of n numbers $x_1, x_2, x_3, \dots x_n$, the arithmetic mean is calculated as:

$$\overline{x} = x_a + \frac{(x_1 - x_a) + (x_2 - x_a) + (x_3 - x_a) + \cdots + (x_n - x_a)}{n}$$

- If $x_1, x_2, x_3, \dots x_n$ be n observations and $f_1, f_2, f_3, \dots f_n$, be their corresponding frequencies, their arithmetic mean is calculated as:

$$\overline{x} = x_a + \frac{(x_1 - x_a) \times f_1 + (x_2 - x_a) \times f_2 + \ldots (x_n - x_a) \times f_n}{f_1 + f_2 + \cdots + f_n}$$

Let us take a few examples:

(1) What is the mean of 123, 134, 128, 139, 141, and 127?

Explanation:

$$\text{Mean} = \frac{123 + 134 + 128 + 139 + 141 + 127}{6} = 132$$

Alternate approach:

Let the assumed mean = 134.

Thus, the deviations are:

- $123 - 134 = -11$
- $134 - 134 = 0$
- $128 - 134 = -6$
- $139 - 134 = 5$
- $141 - 134 = 7$
- $127 - 134 = -7$

$$=> \text{Mean} = 134 + \frac{-11 + 0 - 6 + 5 + 7 - 7}{6} = 134 - \frac{12}{6} = 132$$

(2) What is the mean of the observations shown below?

Observations	Frequencies
11	14
7	8
15	6
13	12

Explanation:

we have

Observations	Frequencies	Observation × Frequency
11	14	$11 \times 14 = 154$
7	8	$7 \times 8 = 56$
15	6	$15 \times 6 = 90$
13	12	$13 \times 12 = 156$
	Total = 40	**Total = 154 + 56 + 90 + 156 = 456**

$=> \text{Mean} = \frac{456}{40} = 11.4$

Alternate approach:

Let the assumed mean be 11.

Thus, we have

Observations	Modified Observation = Observation − Assumed mean	Freq.	Mod. Obs. × Freq.
11	$11 - 11 = 0$	14	$0 \times 14 = 0$
7	$7 - 11 = -4$	8	$(-4) \times 8 = -32$
15	$15 - 11 = 4$	6	$4 \times 6 = 24$
13	$13 - 11 = 2$	12	$2 \times 12 = 24$
		Total = 40	**Total = 16**

$=> \text{Mean} = 11 + \frac{16}{40} = 11.4$

15.2 Median:

Median refers to the middle value of all observations arranged in ascending or descending order.

Thus, we have

- **If n is odd:** Median = Value of the $\left(\frac{n+1}{2}\right)^{th}$ observation.
- **If n is even:** Median = Average of $\left(\frac{n}{2}\right)^{th}$ and $\left(\frac{n}{2}+1\right)^{th}$ observations.

Let us take a few examples:

(1) What is the median of 123, 134, 128, 139, 141 and 126 ?

Explanation:

After arranging data in ascending order, we have

123, 126, 128, 134, 139, 141

Since there are 6 (even) terms, we have

$$\text{Median} = \frac{\left(\frac{6}{2}\right)^{th}\text{ term} + \left(\frac{6}{2}+1\right)^{th}\text{ term}}{2} = \frac{3^{rd}\text{ term} + 4^{th}\text{ term}}{2}$$

$$123, 126, \underset{3^{rd}}{\underline{128}}, \underset{4^{th}}{\underline{134}}, 139, 141$$

$$= \frac{128 + 134}{2} = 131$$

(2) What is the median of the observations shown below?

Observations	Frequencies
11	12
7	8
15	6
13	14

Explanation:

After arranging the observations in ascending order, we have

Observations	Frequencies	Cumulative Frequency
7	8	8
11	12	12 + 8 = 20
13	14	14 + 20 = 34
15	6	6 + 34 = 40
	Total = 40	

Since there are 40 observations, we have

$$\text{Median} = \frac{\left(\frac{40}{2}\right)^{\text{th}} \text{ term} + \left(\frac{40}{2} + 1\right)^{\text{th}} \text{ term}}{2} = \frac{20^{\text{th}} \text{ term} + 21^{\text{st}} \text{ term}}{2}$$

Observations	Frequencies	Cumulative Frequency	
7	8	8	
11 <—	12	12 + 8 = 20 <—	**20th term**
13 <—	14	14 + 20 = 34 <—	**21st term**
15	6	6 + 34 = 40	
	Total = 40		

$$\text{Median} = \frac{11 + 13}{2} = 12$$

15.3 Range:

Range refers to the difference between the values of the maximum and minimum observations.

For example, the range of the set of observations 11, 13, 23, 21, 13, 17, 19 is 23 − 11 = 12

15.4 Standard Deviation:

Standard deviation shows the spread of the observations. If the observations are far away from the mean, the standard deviation would be high; similarly, if the observations are close to the mean, the standard deviation would be low.

For a set of n numbers $x_1, x_2, x_3, \ldots x_n$, the Standard Deviation (s) is calculated as:

If $\overline{x}$ is the mean of the observations:

$$s = \sqrt{\frac{(x_1 - \overline{x})^2 + (x_2 - \overline{x})^2 + \cdots + (x_n - \overline{x})^2}{n}}$$

Let us take an example:

(1) What is the standard deviation of 12, 15, 16 and 17 ?

$$\text{Mean} = \overline{x} = \frac{12 + 15 + 16 + 17}{4} = 15$$

$$=> s = \sqrt{\frac{(12 - 15)^2 + (15 - 15)^2 + (16 - 15)^2 + (17 - 15)^2}{4}}$$

$$= \sqrt{\frac{9 + 0 + 1 + 4}{4}} = \sqrt{3.5} = 1.87$$

A few important rules:

For a set of n numbers: $x_1, x_2, x_3, \ldots x_n$

- If a constant number k is added to (or subtracted from) each term of the above set of observations, we have
  - The Mean increases (or decreases) by k.
  - The Median increases (or decreases) by k.
  - The Range remains unchanged.
  - The Standard Deviation remains unchanged.

- If a constant number k is multiplied (divided) by each term of the above set of observations, we have
    - The Mean is multiplied (divided) by k.
    - The Median is multiplied (divided) by k.
    - The Range is multiplied (divided) by k.
    - The Standard Deviation is multiplied (divided) by k.

- If a new term, whose value is equal to the mean, is introduced in the above set of observations, we have
    - The Mean remains unchanged.
    - Nothing can be concluded about Median; it depends on the distribution.
    - The Range remains unchanged.
    - The Standard Deviation decreases.

- If the terms of a set of observations arranged in ascending or descending order have a constant difference between consecutive terms, i.e., they form an Arithmetic Sequence, we have Mean = Median

Chapter 16

Practice Questions: Statistics

Questions on Average (Arithmetic Mean)

1. If all the elements of a set are multiplied by a constant, and the average (arithmetic mean) remains unchanged, which of the following statements must be true?

   I. Average (arithmetic mean) = 0

   II. The sum of the elements of the set equals 0.

   III. The set contains at least one positive and one negative number.

   (A) I only
   (B) II only
   (C) III only
   (D) I, and II only
   (E) I, II, and III

2. The average (arithmetic mean) of x and y is 80, and the average of x and z is 200; what is the value of $\frac{(z-y)}{8}$?

   (A) 80
   (B) 30
   (C) 0
   (D) -30
   (E) -120

3. The average (arithmetic mean) of six numbers is 7.5. If one of the six numbers is multiplied by 4, the average of the numbers decreases by 0.5. Which of the following numbers among the six is multiplied by 4?

   (A) 2
   (B) 4
   (C) 4.5
   (D) 6
   (E) 6.5

4. The average (arithmetic mean) of six numbers is 7.5. If one of the six numbers is multiplied by 4, the average of the numbers decreases by 0.5, which of the six numbers is multiplied by 4?

   (A) -3
   (B) -2
   (C) -1
   (D) 2
   (E) 3

5. A student's average (arithmetic mean) score on 5 tests is 38. What must be the student's score on the 6th test for the average score on all the 6 tests to be 41?

(A) 40

(B) 44

(C) 50

(D) 54

(E) 56

6. For the first x days, the average (arithmetic mean) rainfall was 32 millimeters per day. If today's rainfall of 56 millimeter increased the average rainfall to 36 millimeters per day, what is the value of x ?

(A) 2

(B) 3

(C) 5

(D) 6

(E) 7

7. A student appeared for a number of tests, with each test a maximum mark of 200. His average score is 70 points higher than his lowest score in any test. However, the median score is 50 points higher than the lowest score in any test. If the range of scores is 160 and the number of tests is an odd number, what is the minimum number of tests that the student appeared for?

(A) 3

(B) 5

(C) 7

(D) 9

(E) 11

8. The average of 7 numbers is 12. The average of the 4 smallest numbers in this set is 8, while the average of the 4 greatest numbers in this set is 20. How much greater is the sum of the 3 greatest numbers than the sum of the 3 smallest numbers?

(A) 32

(B) 46

(C) 48

(D) 80

(E) 84

9. The average age of the members of a club is 32. If 4 members are 25 years old and none of the members is more than 60 years old, which of the following could be the number of members in the club?

I. 5

II. 6

III. 9

(A) Only I

(B) Only II

(C) Only III

(D) Both II and III

(E) I, II and III

10. The average of five distinct integers is 65. If the largest integer is 75, what is the maximum possible value of the smallest integer?

(A) 60

(B) 61

(C) 62

(D) 63

(E) 64

11. Four students contributed to a charity drive, and the average amount contributed by the students was \$20. If no student gave more than \$25 and all contributed different whole-dollar amounts, what is the ratio of the amounts contributed by the highest contributor to the amount contributed by the lowest contributor?

(A) 3

(B) $3\frac{1}{8}$

(C) $3\frac{5}{8}$

(D) 5

(E) $5\frac{1}{2}$

12. The mean of n terms is x. If the first term is increased by 1, the second by 2, the third by 3, and so on, which of the following denotes the new mean?

(A) $x + \frac{n-1}{2}$

(B) $x + \frac{n}{2}$

(C) $x + \frac{n+1}{2}$

(D) $x + n - 1$

(E) $x + n$

13. If the average (arithmetic mean) of seven distinct positive integers is 14, what is the least possible value of the greatest of the seven numbers?

(A) 14
(B) 17
(C) 18
(D) 20
(E) 77

Questions on Median

14. A student scored 43, 47, 39, 27, and 34 in the five tests he took. What could be the score on his sixth test if, after the sixth test, his mean score is equal to his median score?

I. 29
II. 36
III. 56

(A) Only I
(B) Only II
(C) Only III
(D) Only I and III
(E) I, II and III

15. If Set X consists of the following unique integers: $-1, 18, 4, p, 3, 16, -2$, and -26, which of the following CANNOT be the median of Set X?

(A) 1
(B) 1.5
(C) 2
(D) 2.5
(E) 3

16. Which of the following numbers CANNOT be included in the set of the following numbers to decrease the median of the set by 1.5?

Set: $\{28, 18, 22, 21, 7, 13\}$

(A) -7
(B) 10
(C) 11.5
(D) 18

(E) 18.5

17. If two sets, Set X: $\{-2, 23, 4, -8, 6\}$, and Set Y: $\{0, -2, -8, 8, 12, 13, y\}$ have equal medians, what is the value of y ?

(A) 0

(B) 2

(C) 3

(D) 4

(E) 8

18. If the median of X consecutive integers is 50.5, which is the smallest integer among the following?

(A) 51 - X/2

(B) 50.5 - X/2

(C) 50 - X/2

(D) 49.5 - X/2

(E) 49 - X/2

19. Five people have marbles with them, with counts of 23, 45, 51, 66, and 73. If they redistribute the marbles such that each person has a distinct number of marbles and the median value is maximized, what is the range of the number of marbles with them?

(A) 83

(B) 84

(C) 85

(D) 86

(E) 88

20. The heights of students in a class are measured. It is found that, for boys, the average height is $(3x + 1)$ feet, while for girls, the average height is $(6 - 2x)$ feet. If there are a total of $5x$ students, with boys and girls in the ratio $2 : 3$. What is the average height of all students taken together?

(A) 4 feet

(B) $4x$ feet

(C) 20 feet

(D) $20x$ feet

(E) $12x^2$ feet

21. A survey is conducted on the monthly salaries of households in a locality. For each household, the salary of the eldest member is considered. The data is then grouped according to different age groups. For each age group, the median salary is considered, as shown in the table below.

Age-group	Between 15 and 25 years	Between 25 and 35 years	Between 35 and 45 years	Between 45 and 55 years	Between 55 and 65 years
Median Salary	\$12,000	\$24,000	\$36,000	\$38,000	\$28,000

If the data for the age groups 15–25 years and 55–65 years are ignored, which of the following statements would be correct?

I. The mean of the above median salaries would change by less than \$5,000.
II. The median of the above median salaries would change by less than \$5,000.
III. The difference between the new mean and new median of the above median salaries is less than \$5,000.

(A) Only I
(B) Only II
(C) Only III
(D) Both I and II
(E) I, II and III

22. If x and y are positive integers and the median of the numbers $3^{x-1}5^y$, 3^x5^{y-1} and 3^x5^y is 2,025, what is the value of xy ?

(A) 8
(B) 9
(C) 10
(D) 12
(E) 15

Questions on Average and Median

23. The average (arithmetic mean) of the five positive integers $k, m, r, s,$ and t is 16, and $k < m < r < s < t$. If t is 40, what is the greatest possible value of the median of the five integers?

(A) 16
(B) 18
(C) 19
(D) 20
(E) 38

24. Nine students each took a test with 50 questions, each weighted at 6 marks. Their average (arithmetic mean) score was 150, and the median score was 120. Which of the following statements must be true?

I. At least one student had a score of more than 180.
II. At least one student had a score of more than 120 and less than 150.
III. At least one student had a score of less than 120.

(A) Only I
(B) Only II
(C) Only III
(D) Only I and III
(E) Only II and III

25. Nine boys have an average (arithmetic mean) weight of 36 kg and a median weight of 38 kg. If the weight of the heaviest boy is 3 kg less than 3 times the weight of the lightest boy, what is the maximum possible weight, in kg, of the heaviest boy?

(A) 38
(B) 54
(C) 72
(D) 94
(E) 100

26. The number of defects in eight of the nine batches of batteries is 2, 4, 6, 5, 3, 3, 7, and 8, respectively. If the median defect per batch for all the nine batches equals the average (arithmetic mean) defect per batch for all the nine batches, which of the following would be the number of defects in the ninth batch?

(A) 2
(B) 3
(C) 4
(D) 5
(E) 7

27. For the set of numbers $\{x, y, (x+y), (x-4y), xy, 2y\}$, if $y > 6$ and the mean of the set equals $(y+3)$, which of the following denotes the median of the above set?

(A) $x + y$
(B) $\frac{x}{2} + y$
(C) $x + \frac{y}{2}$
(D) $x - 4y$
(E) $\frac{x+y}{2}$

28. Four positive integers, a, b, c and d are such that $a < b < c < d$. The mean of the above numbers is 12, and the median is 7. If one of the numbers is 8, what is the minimum possible value of d ?

(A) 26

(B) 27

(C) 28

(D) 29

(E) 30

29. If the mean of 3^{x-1}, $3^{x-1}5$, and $3^{x}5^{2}$ is 243, what is the median of the above numbers?

(A) 15

(B) 45

(C) 81

(D) 135

(E) 225

Questions on Range

30. Which of the following could be the range of set of integers $\{-43, 76, a, b, c\}$?

I. -129

II. 112

III. 154

(A) I only

(B) II only

(C) III only

(D) I or II only

(E) II or III only

31. A set contains 8 distinct positive integers. If the smallest three numbers decrease by 2 and the largest two numbers increase by 3, which of the following statements is correct?

I. The arithmetic mean (average) will change.

II. The median will change.

III. The range will change.

(A) Only I

(B) Only II

(C) Only III

(D) Both I and II

(E) Both I and III

32. The range of heights for students in Class A is 9 inches, and that for Class B is 6 inches. Which of the following statements alone is sufficient to determine the range of heights for all students in both classes?

I. The tallest student in Class B is 5 inches taller than the tallest student in Class A.

II. The shortest student in Class B is 1 inch shorter than the tallest student in Class A.

III. The shortest student in Class A is 8 inches shorter than the shortest student in Class B.

(A) Only I
(B) Only II
(C) Only III
(D) Both I and II
(E) I, II and III

Questions on Standard deviation

33. The mean score and the standard deviation of the scores in an exam in the ninth grade were 70 and 2, respectively. John, a student in the ninth grade, scored 75 in the exam. The mean score and the standard deviation of the scores in an exam in the tenth grade were 65 and 3, respectively. Bob, a student in the tenth grade, scored 72 in the exam. Which of the following statements must be correct?

I. John's performance was better than Bob's performance in their respective exams.

II. In the ninth grade, there was at least one student who scored more than John.

III. In the tenth grade, there was at least one student who scored less than Bob.

(A) Only I
(B) Only II
(C) Only III
(D) Only I and III
(E) Only II and III

34. Set X: {10, 20, 30, 40}

Which of the following sets' standard deviation is less than Set X's standard deviation?

$P = \{2, 12, 22, 32\}$
$Q = \{-3, 7, 17, 27\}$
$R = \{-6, 4, 14, 24\}$

(A) Set P
(B) Set Q

(C) Set R

(D) Sets P and R

(E) None

35. If Set X is $\{p,\ q,\ r\}$, of the following sets has the same standard deviation as Set X?

(A) $\{p+1,\ q,\ r\}$

(B) $\{p,\ q+1,\ r\}$

(C) $\{p,\ q,\ r+1\}$

(D) $\{p+qr,\ q(1+r),\ r(q+1)\}$

(E) $\{p-1,\ q+1, r\}$

36. $A = \{42, 32, 22, 12, 2\}$
$B = \{95, 90, 85, 80, 75\}$
$C = \{40, 20, 0, -20, -40\}$

Which of the following represents the correct ordering of the standard deviations of the sets A, B, and C, from the largest to the smallest?

(A) ABC

(B) BCA

(C) CAB

(D) ACB

(E) CBA

37. If one of the two numbers P and Q is positive and the other negative, which of the following could be the standard deviation of set $\{P, -8, -12, 4, 0, -4, 8, Q\}$?

I. -2

II. -4

III. 0

(A) I only

(B) II only

(C) I, and II only

(D) I, and III only

(E) None of these

38. Set A consists of 25 numbers with an average (arithmetic mean) of 5 and a standard deviation of 4. If, from the set, two numbers 3 and 5 are swapped with the following pairs of numbers (given in options), which pair would most decrease Set A's standard deviation?

(A) $\{0, 8\}$

(B) $\{5, 3\}$

(C) $\{2, 6\}$

(D) $\{4, 4\}$

(E) $\{1, 7\}$

39. The distribution of test scores of a large number of test takers is symmetric about the mean score μ. If 95 percent of the distribution lies within 2 standard deviation of the mean, what percent of the distribution is more than $(\mu - 2\sigma)$?

(A) 2.5%

(B) 5%

(C) 84%

(D) 95%

(E) 97.5%

40. Which of the following sets has the smallest standard deviation?

(A) $\{-10, -10, -10, -10, -10\}$

(B) $\{-10, -8, -6, -4, -2\}$

(C) $\{-1, 0, 0, 0, 1\}$

(D) $\{-1, -1, 0, 0, 0\}$

(E) $\{-1, 0, 1, 2, 3\}$

Chapter 17

Answer Key: Statistics

(1) D	(9) E	(17) D	(25) C	(33) D
(2) B	(10) B	(18) A	(26) E	(34) E
(3) C	(11) B	(19) C	(27) B	(35) D
(4) C	(12) C	(20) A	(28) D	(36) C
(5) E	(13) B	(21) C	(29) B	(37) E
(6) C	(14) D	(22) C	(30) C	(38) D
(7) A	(15) E	(23) B	(31) C	(39) E
(8) C	(16) E	(24) A	(32) E	(40) A

Chapter 18

Solution: Statistics Practice Questions

Questions on Average (Arithmetic Mean)

1. If all the elements of a set are multiplied by a constant, the mean also gets multiplied by the same constant.

   However the mean remaining unchanged can happen only if mean = 0, so the statement I is true.

   Since sum of all the elements = Mean $\times$ # of elements
   => Sum of all the elements = 0 as the mean = 0, so the statement II is true.

   Statement 3 is also not necessary because the set may contain all zeros.

   The correct answer is option D.

2. We know that arithmetic mean of x and y is 80,

   => $(x + y)/2 = 80$

   => $x + y = 160...(1)$

   Also, we know that arithmetic mean of x and z is 200,

   => $(x + z)/2 = 200$

   => $x + z = 400...(2)$

   To get the value of $(z - y)/8$, we subtract equation (2) from the equation (1) and get,

   => $z - y = 400 - 160 = 240;$

   => $(z - y)/8 = 240/8 = 30$

   The correct answer is option B.

3. Say the three numbers are $a, 7,$ and b with a being the largest and b being the smallest.

   => $(a + 7)/2 = 8$

   => $a + 7 = 16$

   => $a = 9$

Similarly, $(a + 7 + b)/3 = 6$

=> $9 + 7 + b = 18$

=> $b = 2$

So, the numbers are 9, 7, & 2.

So, the mean of the two smallest numbers = $(7 + 2)/2 = 4.5$

The correct answer is option C.

4. Given: Average = 7.5

=> Sum of 6 numbers = $6 \times$ average = $6 \times 7.5 = 45$

Say the number multiplied by 4 is x, so the number becomes $4x$. Or the sum of the numbers increases by $3x$.

=> New sum = $45 + 3x$

Given that new average = $7.5 - 0.5 = 7$

=> New sum = $6 \times$ new average = $6 \times 7 = 42$

=> $45 + 3x = 42$

=> $x = -1$

The correct answer is option C.

Note that the question stem stated that after multiplying a number by 4, the average decreased rather than anticipated—increased, it implies that the number must be negative; thus only options A or B could be true.

5. Sum of scores of 5 tests = $5 \times 38 = 190$;

Say the score on the 6th test is x, so

Sum of scores of 6 tests = $190 + x$;

=> Average score = $\frac{(190 + x)}{6}$;

=> $41 = \frac{(190 + x)}{6}$; given that the average is 41.

=> $x = 56$

The correct answer is option E.

Alternate approach:

You find that to increase the average score from 38 to 41, you need to increase it by 3. Since '3' is an average increase per test, to get the total increase, you must multiply it by the total number of tests, i.e., 6.

=> Total increase = $3 \times 6 = 18$;

=> Desired score in the 6th test = 38 + 18 = 56

6. Total rainfall in first x days = $32x$;

Total rainfall in $(x + 1)$ days = $(32x + 56)$;

=> Average rainfall in $(x + 1)$ days = $\frac{(32x + 56)}{(x + 1)}$

=> $36 = \frac{(32x + 56)}{(x + 1)}$; given that the average rainfall in $(x + 1)$ days = 36 mm

=> $36(x + 1) = (32x + 56)$

=> $36x + 36 = 32x + 56$

=> $4x = 20$

=> $x = 5$ days

The correct answer is option C.

Alternate approach:

Given that increase in average = 36 - 32 = 4;

And difference of average of x days, 32 and today's rainfall, 56 = 56 - 32 = 24;

=> $24 = 4 \times (x + 1)$

=> $x = 5$ days

7. Let the minimum score obtained in any test be m

Thus, mean score = $(m + 70)$

Also, median score = $(m + 50)$

Since the range is 160, the maximum score = $(m + 160)$

Since the number of tests is an odd number, there must be one test having the median score, i.e., $(m + 50)$

Thus, we have three scores with us: m, $(m + 50)$ and $(m + 160)$

The mean of the above three scores = $\left(\frac{m + m + 50 + m + 160}{3}\right) = \frac{3m + 210}{3} = m + 70$

Since the mean mentioned is also $(m + 70)$ and matches with the derived mean, the minimum number of tests is 3.

The correct answer is option A.

8. Since the average of 7 numbers is 12, the sum of the numbers = $7 \times 12 = 84$.

Let the numbers, when arranged in ascending order, be a, b, c, d, e, f and g.

Thus, we have

$a + b + c + d + e + f + g = 84 \ldots$ (i)

Since the average of the 4 smallest numbers in this set is 8, we have

$a + b + c + d = 8 \times 4 = 32 \ldots$ (ii)

Since the average of the 4 greatest numbers in this set is 20, we have

$d + e + f + g = 20 \times 4 = 80 \ldots$ (iii)

(ii) + (iii) – (i):

$d = 28$

=> $a + b + c = 32 - 28 = 4$

$e + f + g = 80 - 28 = 52$

Thus, the required difference = $52 - 4 = 48$

The correct answer is option C.

9. Let the number of members be x.

Thus, the sum of ages of all the members = $32x$

We know that the sum of ages of 4 members = $25 \times 4 = 100$

Thus, the sum of ages of $(x - 4)$ members = $32x - 100$

=> the average of $(x - 4) = \dfrac{32x - 100}{x - 4}$

=> $\dfrac{32x - 100}{x - 4} \leq 60$; we know that no member is more than 60 years of age

=> $32x - 100 \leq 60x - 240$

=> $x \geq 5$

=> x can assume any value 5, 6, 7, 8, 9, ...

The correct answer is option E.

Alternate approach:

Let us use the options to determine the number of members in the club.

- Option A: 5 members:

Sum of ages of 5 members = $32 \times 5 = 160$
Since the average age of 4 members is 25 years, their total age = $25 \times 4 = 100$ years
Thus, the age of the 5th member = $160 - 100 = 60 \leq 60$ (maximum possible age) - Satisfies

- Option B: 6 members:

Sum of ages of 6 members = $32 \times 6 = 192$
Since the average age of 4 members is 25 years, their total age = $25 \times 4 = 100$ years
Thus, the sum of ages of 2 members = $192 - 100 = 92$
Thus, the average age of 2 members = $\frac{92}{2} = 46 \leq 60$ (maximum possible age) - Satisfies

- Option C: 9 members:

Sum of ages of 9 members = $32 \times 9 = 288$

Since the average age of 4 members is 25 years, their total age = $25 \times 4 = 100$ years
Thus, the sum of ages of 5 members = $288 - 100 = 188$
Thus, the average age of 5 members = $\frac{188}{5} = 33.60 \leq 60$ (maximum possible age) - Satisfies

10. Since the average of 5 integers is 65; thus, the total = $65 \times 5 = 325$

The largest integer is 75

Thus, sum of the remaining 4 integers = $325 - 75 = 250$

Thus, the average of the 4 integers = $\frac{250}{4} = 62.5$

The smallest integer will be maximized if these 4 integers are as close to each other as possible.

Since the numbers are distinct integers, let us subtract 0.5 from the average (62.5) for one of the numbers and add 0.5 to the average (62.5) for the other number; giving us $62.5 - 0.5 = 62$ and $62.5 + 0.5 = 63$.

There are still two numbers left.

Let us subtract 1.5 from the average (62.5) for one of them and add 1.5 to the average (62.5) for the other; giving us $62.5 - 1.5 = 61$ and $62.5 + 1.5 = 64$.

Thus, the 4 integers are 61, 62, 63 and 64.

Thus, the maximum possible value of the smallest integer = 61.

The correct answer is option B.

Alternate approach:

Since we need to maximize the smallest number, we need to make the distance between the consecutive numbers minimum in this scenario 1.

Let the smallest number be x; thus the 2nd smallest number = $(x + 1)$, and other numbers $(x + 2)$ and $(x + 3)$.

We have the sum of the above 4 numbers as 250.

$x + (x + 1) + (x + 2) + (x + 3) = 250$

$=> 4x = 244$

$=> x = 61$

11. Since the average contribution by 4 students was \$20, total contribution = \$(20 × 4) = \$80

To minimize the contribution by any student, we need to maximize the contribution made by the other three students. Since the contributions are different, we assume that the contributions made by 3 students are \$25 (maximum), \$24 and \$23.

Thus, contribution (minimum) made by the 4th student = \$(80 − 25 − 24 − 23) = \$8

Thus, the required ratio = $\frac{25}{8} = 3\frac{1}{8}$

The correct answer is option B.

12. Let the n terms be $t_1, t_2, t_3, \ldots t_n$

Since the mean is x, we have

$$\frac{t_1 + t_2 + \cdots + t_n}{n} = x$$

$=> t_1 + t_2 + \cdots + t_n = nx$

The new series is obtained by adding '1' to the 1st term, '2' to the 2nd term, and so on.

Thus, the new terms are:

$(t_1 + 1), (t_2 + 2), (t_3 + 3), \ldots (t_n + n)$

Thus, the sum of the above terms

$= t_1 + t_2 + \cdots + t_n + (1 + 2 + 3 + \cdots + n)$

$= nx + \frac{n(n+1)}{2}$

Thus, the new mean

$= \frac{nx + \frac{n(n+1)}{2}}{n}$

$= x + \frac{n+1}{2}$

The correct answer is option C.

Alternate approach:

Let us assume that there are three terms (1, 2, 3) and their mean = 2; thus $n = 3$ & $x = 2$

After the increment, the terms are: $1 + 1,\ 2 + 2,\ 3 + 3 \Rightarrow 2, 4, 6$

The new average $= \frac{2+4+6}{3} = 4$

By plugging the values of x & n into options, we find that both options C and D give the new average = 4. So we cannot decide which is the correct answer. Sometimes test-makers make such tricky options that if one goes for a hit-and-trial approach, he or she may still not be able to conclude the correct answer.

Since we have zeroed in on options C and D, we need to eliminate one of them, but for that to happen, we will have to choose another set of numbers.

Let us assume that there are three terms (2, 3, 4) and their mean = 3; thus $n = 3$ & $x = 3$

After the increment, the terms are $2 + 1,\ 3 + 2,\ 4 + 3 \Rightarrow 3, 5, 7$

The new average = $\frac{3 + 5 + 7}{3} = 5$

We can still not eliminate C or D as both options still qualify.

Let us take another set:

Let us assume that there are four terms (1, 3, 5, 7) and their mean = 4; thus $n = 4$ & $x = 4$

After the increment, the terms are $1 + 1,\ 3 + 2,\ 5 + 3,\ 7 + 4 \ \Rightarrow\ 2,\ 5,\ 8,\ 11$

The new average = $\frac{2 + 5 + 8 + 11}{4} = 6.5$

Let us plug in $x = 4$ & $n = 4$ in options C and D.

(C) $x + \frac{n + 1}{2} = 4 + \frac{4 + 1}{2} = 6.50$ - Satisfies - Correct answer

(D) $x + n - 1 = 4 + 4 - 1 = 7 \neq 6.50$ - Does not satisfy

13. Let the seven distinct positive integers be $p,\ q,\ v,\ w,\ x,\ y$ and z such that $p > q > v > w > x > y > z$.

Thus, we need to find the least possible value of the largest among the seven, i.e. p.

Let's assume that w, the middle-most number = average of the seven numbers = 14.

Now to maintain the average equals to 14, increase $v,\ q,$ & p by 1, 2, & 3, respectively and decrease $x,\ y,$ & z by 1, 2, & 3, respectively.

Thus, $p = 14 + 3 = 17$ - the least possible value of the greatest of the seven numbers

The correct answer is Option B.

Questions on Median

14. Say the score in the sixth test = n

Thus, the scores in the six tests are = 43, 47, 39, 27, 34, and n, respectively.

Thus, the total score = $43 + 47 + 39 + 27 + 34 + n = 190 + n$

Thus, the mean score = $\frac{190 + n}{6}$

We know that the mean score and the median score are the same.

Working with the statements:

I. $n = 29$:

$$\text{Mean} = \frac{190 + 29}{6} = \frac{219}{6} = 36.5$$

The scores, when arranged in order, are 27, 29, 34, 39, 43, and 47

The median is the average of the two middle terms.

Thus, we have

$$\text{Median} = \frac{\left\{\left(\frac{6}{2}\right)^{\text{th}} \text{term} + \left(\frac{6}{2} + 1\right)^{\text{th}} \text{term}\right\}}{2} = \frac{\left\{3^{\text{rd}} \text{ term} + 4^{\text{th}} \text{ term}\right\}}{2}$$

$$= \frac{34 + 39}{2} = \frac{73}{2} = 36.5$$

Thus, Mean = Median - Satisfies

II. $n = 36$:

$$\text{Mean} = \frac{190 + 36}{6} = \frac{226}{6} = 37.67$$

The scores, when arranged in order, are 27, 34, 36, 39, 43, and 47

The median is the average of the two middle terms.

Thus, we have

$$\text{Median} = \frac{\left\{\left(\frac{6}{2}\right)^{\text{th}} \text{term} + \left(\frac{6}{2} + 1\right)^{\text{th}} \text{term}\right\}}{2} = \frac{\left\{3^{\text{rd}} \text{ term} + 4^{\text{th}} \text{ term}\right\}}{2}$$

$$= \frac{36 + 39}{2} = \frac{73}{2} = 36.5$$

Thus, Mean $\neq$ Median - Does not satisfy

III. $n = 56$:

$$\text{Mean} = \frac{190 + 56}{6} = \frac{246}{6} = 41$$

The scores, when arranged in order, are 27, 34, 39, 43, 47, and 56

The median is the average of the two middle terms.

Thus, we have

$$\text{Median} = \frac{\left\{\left(\frac{6}{2}\right)^{\text{th}} \text{term} + \left(\frac{6}{2}+1\right)^{\text{th}} \text{term}\right\}}{2} = \frac{\left\{3^{\text{rd}} \text{ term} + 4^{\text{th}} \text{ term}\right\}}{2}$$

$$= \frac{39+43}{2} = \frac{82}{2} = 41$$

Thus, Mean = Median - Satisfies

The correct answer is option D.

15. Let us first arrange the elements of the set in ascending order, ignoring p as we do not yet know where it lies.

=> $-26, -2, -1, 3, 4, 16, 18$

It is not necessary to know the exact placement of p in order to find which option could be the median of the set.

When p is included, the set has an even number of elements. Consequently, the median will be the average of the two middle-most terms.

It is important to note that the set consists of "unique integers". So, p cannot be equal to any of the existing terms in the set.

Let us analyze the set on the basis of the placement of p.

(1) Say $p < -26$, so the set would be: $\{p, -26, -2, -1, 3, 4, 16, 18\}$
=> Median = $(-1+3)/2 = 1$

(2) Say $-2 > p > -26$, so the set would be: $\{-26, p, -2, -1, 3, 4, 16, 18\}$
=> Median = $(-1+3)/2 = 1$ (No change in the value of median)

(3) Say $3 > p > -1$, so the set would be: $\{-26, -2, -1, p, 3, 4, 16, 18\}$; p cannot lie between -2, and -1 as p is a unique integer.
=> Median = $(p+3)/2$;

(a) p can have any value: 0, 1, or 2

(b) Median = $(0 + 3)/2 = 1.5$, or $(1 + 3)/2 = 2$, $(2 + 3)/2 = 2.5$

(4) Say $4 > p$, so the set would be: $\{-26, -2, -1, 3, 4, p, 16, 18\}$ or $\{-26, -2, -1, 3, 4, 16, p, 18\}$ or $\{-26, -2, -1, 3, 4, 16, 18, p\}$

=> In either case, Median = $(4 + 3)/2 = 3.5$

So, the possible values of median are: $1, 1.5, 2, 2.5,$ & 3.5.

Thus, the median cannot be 3.

The correct answer is option E.

16. Let us calculate the median for the given set.

First, arrange the set in ascending order: $\{7, 13, 18, 21, 22, 28\}$

=> Median = $(18 + 21)/2 = 19.5$

Desired median = $19.5 - 1.5 = 18$

Since upon including an extra number, the number of elements in the set would be 7, an odd number, and to make sure that 18 (the desired median) lies in the middle, we can only add a new number that is ≤ 18.

The new set may look like the following.

$\#, 7, 13, 18, 21, 22, 28$ or $7, \#, 13, 18, 21, 22, 28$ or $7, 13, \#, 18, 21, 22, 28$. In each case, median is 18.

Thus, 18.5 cannot be added, else the new median would be $\frac{18+18.5}{2} = 18.25 \neq 18$.

The correct answer is option E.

17. First, arrange the set X in an ascending order, we get $-8, -2, 4, 6, 23$

=> Median = 4 (Middle-most term)

=> Median of Set Y = 4

Arranging Set Y in ascending order, we get $-8, -2, 0, 8, 12, 13$, excluding y as we do not know its value. We know that Set Y has an odd number of terms, so there would be a unique term equal to median value (here 4). Since there is no term equal to 4, only y can be 4.

The correct answer is option D.

18. The fact that the median of all consecutive integers is a non-integer implies that the number of consecutive integers, X is an even number.

The median, 50.5 is derived by taking the average of 50 & 51.

The list looks like this,

N1, N2, N3, ..., 50, 51, ...;

Since X is even, X/2 integers will be on the left of the median, and X/2 integers will be on the right of the median, 50.5. Counting of the smaller integer must start from 51, and we must deduct X/2 to reach the smallest; thus, the smallest integer = 51 – X/2.

The correct answer is option A.

Alternate approach:

Since we already concluded that X is an even number, we can assume X.

Say X = 4. So, the list would be:

49, 50, 51, 52; the smallest being 49

Only option A, 51 – X/2 = 51 – 4/2 = 49 is applicable.

19. The numbers of marbles with the 5 people are 23, 45, 51, 66 and 73

Thus, total number of marbles = 23 + 45 + 51 + 66 + 73 = 258.

We need to maximize the median value.

The median of 5 terms $= \left(\frac{5+1}{2}\right)^{\text{th}}$ term = 3rd term.

Let the terms be arranged in ascending order.

Since the median has to be maximized, we need to maximize the terms following the median. Thus, we need to maximize the 3rd, 4th and 5th terms.

Thus, we minimize the 1st and 2nd terms, keeping them at 1 and 2, respectively.

Thus, the number of marbles with the other 3 people = 258 − 1 − 2 = 255.

Since the above 3 people have different numbers of marbles, let the numbers be a, $(a + 1)$ and $(a + 2)$.

Thus, we have

$a + (a + 1) + (a + 2) = 255 => a = 84$

Thus, the numbers of marbles are 1, 2, 84, 85 and 86 (the median is maximized).

Thus, the range = 86 - 1 = 85

The correct answer is option C.

20. Total number of students = $5x$

Since boys and girls are in the ration 2 : 3, we have

Number of boys = $\frac{2}{2+3} \times 5x = 2x$

Number of girls = $5x - 2x = 3x$

Average height of boys = $(3x + 1)$ feet.

Thus, sum of heights of all boys = $2x\,(3x + 1) = 6x^2 + 2x$

Average height of girls = $(6 - 2x)$ feet.

Thus, sum of heights of all girls = $3x\,(6 - 2x) = 18x - 6x^2$

Thus, sum of heights of all students

= Sum of heights of boys + Sum of heights of girls = $20x$

Thus, the average height of all students = $\frac{20x}{5x}$ = 4 feet

The correct answer is option A.

21. In the table given above, the median salary of each age-group is considered, i.e., the salary of one individual is considered a representation of each age-group.

Thus, the mean salary is simply the sum of the above 5 salaries divided by 5.

Similarly, the median salary is the $\left(\frac{5+1}{2}\right)^{\text{th}}$ term, i.e., 3rd term, when the above salaries are arranged in order.

Thus, we have

$$\text{Mean} = \$\left(\frac{12,000 + 24,000 + 36,000 + 38,000 + 28,000}{5}\right) = \$27,600$$

Median: The salaries in ascending order are 12,000, 24,000, 28,000, 36,000 and 38,000:

=> Median = $28,000

After the two age groups are ignored, we have

$$\text{Mean} = \$\left(\frac{24,000 + 36,000 + 38,000}{5}\right) =\approx \$32,667$$

Median: The salaries in ascending order are 24,000, 36,000 and 38,000

=> Median = Middle term = $36,000

Thus, we have

Difference in mean = $32,667 – $27,600 = $5,067 – Statement I is incorrect

Difference in median = $36,000 – $28,000 = $8,000 – Statement II is incorrect

Difference between new mean and median = $36,000 – $32,667 = $3,333 - Statement III is correct

Thus, only Statement III is correct.

The correct answer is option C.

22. We have

$$3^{x-1}5^y = \frac{3^x5^y}{3}$$

$$3^x5^{y-1} = \frac{3^x5^y}{5}$$

Thus, the smallest of the three terms is $\frac{3^x5^y}{5}$ and the largest is 3^x5^y

Thus, the three terms, when arranged in ascending order are $\frac{3^x5^y}{5}$, $\frac{3^x5^y}{3}$ and 3^x5^y

Thus, the median value = $\frac{3^x5^y}{3} = 3^{x-1}5^y$

Thus, we have

$$3^{x-1}5^y = 2,025 = 3^45^2$$

$$=> x - 1 = 4, \text{ i.e., } x = 5 \text{ and } y = 2$$

$$=> xy = 10$$

The correct answer is option C.

Questions on Average and Median

23. Since the average of the five positive integers is 16, we have

$$\frac{k + m + r + s + t}{5} = 16$$

$$=> k + m + r + s + t = 80$$

Since $t = 40$, we have

$$k + m + r + s = 80 - 40 = 40 \ldots \text{(i)}$$

Also, we have

$$k < m < r < s < t$$

$$=> k < m < r < s < 40$$

The median is the middle term when the numbers are arranged in order.

Thus, the median is r.

Since we need to maximize r, and at the same time keep all the numbers positive integers, let's minimize $k = 1$ and $m = 2$.

So, $k = 1,\ m = 2$

Thus, from (i), we have

$r + s = 40 - (1 + 2) = 37$

Since $r < s$, we have

$s = r + 1$

$r = 18,\ s = 19$

Thus, the maximum value of the median is 18.

The correct answer is option B.

24. We know that the average score was 150 and the median score was 120.

Since there are 9 students, the median value would be that for the $\left(\frac{9+1}{2}\right)^{\text{th}} = 5^{\text{th}}$ student (after arranging them in ascending or descending order).

Let the average score of the first 4 students be f, and the average score of the last 4 students be l.

Since the average score is 150, we have

$$\frac{4f + 120 + 4l}{9} = 150$$

$4f + 120 + 4l = 1,350$

$=> 4f + 4l = 1,230$

$=> f + l = 307.50 \ldots$(i)

Since the score of the first four students must be less than or equal to 120, we have

$f \leq 120$

Thus, from (i):

$l \geq 307.50 - 120 = 187.50$

Thus, the average score of the last 4 students is at least 187.50.

Thus, at least one student must have got more than 180 score.

Thus, statement I must be true.

However, the other statements may or may not be correct.

Let's test them.

Statement II: At least one student had a score of more than 120 and less than 150.

The best approach to testing this statement is to think of a scenario wherein the situation depicted above fails.

The statement states that "at least one student had a score of more than 120 and less than 150." To invalidate this, assume that each of the four high scorers scored 300 (the maximum possible score in the test = $50 \times 6 = 300$), and each of the first four students scored x. Needless to say, the fifth score in an ascending order would be 120 (Median score).

After calculating the value of x, if $150 > x > 120$, the statement must be true; otherwise, it must not be true.

$x \times 4 + 120 + 300 \times 4 = 9 \times 150$

$4x + 120 + 1,200 = 1,350$

$4x = 30$

$x = 7.50 \ngtr 120$, thus, Statement II must not be true.

Statement III: At least one student had a score of less than 120.

Like Statement II, the best approach to test this statement is to think of a scenario wherein the situation depicted above fails.

The statement states that "At least one student had a score of less than 120." To invalidate this, assume that each of the first five students scored 120, and each of the next four students scored x.

After calculating the value of x, if $x < 120$, the statement must be true, else must not be true.

$120 \times 5 + x \times 4 = 9 \times 150$

$600 + 4x = 1,350$

$x = 187.5 \nless 120$, thus, the statement must not be true.

The correct answer is option A.

25. Given,

Average wt. of 9 boys = 36 kg;
=> Total wt. of 9 boys = 9×36 = 324 kg

Median wt. of 9 boys = 38 kg
Say the wt. of the lightest boy is x kg
=> Wt. of the heaviest boy = $(3x - 3)$ kg

Since the # of boys is an odd number, a boy with median wt., 38 kg, must be present among 9 boys. As of now we have some information about only 3 boys. If we arrange the boys according to their weights, they would be put up as following:

x, B2, B3, B4, 38, B6, B7, B8, $(3x - 3)$
=> x = Wt of the lightest boy (B1); 38 = Median (B5); & $(3x-3)$ = Wt of the heaviest boy (B9)

Since the question asks us to get the maximum value of $(3x - 3)$, we must try to keep the weights of other 6 boys to the lowest.

=> Since B1 = x is the lowest value, the least value for each of B2, B3, & B4 would be x
=> Similarly, Since Median = B5 = 38, the least value for each of B6, B7, & B8 would be 38

As per the designated values, the arrangement of boys according to their weights is given below.

$x, x, x, x, 38, 38, 38, 38, (3x - 3)$

=> Sum of 9 boys = $x + x + x + x + 38 + 38 + 38 + 38 + (3x - 3)$

=> $324 = 7x + 152 - 3$

=> $7x = 175$

=> $x = 25$ kg

=> Wt of the heaviest boy = $3x - 3 = 3 \times 25 - 25 = 72$ kg

The correct answer is option C.

26. Given that,

Median defect per batch = Average defect per batch; for all the nine batches;

Let us first deduce the median.

By arranging the eight values of defects in ascending order, we get;

$2, 3, 3, 4, 5, 6, 7, 8$

From the options, if the # of defects in the ninth batch is 2, 3, or 4, the median would be 4—Middle-most value.

However if it is either 5, 6, 7 or 8, the median would be 5—Middle-most value, so we are not sure about the value of median, but it would be either 4 or 5.

Let us calculate the average now. It must be either 4 or 5 as it is given that average = median.

Say the # of defects in the ninth batch is x,

=> Average = $(2 + 3 + 3 + 4 + 5 + 6 + 7 + 8 + x)/9 = (38 + x)/9$;

=> $(36 + 2 + x)/9$

=> $4 + \frac{(2+x)}{9}$

$\frac{(2+x)}{9}$ must be 0 or 1; thus x must be -2 or 7.

Since # of defects cannot be negative, so $x = 7$.

The correct answer is option E.

27. The sum of the numbers of the set

$= x + y + (x + y) + (x - 4y) + xy + 2y$

$= 3x + xy = x(y + 3)$

Thus, the mean

$$= \frac{x(y + 3)}{6}$$

Since the mean is $(y + 3)$, we have

$$\frac{x(y + 3)}{6} = y + 3$$

$\Rightarrow x = 6$

We know that: $y > 6$

$\Rightarrow y = 6 + k$, where k is a positive number

Thus, the terms are: $\{x, y, (x + y), (x - 4y), xy,\ 2y\}$

$= \{6,\ (6 + k),\ (12 + k),\ (-18 - 4k),\ (36 + 6k),\ (12 + 2k)\}$

Arranging in ascending order, we have

$\{(-18 - 4k),\ 6,\ (6 + k),\ (12 + k),\ (12 + 2k), (36 + 6k)\}$

Thus, the median of the above 6 terms is the average of the 3rd and 4th terms

$$= \frac{(6 + k) + (12 + k)}{2}$$

$= 9 + k = 3 + (6 + k)$

$= 3 + y = \frac{6}{2} + y$

$= \frac{x}{2} + y$

The correct answer is option B.

Alternate approach:

We have determined that $x = 6$.

Since $y > 6 => y > x$

Thus, the terms in ascending order are:

$\{(x - 4y),\ x, y, (x + y), 2y,\ xy\}$

Thus, the median is the average of the 3rd and 4th terms $= \frac{y + (x + y)}{2} = \frac{x}{2} + y$

28. Since the mean of the 4 numbers is 32, we have

$\frac{a + b + c + d}{4} = 12$

$=> a + b + c + d = 48 \ldots$ (i)

Since the median is 7, and the numbers are in ascending order, we have

$\frac{b + c}{2} = 7$

$=> b + c = 14 \ldots$ (ii)

We know that one of the numbers is 8. Thus, we have the following possibilities:

- $a = 8$: Not possible, since the mean of b and c, both of which are greater than a, is 7
- $b = 8$: Not possible, since $b + c = 14$ and $b < c$
- $c = 8$: This is possible, since from $b + c = 14$; thus, we have $b = 6$.

Thus, to minimize d, we need to maximize a. The maximum value of a can only be 5 since $a < b$.

Thus, we have $d = 48 - (a + b + c) = 48 - (5 + 14) = 29$

The correct answer is option D.

29. Mean of the three terms

$$= \frac{3^{x-1} + 3^{x-1}5 + 3^x 5^2}{3}$$

$$= \frac{3^{x-1}\left(1 + 5 + 3 \times 5^2\right)}{3}$$

$$= 3^{x-2}(1 + 5 + 75)$$

$$= 3^{x-2} \times 3^4$$

$$= 3^{x+2}$$

Thus, we have

$3^{x+2} = 243 = 3^5$

$=> x = 3$

Thus, the three terms are 3^{x-1}, $3^{x-1}5$ and $3^x 5^2$

$= 3^{3-1}$, $3^{3-1}5$ and $3^3 5^2$

Thus, the median is $3^2 5 = 45$.

The correct answer is option B.

Questions on Range

30. We do not know the values of $a, b,$ & c. So,

Minimum value of range $= 76 - (-43) = 119$; if any of $a, b,$ & c is greater than 76 and/or any of $a, b,$ & c is smaller than -43, the range would be greater than 119.

Since range is always positive, option I is ruled out.

Option II is ruled out since its value is less than 119, we know that the minimum value of range should be 119.

The correct answer is option C.

31. The smallest 3 numbers are reduced by 2 each; thus, the total is reduced by $3 \times 2 = 6$.

The largest 2 numbers are increased by 3 each; thus, the total is increased by $2 \times 3 = 6$.

Thus, the sum of the 8 numbers of the set remains unchanged, as the sum is reduced and increased by the same number.

Thus, the mean remains unchanged. - Statement I is incorrect

The median of the set of 8 numbers is the average of the 4th and 5th numbers, once the numbers are arranged in order, say, ascending order.

The smallest three numbers, i.e., 1st, 2nd and 3rd numbers are reduced, and the largest two numbers, i.e., 7th and 8th numbers are increased.

Thus, the 4th and 5th numbers remain unchanged. Also, the order of the numbers remains the same since the smaller numbers become smaller and the larger numbers become larger.

Thus, the median remains unchanged. - Statement II is incorrect

The range is the difference between the largest and the smallest numbers. Since the largest number is increased by 3 and the smallest number is reduced by 2, the range will increase by $3 - (-2) = 5$.

Thus, the range would change. - Statement III is correct.

The correct answer is option C.

32. Since the range of heights of students in Class A is 9 inches, let the heights of the shortest and tallest students in Class A be x and $(x + 9)$ inches, respectively.

Since the range of heights of students in Class B is 6 inches, let the heights of the shortest and tallest students in Class B be y and $(y + 6)$ inches, respectively.

We need to determine the range of heights of all students taken together. Thus, we need to know the relative values of x and y.

- Statement I: The tallest student in Class B is 5 inches taller than the tallest student in Class A

=> $y + 6 = (x + 9) + 5$ => $y = x + 8$

=> The heights of the shortest and the tallest students of Class B are $y = x + 8$ and $y + 6 = x + 14$, respectively

The heights of the shortest and the tallest students in Class A are x and $x + 9$, respectively

Thus, considering all students, the height of the shortest student is x inches, and the height of the tallest student is $(x + 14)$ inches.

Thus, the range is 14 inches.

- Statement II: The shortest student in Class B is 1 inch shorter than the tallest student in Class A.

  => $y = (x + 9) - 1$ => $y = x + 8$

  => The heights of the shortest and tallest students in Class B are $y = x + 8$ and $y + 6 = x + 14$, respectively

  The heights of the shortest and tallest students in Class A are x and $x + 9$, respectively

  Thus, considering all students, the height of the shortest student is x inches and the height of the tallest student is $(x + 14)$ inches.

  Thus, the range is 14 inches.

- Statement III: The shortest student in Class A is 8 inches shorter than the shortest student in Class B.

  => $x = y - 8$ => $y = x + 8$

  => The heights of the shortest and the tallest students of Class B are $y = x + 8$ and $y + 6 = x + 14$, respectively

  The heights of the shortest and tallest students in Class A are x and $x + 9$, respectively.

  Thus, considering all students, the height of the shortest student is (x) inches and the height of the tallest student is $(x + 14)$ inches.

  Thus, the range is 14 inches.

Thus, the range of all students combined can be determined from each of the statements.

The correct answer is option E.

Questions on Standard deviation

33. Let us analyze the performances of John and Bob:

John:

John's score = 75

Mean score of all students = 70

Difference between John's score and the mean score = 75 - 70 = 5

Standard deviation of all scores = 2

Thus, the number of standard deviations John's score is more than the mean score = $\frac{5}{2}$

= 2.5

Bob:

Bob's score = 72

Mean score of all students = 65

Difference between Bob's score and the mean score = 72 - 65 = 7

Standard deviation of all scores = 3

Thus, the number of standard deviations Bob's score is more than the mean score = $\frac{7}{3}$

= 2.33

Thus, John's score is farther (+2.5×SD) from his class's mean score than Bob's score (+2.33×SD) is from his class's mean score. In other words, a relatively greater number of students would be below John's ranking in his class than would be below Bob's ranking in his class.

Thus, John's performance in his exam is better than Bob's. - Statement I is correct.

Statement II:

Since John's score is higher than the mean score, there may or may not be a student scoring higher than John. - Statement II may be correct, but it must not be correct.

Statement III:

Since Bob's score is higher than the mean score, there must be a student who scored less than the mean score and hence scored less than Bob. - Statement III is correct.

The correct answer is option D.

34. Adding or subtracting a constant number from all the elements of a set does not affect the standard deviation (SD) of the set. We see that all the sets—P, Q, and R—can be obtained by adding or subtracting a constant from the elements of the given set X; hence all the sets have equal standard deviations (SD).

The correct answer is option E.

Note that, if you multiply each element of a set by a constant m, the SD would be m times the original SD. Say, each element of the set is multiplied by 3, then the SD would be 3 times the original SD = 3SD.

Similarly, if you divide each element of the set by a constant m, the SD would be $1/m$ times the original SD = SD/$|m|$.

Multiplying with or dividing by a negative number does not affect SD's sign. SD is always a positive number.

35. Adding or subtracting a constant from each element in the set has no effect on the standard deviation; however multiplication and division do change the SD.

Looking at options A, B, C, & E, we find that a constant term is NOT added to or NOT subtracted from each element; this implies that the SD of any of the sets would not be the same as the SD of Set X. So the only option left is: D—correct answer.

Though, at first sight, it does not look lie option D satisfies the criterion, expanding each element will do the trick.

D. $\{p + qr,\ q(1 + r),\ r(q + 1)\}$

- I element: $p + qr$
- II element: $q(1 + r) = q + qr$
- III element: $r(q + 1) = r + qr$

You must have observed that each element is derived by adding a constant qr to each element of Set X. This implies that the SD of the set would be the same as that of Set X.

Hence, the correct option is D.

Alternate approach:

Say, $p = 1,\ q = 2$ & $r = 3$. No need to calculate SD.

Let us analyze the options by plugging in the values of $p,\ q$ & r. The values would be as follows.

A. 2, 2, 3
B. 1, 3, 3
C. 1, 2, 4
D. 7, 8, 9: Correct; as constant 6 is added to $\{p = 1,\ q = 2,\ r = 3\}$, implying same SD.
E. 0, 2, 3

36. It is to be noted that the number of elements in each set is the same: 5 each. So only the deviations w.r.t. the mean would matter and be deciding.

For set A, we find that the deviation of any element from the successive element is 10.
For set B, we find that the deviation of any element from the successive element is 5.
For set C, we find that the deviation of any element from the successive element is 20.

Thus, of the three sets, the deviation (20) in Set C is the greatest; thus, its standard deviation would also be the greatest.

There is no need to calculate SDs for sets; we can conclude that SD for set C would be the largest and SD for set B would be the smallest.

Note that the above logical deduction approach is applicable if there are an equal number of elements in each set and the elements are evenly spaced.

The correct answer is option C.

37. Since SD is always positive, so option I & II are ruled out.

SD for the given set CANNOT be '0' as there are deviations among elements. For SD to be '0', all the elements of a set must be equal.

The correct answer is option E.

38. We know that **SD** = $\sqrt{\frac{(x_1 - \bar{x})^2 + (x_2 - \bar{x})^2 + (x_3 - \bar{x})^2 + ...(x_n - \bar{x})^2}{n}}$; where $\bar{x}$ is arithmetic mean and $x_1, x_2, x_3, ..., x_n$ are elements

Since # of elements are two {3, 5}, we can reduce the formula to

SD = $\sqrt{\frac{(x_1 - \bar{x})^2 + (x_2 - \bar{x})^2}{2}}$

Since the # of elements to be swapped also two, we can deduce

SD $\propto [(x_1 - \bar{x})^2 + (x_2 - \bar{x})^2]$

To get a lesser value of SD, the value of $[(x_1 - \bar{x})^2 + (x_2 - \bar{x})^2]$ must be decreased.

Let us calculate the value of $[(x_1 - \bar{x})^2 + (x_2 - \bar{x})^2]$ for the elements to be swapped: {3, 5}, given that $\bar{x} = 5$.

For {3, 5}, $[(x_1 - \bar{x})^2 + (x_2 - \bar{x})^2] = [(3 - 5)^2 + (5 - 5)^2] = 4$

The option that has the least value of $[(x_1 - \bar{x})^2 + (x_2 - \bar{x})^2]$ would be the answer as it would decrease the SD most!

It is to be noted that the sum of the swapped elements {3, 5}, and the elements to be swapped (all options—options A through E) is 8; this implies that the mean of the set will not get affected; it will remain 5—a very important aspect.

Let us take each option and calculate the values.

(A) {0, 8}: $[(0 - 5)^2 + (8 - 5)^2] = 34 > 4$ (derived value); it would rather increase the SD.

(B) {5, 3}: $[(5 - 5)^2 + (3 - 5)^2] = 4 = 4$ (derived value); it would NOT decrease the SD.

(C) {2, 6}: $[(2 - 5)^2 + (6 - 5)^2] = 10 > 4$ (derived value); like Option A, it would rather increase the SD.

(D) {4, 4}: $[(4-5)^2+(4-5)^2] = 2 < 4$ (derived value); it would decrease the SD—correct answer.

(E) {1, 7}: $[(1-5)^2+(7-5)^2] = 20 > 4$ (derived value); like option A & C, it would rather increase the SD.

The correct answer is option D.

39. Let us understand symmetrical distribution of a characteristic in a large population. It is also called Binomial distribution or Normal distribution.

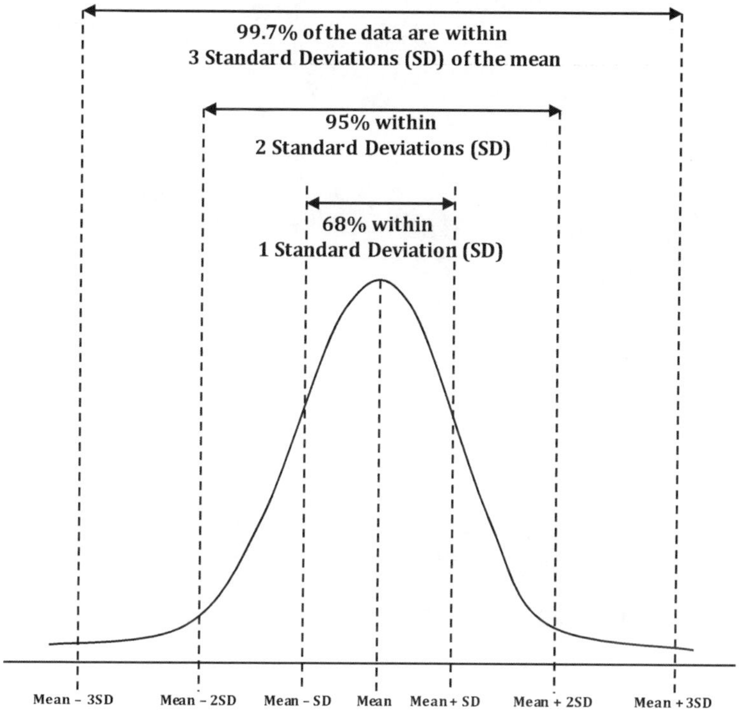

As you see in the figure, there is a bell-shaped curve that is symmetrical about the mean μ. Within the range of $\mu \pm \sigma$, approximately 68% population lies; within the range of $\mu \pm 2\sigma$, approximately 95% of the population lies; and within the range of $\mu \pm 3\sigma$, approximately 99.7% population lies.

Some results:

(1) Within the range of μ & $\mu + \sigma$, approximately 68/2 = 34% population lies

(2) Within the range of μ & $\mu - \sigma$, approximately 68/2 = 34% population lies

(3) Within the range of μ & $\mu + 2\sigma$, approximately 95/2 = 42.5% population lies

(4) Within the range of μ & $\mu - 2\sigma$, approximately 95/2 = 42.5% population lies

(5) Within the range of μ & $\mu + 3\sigma$, approximately 99.7/2 = 49.85% population lies

(6) Within the range of μ & $\mu - 3\sigma$, approximately 99.7/2 = 49.85% population lies

(7) Within the range of μ & ∞, exactly 50% population lies

(8) Within the range of μ & $-\infty$, exactly 50% population lies

(9) Within the range of $\mu - 3\sigma$ & $-\infty$, approximately (100 - 99.7)/2 = 0.15% population lies

(10) Within the range of $\mu + 3\sigma$ & ∞, approximately (100 - 99.7)/2 = 0.15% population lies

(11) Within the range of $\mu - 2\sigma$ & $-\infty$, approximately (100 - 95)/2 = 2.5% population lies

(12) Within the range of $\mu + 2\sigma$ & ∞, approximately (100 ? 95)/2 = 2.5% population lies

(13) Within the range of $\mu - \sigma$ & $-\infty$, approximately (100 - 68)/2 = 16% population lies

(14) Within the range of $\mu + \sigma$ & ∞, approximately (100 ? 68)/2 = 16% population lies

(15) Within the range of $\mu - 3\sigma$ & ∞, approximately (100 - 0.15) = 99.85% population lies
Or, we can also say, 99.85% of the distribution is more than $\mu - 3\sigma$

(16) Within the range of $\mu + 3\sigma$ & ∞, approximately (100 - 0.15) = 99.85% population lies
Or, we can also say, 99.85% of the distribution is less than $\mu + 3\sigma$

(17) Within the range of $\mu - 2\sigma$ & ∞, approximately (100 - 2.5) = 97.5% population lies
Or, we can also say, 97.5% of the distribution is more than $\mu - 2\sigma$

(18) Within the range of $\mu + 2\sigma$ & $-\infty$, approximately (100 - 0.15) = 97.5% population lies
Or, we can also say, 97.5% of the distribution is less than $\mu + 2\sigma$

(19) Within the range of $\mu - \sigma$ & ∞, approximately (100 - 16) = 84% population lies
Or, we can also say, 84% of the distribution is more than $\mu - \sigma$

(20) Within the range of $\mu + \sigma$ & $-\infty$, approximately (100 - 16) = 84% population lies
Or, we can also say, 84% of the distribution is less than $\mu + \sigma$

Coming to our question, we know that within the range of $\mu \pm 2\sigma$, approximately 95% of the population lies; thus, 100 − 95 = 5% lies beyond the range of $\mu \pm 2\sigma$. This means that 5/2 = 2.5% population lies on either side of $\mu \pm 2\sigma$ range. Thus, 100 − 2.5 = 97.5% percent of the distribution is more than $(\mu - 2\sigma)$.

The correct answer is option E.

40. The computation of standard deviation (SD) is not within the scope of the GMAT; however, its analysis is. This question can be answered by analyzing the values.

The standard deviation is the measure of the spread of values in the set with respect to the mean of the set. The farther the values are on either side of the mean, the greater the standard deviation, and vice versa.

The smallest possible value of SD is 0, and that's when all of the numbers in a group are the SAME. We see that only the set $\{-10, -10, -10, -10, -10\}$ has all the numbers the same; thus, its SD = 0, minimum.

Other sets will have SD > 0. We are not asked to choose a set that has the largest SD; had that been the question, the answer would have been option B: $\{-10, -8, -6, -4, -2\}$ because collectively the spread of the numbers in the set w.r.t. its mean (= -6) would be greater than that for any other set.

The correct answer is option A.

Chapter 19

Talk to Us

Have a Question?

Email your questions to *info@manhattanreview.com*. We will be happy to answer you. Your questions can be related to a concept, an application of a concept, an explanation of a question, a suggestion for an alternate approach, or anything else you wish to ask regarding the GMAT.

Please mention the page number when quoting from the book.

GMAC Resources

- *Official Guide*: It is the best resource to prepare for the GMAT. It is a complete GMAT book. It comes with a Diagnostic test, which helps you measure your capability beforehand. It features Verbal, Quantitative, and Data Insights questions types. The book contains an access code to avail GMATPrep Software, Online Question Bank and Exclusive Videos.

- *GMATPrep Software*: If you buy the OG, you get a free online resource from the GMAC—the testmaker. Apart from practice questions and explanation, it also has two genuine Computer Adaptive tests; you can also buy four additional CATs and few practice questions upon the payment.

Best of luck!

Happy Learning!

Professor Dr. Joern Meissner
& The Manhattan Review Team

Made in United States
Orlando, FL
04 November 2026

90202235R00258